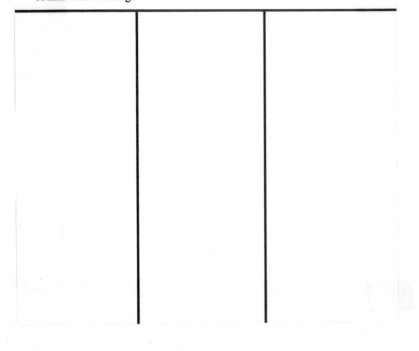

MAGNETOHYDRODYNAMIC TURBULENCE

This book presents an introduction to, and state-of-the-art account of, magnetohydrodynamic (MHD) turbulence, an active field both in general turbulence theory and in various areas of astrophysics. The book starts by introducing the MHD equations and certain useful approximations. The transition to turbulence is then discussed, including the problem of finite-time singularities of the ideal equations and the excitation of instabilities. The second part of the book deals with incompressible MHD turbulence, the macroscopic aspects connected with the various self-organization processes, the phenomenology of the turbulence spectra, two-point closure theory, and intermittency. The third part considers two extensions: two-dimensional turbulence and compressible (in particular, supersonic) turbulence. Because of the similarities in the theoretical approach, these chapters start with a brief account of the corresponding methods developed in hydrodynamic turbulence. The final part of the book is devoted to three astrophysical topics: turbulence in the solar wind, in accretion disks, and in the interstellar medium. This book is suitable for graduate students and researchers working in turbulence theory, plasma physics, and astrophysics.

DIETER BISKAMP received his Ph.D. from the University of Munich. Following a postdoctoral period at the Max Planck Institute for Astrophysics, he went on to work at the Space Research Institute in Frascati. In 1972, he became a Senior Research Scientist at the Max Planck Institute for Plasma Physics. From 1981 to 1995 he was head of the General Theory Group and subsequently head of the Nonlinear Plasma Dynamics Group, a position he held until 2001. In 1979 he was visiting Professor at the University of Texas and in 1995 COE visiting Professor at the National Institute for Fusion Science in Nagoya. He currently works as a consultant at the Center for Interdisciplinary Plasma Science at the Max Planck Institute for Extraterrestrial Physics. He is the author of two previous books, *Nonlinear Magnetohydrodynamics* and *Magnetic Reconnection in Plasmas.*

MAGNETOHYDRODYNAMIC TURBULENCE

DIETER BISKAMP
Center for Interdisciplinary Plasma Science, Garching

CAMBRIDGE
UNIVERSITY PRESS

PUBLISHED BY THE PRESS SYNDICATE OF THE UNIVERSITY OF CAMBRIDGE
The Pitt Building, Trumpington Street, Cambridge, United Kingdom

CAMBRIDGE UNIVERSITY PRESS
The Edinburgh Building, Cambridge CB2 2RU, UK
40 West 20th Street, New York, NY 10011-4211, USA
477 Williamstown Road, Port Melbourne, VIC 3207, Australia
Ruiz de Alarcón 13, 28014 Madrid, Spain
Dock House, The Waterfront, Cape Town 8001, South Africa

http://www.cambridge.org

First published 2003

Printed in the United Kingdom at the University Press, Cambridge

Typeface Times 10/13 pt *System* LaTeX 2_ε [TB]

A catalog record for this book is available from the British Library

Library of Congress Cataloging in Publication data

Biskamp, D.
Magnetohydrodynamic turbulence / Dieter Biskamp.
p. cm.
Includes bibliographical references and index.
ISBN 0 521 81011 6
1. Magnetohydrodynamic instabilities. 2. Turbulence. I. Title.
QC718.5.M36 B55 2003
538′.6–dc21 2002034948

ISBN 0 521 81011 6 hardback

To Robert H. Kraichnan

Contents

Preface

Turbulence in electrically conducting fluids is necessarily accompanied by magnetic-field fluctuations, which will, in general, strongly influence the dynamics. It is true that, in our terrestrial world, conducting fluids in turbulent motion are rare. In astrophysics, however, material is mostly ionized and strong turbulence is a widespread phenomenon, for instance in stellar convection zones and stellar winds and in the interstellar medium. Turbulent magnetic fields are therefore expected to play an important role. Despite the fact that, on a microscopic level, astrophysical plasmas exhibit rather diverse properties, a unified macroscopic treatment in the framework of magnetohydrodynamics (MHD) to describe the most important magnetic effects is appropriate. Hence there is much interest in MHD turbulence in the astrophysical community. Considerable interest comes also from the side of pure theory, where MHD turbulence introduces new concepts into turbulence theory, as the large number of articles on this topic in the literature shows. However, to date no monograph on MHD turbulence seems to have been written. I therefore believe that a treatise both introducing the field and reviewing the current state of the art could be welcome.

The book consists of four major parts: an introductory part, Chapters 2 and 3, discusses the MHD model and the transition to turbulence; the second part, Chapters 4–7, focusses on the theory of incompressible turbulence; the third part, Chapters 8 and 9, deals with two important extensions, two-dimensional turbulence, which arises in the presence of a strong magnetic field, and, in a sense the opposite case, compressible, in particular supersonic, turbulence; and finally a part concerning applications, Chapters 10–12, treating three areas in which MHD turbulence is observed, or expected to be excited, namely turbulence in the solar wind, in accretion disks, and in the interstellar medium. A book on MHD turbulence is also a book on hydrodynamic turbulence, using and generalizing methods developed for the latter, which the reader will find in the first parts of most chapters. The chapters dealing with the applications contain a general introduction to each field and may be read independently of the rest. Apart from elementary fluid dynamics no particular expertise is required, though some knowlege of plasma physics can sometimes be helpful. I hope that the book will be suitable for those just entering the field and also interesting for researchers in the field.

It is a pleasure to express my gratitude to the many colleagues with whom I enjoyed very fruitful discussions on the various topics of this book. In particular, I would like to thank Axel Brandenburg, Rainer Grauer, Eckart Marsch, Wolf-Christian Müller, Hélène Politano, Annick Pouquet, Rainer Schwenn, and Rudolf Treumann. I am grateful to Barbara Mori for preparing the figures included in the book. I should also like to acknowledge the hospitality and financial support provided by the Center for Interdisciplinary Plasma Science, where most of this book was written, with special thanks to Gregor Morfill. Finally, I want to thank Steven Holt for his excellent copy-editing of the manuscript.

Dieter Biskamp

1

Introduction

Turbulence is a ubiquitous phenomenon. Wherever fluids are set into motion turbulence tends to develop, as everyday experience shows us. When the fluid is electrically conducting, the turbulent motions are accompanied by magnetic-field fluctuations. However, conducting fluids are rare in our terrestrial world, where electrical conductors are usually solid. One of the rare examples of a fast-moving conducting fluid, which has been of some practical importance and concern and to which authors of theoretical studies sometimes referred, is, or better, was the flow of liquid sodium in the cooling ducts of a fast-breeder reactor. It is therefore not surprising that, in contrast to the broad scientific and technical literature on ordinary, i.e., hydrodynamic, turbulence, magnetic turbulence has not received much attention.

The most natural conducting fluid is an ionized gas, called a plasma. It is true that laboratory plasmas, which are confined by strong magnetic fields, notably in nuclear-fusion research, exhibit little dynamics, except in short disruptive pulses. Only the reversed-field pinch, a toroidal plasma discharge of relatively high plasma pressure, exhibits continuous magnetic activity, such that it is sometimes considered more as a convenient device for studying magnetic turbulence rather than as a particularly promising approach to controlled nuclear fusion.

Plasmas are, however, abundant in the extraterrestrial world. It is said that 99% of all material in the universe exists in the plasma state. This does not, however, mean that the plasma properties are always important. In fact, for understanding stellar evolution, during which conditions are mostly quasi-static, or galaxy formation, which is dominated by gravitational forces, the specific plasma properties and the presence of magnetic fields are not crucial and have usually been neglected, atomic and nuclear properties being more important. Only in certain processes have magnetic fields long been acknowledged to play a dominant part, such as in the dynamics of stellar atmospheres and in the generation of cosmic rays. In recent times, however, it appears that the omnipresence and active role of magnetic fields has also been recognized in those astrophysical objects which had formerly been treated as essentially nonmagnetic neutral fluids, for instance in accretion disks and in the interstellar medium. Here magnetic fluctuations provide essential clues regarding the intrinsic transport properties, namely transport of angular momentum in the former and gravitational collapse in the latter. These applications have led to a remarkable revival of interest in magnetic turbulence.

Though magnetic-field fluctuations occur on all scales, down to the smallest plasma modes, magnetic fields are most important at macroscales, i.e., mean wavelengths exceeding the internal plasma scale lengths, such as the ion gyroradius. In this regime magnetohydrodynamics (MHD) provides the appropriate framework, to which this book is restricted. This does not mean that the classical condition for a fluid approach, namely the smallness of the mean free path between interparticle collisions compared with gradient scales, must be satisfied. Indeed, in many dilute plasmas collisions are very rare, but there are other, collective, processes that play a similar dissipative role giving rise to effective (often called anomalous) transport coefficients. Also, in weakly ionized gases magnetic diffusion is not governed by classical resistivity, the friction between ions and electrons, but by ambipolar diffusion, the friction between ions and neutral species. Hence the huge values of the Reynolds numbers and other parameters characterizing a turbulent system, which are typical for astrophysical systems when they are calculated with the classical transport coefficients, should not be taken too seriously. In any case, the dissipation processes, independently of their nature, serve only as energy sinks, which cut off the spectrum of turbulent fluctuations at small scales but do not affect the main turbulence scales.

Since MHD turbulence is related to hydrodynamic turbulence, by following similar equations one may apply, and generalize, the formalism developed for the latter. Hence this book deals necessarily also with hydrodynamic turbulence. Turbulence theory has developed along two rather different lines, one oriented toward technical applications, the other focussing on the intrinsic turbulence properties. The practical importance of the first line is obvious – in popular view, turbulence is, indeed, considered more a technical problem than a physical phenomenon. The second line, which is naturally more interesting to a physicist, is characterized by certain approximations made for algebraic convenience as well as for conceptual clarity, which move the focus away from the practical aspects of turbulence. This is also reflected in the various treatises on turbulence theory published during the past few decades, for instance by Leslie (1973), by Lesieur (1997), and by Frisch (1995), which concentrate more on the formal developments in homogeneous incompressible turbulence theory. While the classical books, especially the two volumes by Monin and Yaglom (1975), contain also major parts dealing with observational results, in more recent works the emphasis has increasingly shifted toward discussion of results of computer simulations, which can be compared more directly with the theoretical models.

In fact, numerical computations have become an indispensable tool in turbulence research. Since statistical fluid theory is notoriously difficult to deal with analytically, a numerical treatment of the time-dependent primitive equations, now generally called numerical simulation, providing an exact solution (within controlled numerical-discreteness effects), is often the only method by which to check the validity of analytical modeling. The numerical scaling laws obtained by varying the values of parameters may be considered *the* solution of a turbulence problem, which should subsequently only be "understood" by invoking a simple intuitive model, or mechanism. The argument often given, namely that numerically attainable Reynolds numbers are simply too small to be useful for understanding real, high-Reynolds-number turbulence is gradually becoming academic, as progress in computer technology allows use of ever larger computational grids. To illustrate the progress in computer power, only 15 years ago two-dimensional (2D) simulations on a grid of 1024^2 points were considered the state of the art, whereas present-day supercomputers can handle the same linear resolution in three dimensions (3D), an increase by a factor of 10^3. Naturally the development of efficient numerical methods and codes and their exploitation has become a major activity for many turbulence theorists. For MHD turbulence numerical simulations play an even greater role than they do for hydrodynamic turbulence, since laboratory experiments are practically impossible and astrophysical systems, in particular solar-wind turbulence, the most important system of high-Reynolds-number MHD turbulence accessible to *in situ* measurements, are too complex to be directly comparable with theoretical results. We shall therefore frequently refer to such numerical studies.

Initially, interest in MHD turbulence focussed on the dynamo problem, notably in Batchelor's early paper (1950) and in the famous article by Steenbeck *et al.* (1966) reviewed later on in Moffatt's book (1978). A milestone in the fundamental scaling theory was the introduction of the Alfvén effect proposed independently by Iroshnikov (1964) and Kraichnan (1965b). This describes small-scale turbulent fluctuations as weakly interacting Alfvén waves propagating along the large-scale field. Because of the reduction of the corresponding spectral transfer the energy spectrum was predicted to be somewhat flatter, $k^{-3/2}$ instead of the Kolmogorov spectrum $k^{-5/3}$. This led to a long-standing debate about which process actually dominates the turbulence dynamics. From the theory side the importance of the Elsässer fields as basic dynamic variables, which incorporate the Alfvén-wave properties and naturally describe aligned states, seems to point to the fundamental role of the Alfvén effect. From the observation side, however, the energy spectrum of solar-wind turbulence was found to be clearly closer to a Kolmogorov law. The solution of this paradox lies in the intrinsic anisotropy of MHD turbulence emphasized by Goldreich and Sridhar (1995), who showed that the spectrum is more strongly developed perpendicularly to the local magnetic field, where the Alfvén effect is not operative. Hence a Kolmogorov-like transfer dynamics should dominate, which is also corroborated by results of recent numerical studies of 3D MHD turbulence.

There are, however, self-organization processes in MHD turbulence that have no hydrodynamic counterpart. These processes originate from certain selective decay properties arising from the presence of several ideal invariants with different decay rates, which have been studied intensively primarily by Montgomery and Matthaeus and their collaborators. Conservation of cross-helicity leads to highly correlated, or aligned, states, while conservation of magnetic helicity gives rise to the formation of force-free magnetic configurations as was first shown by Woltjer (1958) and generalized by Taylor (1974). A further facet of the latter process is the inverse cascade of the magnetic helicity, the excitation of increasingly larger magnetic scales, which is the basis of the nonlinear dynamo effect as noted by Pouquet *et al.* (1976).

The book consists of four parts. Chapters 2 and 3 discuss the properties of the MHD model and the transition from a smooth to a turbulent state; Chapters 4–7 deal with the various aspects of fully developed incompressible turbulence; Chapters 8 and 9 treat two extensions, namely 2D turbulence, which corresponds to the limit of a strong magnetic field, and, in a sense the opposite limit, supersonic turbulence. The last part, Chapters 10–12, considers three astrophysical applications, namely turbulence in the solar wind, in accretion disks, and in the interstellar medium.

Chapter 2 introduces the MHD theory and discusses special approximations, such as incompressibility and the Boussinesq approximation. The MHD equations exhibit a number of ideal conservation laws reflecting the constraints on the turbulence dynamics which are relaxed only by dissipative effects. We then briefly outline the properties of static equilibrium configurations, either magnetic or gravitational, and the different types of linear waves, which arise due to magnetic tension, pressure, and stratification. Finally the MHD equations are formulated in terms of the Elsässer fields, which are particularly well suited to MHD turbulence.

Chapter 3 deals with the transition to turbulence, namely how random motions are generated from a smooth flow. We first discuss the character of the singular solutions developing in the ideal equations, in particular the generation of finite-time singularities. This problem has aroused considerable interest, though its connection to the real, dissipative turbulence is somewhat loose. To date no general mathematical answer has been found, but there are strong indications that the hydrodynamic Euler equations do, indeed, give rise to finite-time singularities, if the initial state is not too symmetric, whereas in the MHD case no finite-time singularities seem to exist. This difference can be pinned down to the structure of the solution at the location of the singularity, filamentary in hydrodynamics and sheet-like in MHD. The development of small-scale random fluid motions is described by the effect of some instability of the quasi-singular structures of the ideal solution. When the velocity shear, pressure gradient, and current density exceed certain thresholds, the fluid becomes Kelvin–Helmholtz unstable, Rayleigh–Taylor unstable or unstable against tearing. We discuss these instabilities and their nonlinear evolution in some detail.

In Chapters 4–7 we consider incompressible turbulence. Chapter 4 focusses on macroscopic properties. We first discuss the Reynolds equations. Here the effects of the small-scale fluctuations are contained in the turbulent Reynolds and Maxwell tensors in the momentum equation and the electromotive force and turbulent resistivity in the induction equation, for which phenomenological expressions based on the mixing-length concept are used. Self-organization processes in MHD turbulence are caused by selective decay. Since magnetic helicity, and, to a lesser extent, also cross-helicity, decay much more slowly than does the turbulence energy, the decay of the latter leads to relaxed states, depending on the initial state either a linear force-free magnetic configuration or an Alfvénic state, in which the velocity and magnetic field are aligned. The energy-decay law is a characteristic property of the turbulence. For finite magnetic helicity the decay is controlled by selective decay, in particular kinetic energy decays more rapidly than does magnetic energy, $E^K \sim t^{-1}$ and $E^M \simeq E \sim t^{-0.5}$. If the magnetic helicity is negligible, the turbulence remains macroscopically self-similar during the decay, $E^K \sim E^M$, and the energy decay is faster, $E \sim t^{-1}$, since the turbulence is less constrained.

High-Mach-number turbulence carries a wide range of spatial scales, which exhibit characteristic scaling properties, notably the internal-range energy spectrum examined in Chapter 5. The scaling behavior becomes particularly transparent if the turbulence is not affected by the inhomogeneity of the global system, but can be considered statistically homogeneous and, possibly, also isotropic, which is the framework of most theories dealing with the intrinsic turbulence properties. To understand the spectral-transfer processes it is helpful to look at the statistical equilibrium states of the nondissipative system truncated in Fourier space, which are called absolute equilibrium states, from which the preferential spectral-transfer, or cascade, direction of a spectral quantity can be read off directly, for instance the inverse cascade of the magnetic helicity. We then derive, in a phenomenological way, the energy spectrum in dissipative MHD turbulence by assuming that there is either a local, Kolmogorov-type, transfer process leading to a $k^{-5/3}$ spectrum or a spectral transfer controlled by the Alfvén effect, which gives rise to the IK spectrum $k^{-3/2}$. In accounting for the inherent spectral anisotropy the latter is essentially ruled out. Numerical simulations of 3D MHD turbulence corroborate the Goldreich–Sridhar concept of a Kolmogorov inertial-range spectrum, which is independent of the amount of magnetic helicity present. Only at small wavenumbers, for which the inverse cascade of the magnetic helicity enhances the magnetic-energy spectrum, is there a difference between helical and nonhelical turbulence.

Two-point closure theory, which we treat in Chapter 6, gives a description of turbulence derived from the basic fluid equations instead of purely phenomenological arguments. Though closure theory cannot treat higher-order correlations correctly because of the basic quasi-normal approximation, it provides, in principle, a self-consistent dynamical theory of the evolution of the two-point correlations, the various spectral quantities in Fourier space. In practice the equations are made tractable by additional phenomonological assumptions, which lead to the EDQNM model which is usually considered. We discuss the MHD closure theory, in particular the special cases of helical and of correlated turbulence.

Chapter 7 deals with the higher-order statistics of turbulence in order to obtain a more detailed picture of the spatial structure. Fluid turbulence is not strictly self-similar, as small-scale eddies are increasingly sparsely distributed, a property which is called intermittency. To familiarize the reader with the general concept of intermittency, we first present some examples demonstrating the difference between self-similar and intermittent behavior. Structure functions $S^{(n)}(l)$, in particular the set of scaling exponents ζ_n, $S^{(n)} \sim l^{\zeta_n}$, describe the statistical distribution of the turbulent structures. In many turbulent systems, however, for which Reynolds numbers are rather modest, the scaling range of the structure functions, especially for higher orders, is too short to yield clear values of the exponents. The scaling range is often significantly broadened when one considers $S^{(n)}(S^{(3)})$ instead of $S^{(n)}(l)$, a property called extended self-similarity, which yields surprisingly accurate values of the relative scaling exponents ζ_n/ζ_3. Third-order structure functions, in turn, satisfy some exact relations derived directly from the fluid equations, Kolmogorov's four-fifths law in hydrodynamic turbulence, Yaglom's four-thirds law for an advected turbulent scalar field, and a similar relation for a third-order structure function of the Elsässer fields in MHD turbulence. Since no further exact relations seem to exist and no approximate scheme is known to date from which to obtain further information about the scaling properties directly from the fluid equations, the only viable approach consists of phenomenological modeling using some physical picture of the turbulence dynamics. We discuss in some detail two different models, the log-normal model and the log-Poisson model, the latter of which gives good agreement with experimental measurements in hydrodynamic turbulence and, *mutatis mutandis*, reproduces equally well the results for MHD turbulence obtained from numerical simulations. By varying the strength of a mean magnetic field B_0 simulations can also be made to show the transition from globally isotropic 3D MHD turbulence for $B_0 = 0$ to 2D turbulence for $B_0 \to \infty$.

Two-dimensional turbulence is considered in Chapter 8. We first discuss the hydrodynamic case, which has attracted much interest. In the presence of two ideal invariants, the energy and the enstrophy, the turbulence dynamics is dominated by a self-organization process, the buildup of large-scale structures due to the inverse cascade of the energy. An analogous process occurs in 2D MHD turbulence, for which now the energy exhibits a direct cascade but the mean-square magnetic potential exhibits an inverse cascade, which leads to the formation of large-scale magnetic structures. The free decay of turbulence proceeds in a macroscopically self-similar way with the asymptotic energy-decay law $E \sim t^{-1}$. The spatial scaling properties are different from those observed in 3D MHD turbulence, scaling exponents being generally lower, in particular $\zeta_3 < 1$, and the energy spectrum is flatter than the Kolmogorov spectrum, roughly consistent with the IK spectrum $k^{-3/2}$. Scaling properties are, however, not uniform in the sense that all structure functions of the same order have the same scaling exponent, since the exponent of the energy flux, a particular type of third-order structure function, is again unity.

The first part in Chapter 9 is devoted to compressible, in particular supersonic, turbulence, which is important in view of many astrophysical applications. Turbulence consists of both eddy motions and shock waves, and, as a rule of thumb, the inertial-range scales are dominated by eddy motions – the energy spectrum follows a Kolmogorov law –, while dissipation occurs mainly through shock waves. Also density fluctuations exhibit a Kolmogorov spectrum, but their spatial distribution is highly intermittent. Except for the case of a strong mean field, the dynamics is dominated by the supersonic flows, while magnetic effects are less important, the field being mainly advected, forming filamentary structures similar to those of the density. The second part of the chapter deals with turbulent convection in stratified systems. We consider first turbulence in the Boussinesq approximation, for which simple spectral laws can be derived, in particular for passive-scalar turbulence, and then add compressibility and magnetic fields. Because of the complexity of these systems only fully dynamic simulations can provide reliable information.

In the last three chapters dealing with some applications we restrict our consideration to astrophysical topics, since in laboratory plasma devices MHD turbulence usually does not occur and the only exception, the reversed-field pinch, has already been reviewed in several treatises (e.g., Biskamp, 1993a). In Chapter 10 we consider the turbulence in the solar wind. This is the only system of high-Reynolds-number MHD turbulence which is accessible to *in situ* observations. Much data has been accumulated from several spacecraft sampling the interplanetary space at various radii and latitudes. Here many properties of homogeneous turbulence theory are recovered. In spite of the large Mach number of the solar wind, compression effects are not dominant in the turbulence. In general inhomogeneities due to radial expansion of the wind and mixing of different types of wind by solar rotation complicate conditions considerably, leaving important observations unexplained.

Chapter 11 treats accretion disks, a widespread phenomenon in astrophysics, in which magnetic turbulence should be present, since it is the only conceivable mechanism for transport of angular momentum, the very agent of accretion. We first give a brief introduction to the main properties of accretion disks before considering more closely the origin of the presumed turbulence. Since in the disk material rotates essentially in Keplerian orbits, the system is hydrodynamically shear-flow stable, even nonlinearly. Turbulence can, however, be excited through the Balbus–Hawley instability when we allow the presence of a weak but finite magnetic field. Numerical simulations indicate that this mechanism may account for the observed accretion rates.

The last chapter deals with the interstellar medium, where the presence of turbulence can be observed directly. As in the previous two chapters, we first give a general overview of the properties of the interstellar medium, in particular its densest parts, the molecular clouds. Observations indicate the occurrence of highly supersonic irregular flows. These act as an effective pressure preventing rapid gravitational collapse and thus explaining the observed long lifetimes and low star-formation rates. The role of the magnetic field, which is random, at least on the larger scales, is still being debated, but it appears that it provides an additional stabilizing effect against contraction. Another interesting feature is the coupling between the magnetic field and the gas, which is only very weakly ionized; while classical resistivity would give rise to almost perfect coupling, in reality the coupling is rather loose because of ambipolar diffusion.

In a book spanning various rather different fields, the individual topics cannot all be treated in depth. We have therefore included numerous references to more specialized reviews and also, of course, the appropriate citations of the original work. However, the number of these citations had to be restricted in order not to diminish the readability of the text. Concerning notation, I have tried as far as possible to avoid denoting different things by the same symbol, giving an explicit warning in the few exceptional cases dictated by traditional notation. I use cgs units in the original equations, because they are still rather common in astrophysics and because of personal preference, but, whenever convenient, introduce suitable normalizations in order to write the equations in nondimensional forms.

2
Magnetohydrodynamics

Magnetohydrodynamics, or MHD in short, describes the macroscopic behavior of an electrically conducting fluid – usually an ionized gas called a plasma –, which forms the basis of this book. By macroscopic we mean spatial scales larger than the intrinsic scale lengths of the plasma, such as the Debye length λ_D and the Larmor radii ρ_j of the charged particles.[1] In this chapter we first derive, in a heuristic way, the dynamic equations of MHD and discuss the local thermodynamics (Section 2.1). Since most astrophysical systems rotate more or less rapidly, it is useful to write the momentum equation also in a rotating reference frame, where inertial forces appear (Section 2.2). Then some convenient approximations are introduced, in particular incompressiblity and, for a stratified system, the Boussinesq approximation (Section 2.3). In MHD theory the ideal invariants, i.e., integral quantities that are conserved in an ideal (i.e., nondissipative) system, play a crucial role in turbulence theory; these are the energy, the magnetic helicity, and the cross-helicity (Section 2.4). Though this book deals with turbulence, it is useful to obtain an quick overview of magnetostatic equilibrium configurations, which are more important in plasmas than stationary flows are in hydrodynamics (Section 2.5). Also the zoology of linear modes, the small-amplitude oscillations about an equilibrium, is richer than that in hydrodynamics (Section 2.6). Finally, in Section 2.7 we introduce the Elsässer fields, which constitute the basic dynamic quantities in MHD turbulence. In this chapter we write the equations in dimensional form, using Gaussian units, to emphasize the physical meaning of the various terms. At the end nondimensionalization in terms of the Alfvén time is introduced, which will be used throughout most of the following.

[1] The fluid approximation also requires that the mean free path is smaller than the gradient scales. For motions perpendicular to the magnetic field, however, the Larmor radius assumes the role of the mean free path.

2.1 MHD equations

2.1.1 Dynamic equations

The momentum equation in MHD can be obtained in a simple heuristic way by considering the forces acting on a fluid element δV with the mass $\rho\,\delta V$, where ρ is the mass density. These are the following.

(a) The Lorentz force. In an electromagnetic field $\boldsymbol{E}$, $\boldsymbol{B}$ a particle of charge q_j is subjected to the Lorentz force $q_j(\boldsymbol{E} + \boldsymbol{v}_j \times \boldsymbol{B}/c)$. The force on a macroscopic fluid element is the sum of the forces acting on the individual particles $\delta q\,\boldsymbol{E} + \delta\boldsymbol{J} \times \boldsymbol{B}/c$, where δq is the net charge and $\delta\boldsymbol{J}$ the electric current carried by the fluid element. Since in most fluids of interest electrostatic fields enforce charge neutrality over macroscopic scales, $\delta q \simeq 0$ (which is called quasi-neutrality and does *not* imply vanishing of the electrostatic field), only the magnetic part of the Lorentz force on the fluid element contributes,

$$\delta V \, \frac{1}{c} \boldsymbol{j} \times \boldsymbol{B}, \tag{2.1}$$

with $\delta\boldsymbol{J} = \boldsymbol{j}\,\delta V$, where $\boldsymbol{j}$ is the current density.

(b) The (thermal) pressure force. We assume that conditions close to local thermodynamic equilibrium pertain; hence the pressure tensor is isotropic, $p_{ij} = p\delta_{ij}$, exerting the force

$$-\oint p\, d\boldsymbol{S} = -\delta V \, \nabla p, \tag{2.2}$$

where the integral is taken over the surface of the fluid element.

(c) The gravitational force[2]

$$\delta V \, \rho \boldsymbol{g}, \tag{2.3}$$

$\boldsymbol{g}$ can be written in terms of the potential ϕ_g, $\boldsymbol{g} = -\nabla\phi_g$.

(d) The viscous force. Similarly to the pressure force (2.2), viscosity acts on the surface of the volume element,

$$\oint \boldsymbol{\sigma}^{(\mu)} \cdot d\boldsymbol{S} = \delta V \, \nabla \cdot \boldsymbol{\sigma}^{(\mu)}, \tag{2.4}$$

[2] In a magnetized plasma gravity is often negligible compared with the Lorentz force. The latter contains a contribution that acts as an effective gravity, $\rho\boldsymbol{g}_{\rm eff} = -2\boldsymbol{\kappa} p$, where $\boldsymbol{\kappa} = \boldsymbol{e}_B \cdot \nabla \boldsymbol{e}_B$ is the field-line curvature, $\boldsymbol{e}_B = \boldsymbol{B}/B$.

where $\boldsymbol{\sigma}^{(\mu)} = \{\sigma_{ij}^{(\mu)}\}$ is the viscous-stress tensor,[3]

$$\sigma_{ij}^{(\mu)} = \mu\left[(\partial_i v_j + \partial_j v_i) - \tfrac{2}{3}\delta_{ij}\,\nabla\cdot\boldsymbol{v}\right], \tag{2.5}$$

and μ is the viscosity, which we assume to be constant. We will later often use the kinematic viscosity $\nu = \mu/\rho$. The first term inside the brackets of (2.5),

$$w_{ij} = \partial_i v_j + \partial_j v_i, \tag{2.6}$$

is called the strain, or deformation, tensor.

On adding up these force contributions we obtain the momentum equation, or equation of motion,

$$\rho\,\frac{d\boldsymbol{v}}{dt} \equiv \rho(\partial_t + \boldsymbol{v}\cdot\nabla)\boldsymbol{v} = -\nabla p + \frac{1}{c}\,\boldsymbol{j}\times\boldsymbol{B} + \rho\boldsymbol{g} + \mu(\nabla^2\boldsymbol{v} + \tfrac{1}{3}\,\nabla\nabla\cdot\boldsymbol{v}). \tag{2.7}$$

The current density $\boldsymbol{j}$ is related to the magnetic field by Ampère's law

$$\nabla\times\boldsymbol{B} = \frac{4\pi}{c}\boldsymbol{j}, \tag{2.8}$$

the displacement current in the full version of Maxwell's equation being omitted. This is consistent with neglecting space charges, such that the continuity equation for the charge density is reduced to $\nabla\cdot\boldsymbol{j} = 0$. On substituting for $\boldsymbol{j}$, the Lorentz force can also be written in the following form:

$$\frac{1}{c}\,\boldsymbol{j}\times\boldsymbol{B} = -\frac{1}{8\pi}\,\nabla B^2 + \frac{1}{4\pi}\,\boldsymbol{B}\cdot\nabla\boldsymbol{B} = -\nabla\cdot\mathcal{T}^M, \tag{2.9}$$

where $\mathcal{T}^M = \{T_{ij}^M\}$ is the magnetic stress tensor,

$$T_{ij}^M = \frac{1}{8\pi}B^2\delta_{ij} - \frac{1}{4\pi}B_iB_j. \tag{2.10}$$

The first term in T_{ij}^M acts as an isotropic pressure, which may be combined with the thermal pressure to give the total pressure

$$P = p + B^2/(8\pi). \tag{2.11}$$

The ratio

$$\beta = 8\pi p/B^2 \tag{2.12}$$

[3] In many treatises on fluid dynamics, the stress tensor is defined as the sum of all surface force contributions, $S_{ij} = -p\delta_{ij} + \sigma_{ij}$, where σ_{ij} is called the deviatoric stress tensor, $\sum_i \sigma_{ii} = 0$, comprising the viscous-stress tensor and possibly other contributions (see, for instance, Batchelor, 1967).

is an important parameter characterizing the strength of the magnetic field in a plasma.

The dynamics of the magnetic field follows from Faraday's law

$$\partial_t \boldsymbol{B} = -c\,\nabla \times \boldsymbol{E}, \tag{2.13}$$

where the electric field is determined by Ohm's law. In the restframe of the fluid element one has $\boldsymbol{E} = \boldsymbol{j}/\sigma$, with σ the electrical conductivity. In the laboratory reference frame, in which the fluid equations are usually considered, the fluid element is moving at the speed $\boldsymbol{v}$ and the electric field is obtained by a Galilean transformation, $\boldsymbol{E} \rightarrow \boldsymbol{E}' = \boldsymbol{E} + \boldsymbol{v} \times \boldsymbol{B}/c$; hence

$$\boldsymbol{E} + \frac{1}{c}\boldsymbol{v} \times \boldsymbol{B} = \frac{1}{\sigma}\boldsymbol{j}, \tag{2.14}$$

which is called the generalized version of Ohm's law (generalized for a conducting fluid in motion). Substituting $\boldsymbol{E}$ into Faraday's law and assuming that we have uniform conductivity yields the advection–diffusion equation for the magnetic field,

$$\partial_t \boldsymbol{B} - \nabla \times (\boldsymbol{v} \times \boldsymbol{B}) = \eta\,\nabla^2 \boldsymbol{B}, \tag{2.15}$$

which we will also call Faraday's law or the induction equation. Here $\eta = c^2/(4\pi\sigma)$ is the magnetic diffusivity which, following convention, we shall also call resistivity. Note that we have always $\nabla \cdot \boldsymbol{B} = 0$. Conservation of mass leads to the continuity equation for the mass density. The change of the mass in volume element δV is given by the flow of mass into, or out of, the element,

$$\frac{d}{dt}\int_{\mathscr{V}} \rho\, dV = -\oint \rho \boldsymbol{v} \cdot d\boldsymbol{S}.$$

Hence, by use of Gauss' theorem,

$$\partial_t \rho + \nabla \cdot \rho \boldsymbol{v} = \frac{d\rho}{dt} + \rho\,\nabla \cdot \boldsymbol{v} = 0. \tag{2.16}$$

For incompressible flows, $\nabla \cdot \boldsymbol{v} = 0$, to which the main part of this book is restricted, the mass density of a fluid element is constant. Conditions for incompressiblity to hold are discussed in Section 2.2.

Finally, we have to find the dynamic equation for the pressure p. For conditions close to local thermodynamic equilibrium the pressure is coupled to the density ρ and temperature T by the equation of state. Since the applications discussed in this book deal mostly with dilute plasmas, we may assume the validity of the ideal-gas law, which for a simple plasma consisting of electrons and one species of singly charged ions reads[4]

$$p = 2(\rho/m_i)k_B T = 2nk_B T, \tag{2.17}$$

where m_i is the ion mass, $n = n_i = n_e$ the particle number density, $\rho \simeq nm_i$, and k_B the Boltzmann constant. Since conduction of heat is a diffusive process, it can be neglected at sufficiently large scales, such that the change of state in a fluid element is adiabatic,

$$d(p\rho^{-\gamma})/dt = 0, \tag{2.18}$$

which is equivalent to an isentropic flow $ds/dt = 0$, where the entropy of an ideal gas is

$$s = c_v \ln(p\rho^{-\gamma}). \tag{2.19}$$

Equation (2.18) is usually written in the form

$$\partial_t p + \boldsymbol{v} \cdot \nabla p + \gamma p \nabla \cdot \boldsymbol{v} = 0. \tag{2.20}$$

The parameter γ is the adiabatic exponent, the ratio of the specific heats $\gamma = c_p/c_v$, $\gamma = \frac{5}{3}$ in the case of a simple gas. To account for local heating and cooling or effects arising from heat conduction, one often introduces a more general polytropic relation $p(\rho) \sim \rho^\gamma$ with some exponent γ smaller than the adibatic value.[5] The limit $\gamma = 1$ describes isothermal pressure variations corresponding to infinite heat conduction; the opposite limit $\gamma \to \infty$ describes incompressible fluid dynamics, $\nabla \cdot \boldsymbol{v} = 0$. In the latter case the pressure can be eliminated from the dynamic equations by taking the curl of the equation of motion, as will be discussed in Section 2.3.

If the heat conduction is not negligible, one has to consider the equation for the internal energy $u = (p/\rho)(\gamma - 1)$ or, more conventionally, for the temperature,

$$\partial_t T + \boldsymbol{v} \cdot \nabla T + (\gamma - 1) T \nabla \cdot \boldsymbol{v} = (\gamma - 1)\kappa \nabla^2 T, \tag{2.21}$$

where κ is the heat diffusivity. For a plasma embedded in a magnetic field, the isotropic diffusion term in (2.21) should be regarded only as a rather coarse phenomenological *Ansatz* in view of the fact that the complicated heat-conduction processes differ strongly between the directions parallel and perpendicular to the magnetic field.

[4] In a hot dilute plasma electron and ion temperatures are often very different; hence, more generally, $p = p_i + p_e = nk_B(T_i + T_e)$.

[5] Concerning nomenclature, one speakes of barotropic pressure in the general case $p(\rho)$, whereas polytropic pressure refers to the special *Ansatz* $p \propto \rho^\gamma$.

In a self-consistent treatment not only heat conduction but also the heating by viscous and Ohmic dissipation have to be accounted for in the energy equation (2.21),

$$\rho(\partial_t + \boldsymbol{v} \cdot \nabla)u + p\,\nabla \cdot \boldsymbol{v} = -\nabla \cdot \boldsymbol{q} + \boldsymbol{\sigma}^{(\mu)} : \nabla \boldsymbol{v} - \frac{1}{\sigma}\boldsymbol{j}^2 \tag{2.22}$$

with the heat flux $\boldsymbol{q} = -\kappa\rho\nabla T$ (the colon denotes the dyadic product).

The dissipation effects introduced above have the form of diffusion processes with the kinematic viscosity ν, and the magnetic and heat diffusivities η and κ. In numerical studies of turbulence it is, however, often practical to use higher-order diffusion operators, or hyperdiffusion,

$$\nabla^2 \to \nabla^{2\alpha}, \quad \alpha > 1, \tag{2.23}$$

in order to concentrate dissipation more strongly at the smallest scales and thus avoid dissipative pollution of the inertial range. The exponent α is called the *dissipativity*. While in some cases such a term represents a physical process, for instance $-\eta_2(\nabla^2)^2\boldsymbol{B}$ added to Ohm's law would describe the effect of electron viscosity, more often than not the choice of $\alpha > 1$, even $\alpha \gg 1$, is made solely for numerical convenience.

2.1.2 The rotating reference frame

Most astrophysical objects, such as planets, stars, accretion disks, and galaxies, are rotating. The origin of the rotation is the conservation of angular momentum during gravitational condensation. To describe these systems it is often useful to consider the dynamics in a corotating reference frame. Consider a coordinate system rotating with the angular frequency $\boldsymbol{\Omega}$. A fluid element at point $\boldsymbol{r}$ with the velocity $\boldsymbol{v}$ in the laboratory system has the velocity $\boldsymbol{u}$ in the rotating system, $\boldsymbol{v} = \boldsymbol{u} + \boldsymbol{\Omega} \times \boldsymbol{r}$. In general, the time derivative of a vector consists of the time derivative in the rotating frame and the motion of the coordinate system,

$$\boldsymbol{v} \equiv \frac{d\boldsymbol{r}}{dt} = \frac{d\boldsymbol{r}}{dt'} + \boldsymbol{\Omega} \times \boldsymbol{r} \tag{2.24}$$

and

$$\begin{aligned}\frac{d\boldsymbol{v}}{dt} &= \frac{d\boldsymbol{v}}{dt'} + \boldsymbol{\Omega} \times \boldsymbol{v} \\ &= \frac{d\boldsymbol{u}}{dt'} + 2\boldsymbol{\Omega} \times \boldsymbol{u} + \boldsymbol{\Omega} \times (\boldsymbol{\Omega} \times \boldsymbol{r}).\end{aligned} \tag{2.25}$$

On substituting for $d\boldsymbol{v}/dt$ from (2.7) the momentum equation in the rotating frame reads

$$\rho(\partial_{t'} + \boldsymbol{u}\cdot\nabla)\boldsymbol{u} = -\nabla p + \frac{1}{c}\boldsymbol{j}\times\boldsymbol{B} + \rho\boldsymbol{g} + \rho\Big[2\boldsymbol{u}\times\boldsymbol{\Omega} + \boldsymbol{\Omega}\times(\boldsymbol{r}\times\boldsymbol{\Omega})\Big] + \mu\,\nabla^2\boldsymbol{u}. \quad (2.26)$$

The first term in the bracketed expression in (2.27) is the Coriolis force. The second is the centrifugal force, which can be written as a gradient,

$$\boldsymbol{\Omega}\times(\boldsymbol{r}\times\boldsymbol{\Omega}) = \nabla\tfrac{1}{2}(\boldsymbol{r}\times\boldsymbol{\Omega})^2, \quad (2.27)$$

and can therefore be included in the pressure or the gravitational potential. The remaining dynamic equations for $\boldsymbol{B}$ and ρ are not changed by the transformation to a rotating coordinate system.

2.2 Incompressibility and the Boussinesq approximation

In most treatises on turbulence theory it is assumed that incompressibility of the fluid holds; $d\rho/dt = 0$ or, equivalently, $\nabla\cdot\boldsymbol{v} = 0$. This allows one to eliminate the pressure by taking the curl of the equation of motion (2.7). Assuming that one has a uniform density $\rho = \rho_0$, one finds the equation for the vorticity $\boldsymbol{\omega} = \nabla\times\boldsymbol{v}$,

$$\partial_t\boldsymbol{\omega} + \boldsymbol{v}\cdot\nabla\boldsymbol{\omega} - \boldsymbol{\omega}\cdot\nabla\boldsymbol{v} = \frac{1}{c\rho_0}(\boldsymbol{B}\cdot\nabla\boldsymbol{j} - \boldsymbol{j}\cdot\nabla\boldsymbol{B}) + \nu\,\nabla^2\boldsymbol{\omega}. \quad (2.28)$$

The velocity is calculated from $\boldsymbol{\omega}$ by solving Poisson's equation

$$\nabla^2\boldsymbol{v} = -\nabla\times\boldsymbol{\omega}. \quad (2.29)$$

Also the pressure p and the total pressure P, (2.11), which are no longer independent dynamic variables, can be computed *a posteriori* from Poisson's equation obtained by taking the divergence of the momentum equation (2.7),

$$\nabla^2 P = -\nabla\cdot[\rho_0\boldsymbol{v}\cdot\nabla\boldsymbol{v} - \boldsymbol{B}\cdot\nabla\boldsymbol{B}/(4\pi)]. \quad (2.30)$$

Let us briefly discuss the conditions for incompressibility to hold. As a rule, the dynamics of the fluid can be assumed incompressible if the fluid motion $\boldsymbol{v}$ is slow compared with the (fastest) compressible wave propagating in the direction of $\boldsymbol{v}$. In addition, the derivative $\partial_t\boldsymbol{v}$ must be sufficiently small, i.e., the frequency must be small compared with that of the compressible wave, such that the ∂_t terms do not exceed the advective terms $\boldsymbol{v}\cdot\nabla$. Consider first a neutral fluid or a high-β plasma. From the pressure equation (2.20) we have

$$\gamma p\,\nabla\cdot\boldsymbol{v} \simeq -\boldsymbol{v}\cdot\nabla p. \quad (2.31)$$

On substituting for ∇p using the momentum equation (2.7), $\nabla p \simeq -\rho \boldsymbol{v} \cdot \nabla \boldsymbol{v}$, one finds

$$\nabla \cdot \boldsymbol{v} \simeq \frac{\rho}{\gamma p} \boldsymbol{v} \cdot \nabla\left(\tfrac{1}{2}v^2\right) \sim \mathsf{M}_\mathrm{s}^2 \frac{v}{L}, \tag{2.32}$$

where

$$\mathsf{M}_\mathrm{s} = v/c_\mathrm{s} \tag{2.33}$$

is the (acoustic or sonic) Mach number, $c_\mathrm{s} = \sqrt{\gamma p/\rho}$ is the speed of sound, and L is a typical gradient scale. In a plasma embedded in a strong magnetic field we have to distinguish between perpendicular and parallel motions. While for the latter (2.32) applies, for the former Faraday's law (2.15) gives the relation

$$B^2\, \nabla \cdot \boldsymbol{v} \simeq -\tfrac{1}{2} \boldsymbol{v} \cdot \nabla B^2.$$

At low β, the magnetic pressure dominates in the equation of motion. On multiplying (2.7) by $\boldsymbol{v}$, we have $\rho \boldsymbol{v} \cdot \nabla v^2 \simeq -\boldsymbol{v} \cdot \nabla B^2/(4\pi)$; hence

$$\nabla \cdot \boldsymbol{v} \simeq \frac{4\pi\rho}{B^2} \boldsymbol{v} \cdot \nabla\left(\tfrac{1}{2}v^2\right) \sim \mathsf{M}_\mathrm{A}^2 \frac{v}{L}. \tag{2.34}$$

Here

$$\mathsf{M}_\mathrm{A} = v/v_\mathrm{A} \tag{2.35}$$

is called the Alfvén Mach number, where $v_\mathrm{A} = B/\sqrt{4\pi\rho_0}$ is the Alfvén velocity, the phase velocity of the Alfvén wave, which will be introduced in Section 2.5.

We thus find that the fluid motion is incompressible if $\mathsf{M}_\mathrm{s} \ll 1$ or, for motions perpendicular to a strong magnetic field in a low-β plasma, $\mathsf{M}_\mathrm{A} \ll 1$. In the latter system it often happens, in particular in laboratory plasmas, that perpendicular motions are incompressible, whereas finite compressibility is important for parallel motions.

When gravity is sufficiently strong, the density is no longer homogeneous. Instead the pressure $p_0(\boldsymbol{x})$ and density $\rho_0(\boldsymbol{x})$ are in hydrostatic equilibrium,

$$\nabla p_0 = \boldsymbol{g}\rho_0, \tag{2.36}$$

which has the typical scale length $L_g = p_0/(g\rho_0)$. Here we concentrate on the case in which the equilibrium magnetic field is negligible; the opposite case of magnetostatic equilibrium is considered in Section 2.5. Thermal convection and turbulence in a stratified medium are conveniently described in terms of the *Boussinesq approximation*. By restricting our consideration to a layer of width L small compared with the equilibrium gradient scale, $L \ll L_g$, we are dealing again with a quasi-homogeneous system, for which the deviations of the dynamic variables from the average equilibrium values p_0, ρ_0, and $\boldsymbol{B}_0$ in this layer are small. One expands the momentum equation to first order in these deviations $\widetilde{\boldsymbol{v}} = \boldsymbol{v}$, $\widetilde{p}$, $\widetilde{\rho}$, $\widetilde{\boldsymbol{B}}$, which are of the order L/L_g. Note, however, that these quantities, though they are small, are still finite; hence the Boussinesq approximation does *not* imply linearization of the equation of motion. Since the velocity is small, the fluid motions can be assumed incompressible. In the Boussinesq approximation the equation of motion (2.7) or, more generally (2.26) including inertial forces in a rotating reference frame, reads

$$\partial_t \boldsymbol{v} + \boldsymbol{v}\cdot\nabla\boldsymbol{v} = -\frac{1}{\rho_0}\nabla\widetilde{p} + \frac{1}{c\rho_0}\widetilde{\boldsymbol{j}}\times(\boldsymbol{B}_0 + \widetilde{\boldsymbol{B}}) + \frac{1}{\rho_0}\boldsymbol{g}\widetilde{\rho} + 2\boldsymbol{v}\times\boldsymbol{\Omega} + \nu\nabla^2\boldsymbol{v}, \tag{2.37}$$

together with the incompressibility condition $\nabla\cdot\boldsymbol{v} = 0$. On taking the curl we obtain the Boussinesq equation in the vorticity equation

$$\partial_t\boldsymbol{\omega} + \boldsymbol{v}\cdot\nabla\boldsymbol{\omega} - \boldsymbol{\omega}\cdot\nabla\boldsymbol{v} = \frac{1}{c\rho_0}[\boldsymbol{B}_0\cdot\nabla\widetilde{\boldsymbol{j}} - \widetilde{\boldsymbol{j}}\cdot\nabla(\boldsymbol{B}_0 + \widetilde{\boldsymbol{B}})] + \frac{1}{\rho_0}\nabla\widetilde{\rho}\times\boldsymbol{g} + 2\boldsymbol{\Omega}\cdot\nabla\boldsymbol{v} + \nu\nabla^2\boldsymbol{\omega}. \tag{2.38}$$

The density fluctuation is usually replaced by the temperature fluctuation. On considering the fluctuations resulting from the gravitational force by perturbing the equilibrium (2.36), we find

$$\frac{\widetilde{p}}{p_0} \sim \frac{L}{L_g}\frac{\widetilde{\rho}}{\rho_0} \ll \frac{\widetilde{\rho}}{\rho_0}. \tag{2.39}$$

Hence the pressure fluctuation is negligible. From the ideal-gas law (2.17) we find

$$\frac{\widetilde{p}}{p_0} = \frac{\widetilde{\rho}}{\rho_0} + \frac{\widetilde{T}}{T_0}, \tag{2.40}$$

such that

$$\widetilde{\rho}/\rho_0 \simeq -\widetilde{T}/T_0. \tag{2.41}$$

For a general medium, in particular a dense liquid, this equation is replaced by

$$\widetilde{\rho}/\rho_0 = -\alpha_p\widetilde{T}, \tag{2.42}$$

with $\alpha_\rho = -(1/\rho)(\partial\rho/\partial T)_p$ the thermal-expansion coefficient. For incompressible motions the temperature fluctuation follows the equation

$$\partial_t\widetilde{T} + \boldsymbol{v}\cdot\nabla(\widetilde{T} + T_0) = \kappa\nabla^2\widetilde{T}. \tag{2.43}$$

The mean temperature gradient can be accounted for by appropriate boundary conditions of $\widetilde{T}$, such that (2.43) assumes the simple form of a scalar advection–diffusion equation:

$$\partial_t \widetilde{T} + \boldsymbol{v} \cdot \nabla \widetilde{T} = \kappa \, \nabla^2 \widetilde{T}. \tag{2.44}$$

The diffusion coefficient κ is called the thermal diffusivity.

If $B = g = \Omega = 0$, (2.37) reduces to the *Navier–Stokes equation* for incompressible hydrodynamics,

$$\partial_t \boldsymbol{v} + \boldsymbol{v} \cdot \nabla \boldsymbol{v} = -(1/\rho_0) \nabla p + \nu \, \nabla^2 \boldsymbol{v}, \quad \nabla \cdot \boldsymbol{v} = 0, \tag{2.45}$$

or, on taking the curl,

$$\partial_t \boldsymbol{\omega} + \boldsymbol{v} \cdot \nabla \boldsymbol{\omega} - \boldsymbol{\omega} \cdot \nabla \boldsymbol{v} = \nu \, \nabla^2 \boldsymbol{\omega}. \tag{2.46}$$

This equation can also be written in the form

$$\frac{d\boldsymbol{\omega}}{dt} = \boldsymbol{\omega} \cdot \nabla \boldsymbol{v} + \nu \, \nabla^2 \boldsymbol{\omega}. \tag{2.47}$$

Hence $\boldsymbol{\omega} \cdot \nabla \boldsymbol{v}$ is the source of vorticity, which is proportional to the velocity gradient along the vorticity and hence is called the vorticity-stretching term. In the ideal limit $\nu = 0$ the Navier–Stokes equation is called *Euler's equation*.

Turbulence is often discussed in the 2D approximation, which is particularly suitable in the case of strongly magnetized plasmas. We therefore write down also the 2D forms of the incompressible MHD equations. Take z as the ignorable coordinate. Because of the condition of vanishing divergence the velocity in the x, y-plane is usually described in terms of the streamfunction ϕ and the magnetic field in terms of A_z, the component of the vector potential in the direction of the ignorable coordinate, or the flux function $\psi = -A_z$,

$$\boldsymbol{v} = \boldsymbol{e}_z \times \nabla \phi, \quad \boldsymbol{B} = \nabla \times A_z \boldsymbol{e}_z = \boldsymbol{e}_z \times \nabla \psi, \tag{2.48}$$

and the vorticity and the current density have only z-components,

$$\omega = \omega_z = \nabla^2 \phi, \quad j = j_z = \frac{c}{4\pi} \nabla^2 \psi. \tag{2.49}$$

Thus in 2D the momentum equation (2.38) becomes (omitting gravity and rotation)

$$\partial_t \omega + \boldsymbol{v} \cdot \nabla \omega = [1/(c\rho_0)] \boldsymbol{B} \cdot \nabla j + \nu \, \nabla^2 \omega, \tag{2.50}$$

while the magnetic-field equation (2.15) reduces to a scalar advection–diffusion equation,

$$\partial_t \psi + \boldsymbol{v} \cdot \nabla \psi = \eta \, \nabla^2 \psi. \tag{2.51}$$

The velocity and magnetic-field components in the z-direction decouple from the dynamics in the x, y-plane. They need not be constant or zero and can be computed *a posteriori*.

2.3 Conservation laws

MHD theory exhibits a number of conservation equations. These play an important role, since they reflect the constraints imposed on the dynamics of the fluid, which are broken only by dissipative effects. There are two classes of invariants, those involving the fluid variables density, velocity, and internal energy, and those involving only the magnetic field.

2.3.1 *Fluid invariants*

Conservation of mass is described by the continuity equation (2.16). To discuss the momentum-conservation law we write the momentum equation (2.7) in conservation form,

$$\partial_t \rho \boldsymbol{v} = \nabla \cdot \mathcal{T} + \rho \boldsymbol{g}, \tag{2.52}$$

where $\mathcal{T} = \{T_{ij}\}$ is the total stress tensor, which can be written in the form

$$\begin{aligned} T_{ij} &= -\left(p + \frac{B^2}{8\pi}\right)\delta_{ij} - \left(\rho v_i v_j - \frac{B_i B_j}{4\pi}\right) + \mu\left(\partial_i v_j + \partial_j v_i - \tfrac{2}{3}\delta_{ij}\, \nabla \cdot \boldsymbol{v}\right) \\ &= -P\delta_{ij} + R_{ij} + \sigma_{ij}^{(\mu)}. \end{aligned} \tag{2.53}$$

The second term R_{ij} is the sum of the Reynolds stress tensor $-\rho v_i v_j$ and the Maxwell tensor $B_i B_j/(4\pi)$. A special variant of (2.52) in rotating systems is the conservation equation of the angular momentum $\rho r v_\phi$ in cylindrical coordinates (r, ϕ, z). Multiplying the azimuthal component of (2.52) by r and rearranging terms gives

$$\begin{aligned} \partial_t(\rho r v_\phi) = &-\nabla \cdot r\left[\rho v_\phi \boldsymbol{v} - \frac{B_\phi \boldsymbol{B}}{4\pi} + \left(p + \frac{B^2}{8\pi}\right)\boldsymbol{e}_\phi\right] \\ &- \nabla \cdot r\mu\left(r\, \nabla \frac{v_\phi}{r} + \tfrac{1}{3}(\nabla \cdot \boldsymbol{v})\boldsymbol{e}_\phi\right). \end{aligned} \tag{2.54}$$

On integrating (2.52) over a volume V one obtains the integral form of the momentum-conservation equation,

$$\frac{d}{dt}\int_V \rho \boldsymbol{v}\, dV = \oint_S \mathcal{T} \cdot d\boldsymbol{S} + \int_V \rho \boldsymbol{g}\, dV, \tag{2.55}$$

i.e., the change of the total momentum is due to the stresses acting on the surface of the volume and to volume, or body, forces. Since the magnetic field often extends beyond the boundary of the fluid, the momentum is not conserved in this case, even in the absence of body forces, because of the surface stresses.

Multiplying the momentum equation (2.7) by $\rho\boldsymbol{v}$ using $\boldsymbol{g} = -\nabla\phi_g$, and the magnetic-field equation (2.15) by $\boldsymbol{B}$, and adding the equations to (2.22) gives the equation for the total energy density:

$$\partial_t\left(\rho(\tfrac{1}{2}v^2 + u + \phi_g) + \frac{1}{8\pi}B^2\right) + \nabla\cdot\boldsymbol{F}^E = 0, \qquad (2.56)$$

with the energy flux

$$\boldsymbol{F}^E = (\tfrac{1}{2}v^2 + h + \phi_g)\rho\boldsymbol{v} - \boldsymbol{\sigma}^{(\mu)}\cdot\boldsymbol{v} + \boldsymbol{q} + \frac{c}{4\pi}\boldsymbol{E}\times\boldsymbol{B}, \qquad (2.57)$$

where h is the enthalpy,

$$h = u + \frac{p}{\rho} = \frac{\gamma}{\gamma-1}\frac{p}{\rho}. \qquad (2.58)$$

Integrating the energy equation (2.56) over V leads to the energy-conservation relation

$$\frac{dE}{dt} = -\oint_S d\boldsymbol{S}\cdot\boldsymbol{F}^E. \qquad (2.59)$$

Hence in an isolated system the total energy

$$E = \int dV\left(\rho(\tfrac{1}{2}\rho v^2 + u + \phi_g) + \frac{1}{8\pi}B^2\right) \qquad (2.60)$$

is conserved.[6] It should also be noted that the Coriolis force in (2.27), being perpendicular to $\boldsymbol{v}$, does not change the energy. In the incompressible limit the internal energy or the pressure is no longer an independent dynamic variable. (Formally speaking, it drops out of the expression for the total energy, since $\rho u = p/(\gamma - 1) \to 0$ in the limit $\gamma \to \infty$.) The energy law is now obtained from the momentum and magnetic-field equations only,

$$\frac{dE}{dt} = -\oint_S d\boldsymbol{S}\cdot\boldsymbol{F}^E - D^E. \qquad (2.61)$$

Here the energy is the sum of kinetic, potential, and magnetic energies,

$$E = \int dV\left(\frac{1}{2}\rho v^2 + \rho\phi_g + \frac{1}{8\pi}B^2\right), \qquad (2.62)$$

[6] Following convention in MHD turbulence theory, we denote the energy by E, since no confusion with the electric field arises, since the latter is eliminated by use of Ohm's law.

$\boldsymbol{F}^E$ contains no dissipative terms,

$$\boldsymbol{F}^E = (\tfrac{1}{2}\rho v^2 + p + \rho\phi_g)\boldsymbol{v} + \frac{1}{4\pi}\boldsymbol{B} \times (\boldsymbol{v} \times \boldsymbol{B}), \tag{2.63}$$

and D^E comprises the energy dissipation,

$$D^E = \int dV \left(\frac{1}{\sigma}\boldsymbol{j}^2 + \mu\boldsymbol{\omega}^2\right). \tag{2.64}$$

Note that, in the incompressible limit, the energy equation is no longer written in conservation form but contains an explicit dissipative energy sink.

A further important conserved quantity in MHD theory is the *cross-helicity*

$$H^C = \int \boldsymbol{v} \cdot \boldsymbol{B}\, dV. \tag{2.65}$$

Multiplying (2.7) by $\boldsymbol{B}/\rho$ and (2.15) by $\boldsymbol{v}$, adding and integrating over a volume V gives

$$\frac{dH^C}{dt} = -\oint_S \boldsymbol{F}^C \cdot d\boldsymbol{S} - D^C \tag{2.66}$$

where

$$\boldsymbol{F}^C = \boldsymbol{v} \times (\boldsymbol{v} \times \boldsymbol{B}) + \left(\phi_g + \frac{\gamma}{\gamma - 1}\frac{p}{\rho}\right)\boldsymbol{B} \tag{2.67}$$

is the cross-helicity flux and

$$D^C = (\nu + \eta)\int_V dV \sum_{i,j} \partial_i B_j\, \partial_i v_j \tag{2.68}$$

gives the dissipative effects on H^C. In (2.66) a polytropic pressure law, $p \sim \rho^\gamma$, and uniform kinematic viscosity $\nu = \mu/\rho$ were assumed. For incompressible motions the r.h.s. of (2.68) reduces to $(\nu + \eta)\int dV\, \boldsymbol{j} \cdot \boldsymbol{\omega}$.

2.3.2 Magnetic invariants

While E and H^C involve both fluid and magnetic contributions, the induction law (2.15) itself implies the invariance of two important purely magnetic quantities, magnetic flux and magnetic helicity. The magnetic flux is defined as the surface integral

$$\Phi = \int_S \boldsymbol{B} \cdot d\boldsymbol{S} \tag{2.69}$$

across a surface $S(t)$ bounded by a closed curve $l(t)$, which is moving with the plasma. On integrating (2.15) over the surface S and applying Stokes' theorem we find

$$\int_S \partial_t \boldsymbol{B} \cdot d\boldsymbol{S} = \oint_l (\boldsymbol{v} \times \boldsymbol{B}) \cdot d\boldsymbol{l} - \frac{c}{\sigma} \oint_l \boldsymbol{j} \cdot d\boldsymbol{l}.$$

The first term on the r.h.s. is just the flux through the change dS of the surface during the time interval dt,

$$\oint (\boldsymbol{v} \times \boldsymbol{B}) \cdot d\boldsymbol{l}\, dt = -\oint \boldsymbol{B} \cdot (\boldsymbol{v} \times d\boldsymbol{l})\, dt = \int_{dS} \boldsymbol{B} \cdot d\boldsymbol{S},$$

such that the total time derivative becomes

$$\frac{d\Phi}{dt} = \int_S \partial_t \boldsymbol{B} \cdot d\boldsymbol{S} + \oint_l \boldsymbol{B} \cdot (\boldsymbol{v} \times d\boldsymbol{l}) = -\frac{c}{\sigma} \oint_l \boldsymbol{j} \cdot d\boldsymbol{l}. \tag{2.70}$$

Hence the magnetic flux is conserved for infinite conductivity $\sigma \to \infty$.

Sweeping the boundary curve l along the field defines a flux tube. In this way flux conservation gives a well-defined physical meaning to the picture of field lines frozen into the plasma, which is often used to illustrate the tight coupling between fluid and magnetic field: Field lines are flux tubes of infinitesimal diameter. In the absence of resistivity they preserve their individuality; they cannot be 'broken' but can only be swirled around by the fluid. Nevertheless, it is well known that most macroscopic processes in plasmas do not conserve the field topology but involve (cutting and) reconnection of field lines. Understanding the efficiency of these reconnection processes even for very small resistivity (or some other nonideal effect in Ohm's law) is a major challenge in plasma theory (see, e.g., Biskamp, 2000a). It may be noted that a similar conservation law arises in hydrodynamics, where Euler's equation, the ideal limit of the Navier–Stokes equation (2.46), can be written in the same form as the ideal magnetic-field equation (2.15),

$$\partial_t \boldsymbol{\omega} - \nabla \times (\boldsymbol{v} \times \boldsymbol{\omega}) = 0. \tag{2.71}$$

Hence the vorticity flux $\int \boldsymbol{\omega} \cdot d\boldsymbol{S} = \oint \boldsymbol{v} \cdot d\boldsymbol{l}$, usually called circulation, is conserved.

Since, as has been discussed, the magnetic field is moved around with the plasma, its structure may become very complicated. A convenient measure of this complexity, essentially the twist and linkage of the field lines, is the *magnetic helicity* H^M,

$$H^M = \int_V \boldsymbol{A} \cdot \boldsymbol{B}\, dV, \tag{2.72}$$

where $\boldsymbol{A}$ is the vector potential, $\boldsymbol{B} = \nabla \times \boldsymbol{A}$. H^M is formally similar to the kinetic helicity $H^K = \int \boldsymbol{v} \cdot \boldsymbol{\omega}\, dV$, which is conserved in ideal hydrodynamics, but, in contrast to the latter case, the integrand of H^M has no direct physical meaning, since it depends on the gauge. We therefore have to make sure that H^M is gauge invariant. On performing a gauge transformation $\boldsymbol{A}' = \boldsymbol{A} + \nabla\chi$, we find

$$H^{M'} - H^M = \int_V \boldsymbol{B} \cdot \nabla\chi\, dV = \oint_S \chi \boldsymbol{B} \cdot d\boldsymbol{S}. \tag{2.73}$$

Hence for gauge invariance the normal component B_n at the boundary surface must vanish, since the gauge function χ is arbitrary. Only in special cases such as cases with periodic boundary conditions is finite B_n permitted. However, many magnetic configurations of interest in astrophysics are either open with field lines extending to infinity, for instance in the solar wind, or are bounded by surfaces crossed by field lines such as coronal loops, which are bounded by the photosphere. In both cases the helicity defined in (2.73) is not gauge invariant. An alternative expression has been proposed by Finn and Antonsen (1985) (see also Berger and Field, 1984)

$$H^M_{\text{alt}} = \int_V dV\, (\boldsymbol{A} + \boldsymbol{A}_0) \cdot (\boldsymbol{B} - \boldsymbol{B}_0), \tag{2.74}$$

where $\boldsymbol{B}_0 = \nabla \times \boldsymbol{A}_0$ is some reference field to be chosen suitably. In an open system $\boldsymbol{B}_0$ may be a static field with the same asymptotic properties as $\boldsymbol{B}$, whereas in a bounded system the normal components of $\boldsymbol{B}_0$ and $\boldsymbol{B}$ should be equal. It is easy to show that H^M_{alt} is indeed gauge invariant, even under separate gauge transformations of $\boldsymbol{A}$ and $\boldsymbol{A}_0$.

We now consider the conservation law for the magnetic helicity. Because of gauge invariance we may choose the gauge such that the scalar potential vanishes, $\boldsymbol{E} = -\partial_t \boldsymbol{A}/c$. On applying Faraday's law (2.13) we find

$$\begin{aligned}\int \partial_t(\boldsymbol{A} \cdot \boldsymbol{B})\, dV &= \int (\boldsymbol{B} \cdot \partial_t \boldsymbol{A} + \boldsymbol{A} \cdot \partial_t \boldsymbol{B})\, dV \\ &= -2c \int \boldsymbol{E} \cdot \boldsymbol{B}\, dV + c \oint (\boldsymbol{A} \times \boldsymbol{E}) \cdot d\boldsymbol{S}.\end{aligned} \tag{2.75}$$

On inserting Ohm's law (2.14) and using the boundary condition $B_n = 0$, the second term on the r.h.s. of (2.75) becomes $-\oint (\boldsymbol{A} \cdot \boldsymbol{B}) \boldsymbol{v} \cdot d\boldsymbol{S}$. Since

$$\oint (\boldsymbol{A} \cdot \boldsymbol{B}) \boldsymbol{v} \cdot d\boldsymbol{S}\, dt = \int_{dV} \boldsymbol{A} \cdot \boldsymbol{B}\, dV$$

is the change of H^M due to the change dV of the volume, we find

$$\frac{dH^M}{dt} = \int \partial_t(\boldsymbol{A}\cdot\boldsymbol{B})\,dV + \oint(\boldsymbol{A}\cdot\boldsymbol{B})\boldsymbol{v}\cdot d\boldsymbol{S} = -\frac{2c}{\sigma}\int dV\,\boldsymbol{j}\cdot\boldsymbol{B}. \qquad (2.76)$$

Hence the helicity is conserved in the ideal limit $\sigma \to \infty$. In a similar way one shows that H^M_{alt} is conserved.

As will be discussed in Chapters 4 and 5, the quadratic quantities E, H^C, and H^M play an important role in turbulence theory. Since the intrinsic turbulence properties are conveniently studied in the homogeneous-turbulence approximation with periodic boundary conditions, the surface terms in the conservation equations (2.59), (2.66), and (2.75) vanish and E, H^C, and H^M are conserved if dissipation is neglected. These quantities are therefore called *ideal invariants*. It is true that, in general, dissipative effects are not negligible at all in a turbulent fluid, even for nominally very small values of the dissipation coefficients, for instance the energy-dissipation rate remains finite in the limit $\mu \to 0$. Nonetheless, the behavior of the ideal system, for which the dissipation coefficients are set equal to zero, reveals important features of the cascade dynamics in real dissipative turbulence as discussed in Section 5.2.

Let us finally give the ideal invariants in 2D incompressible MHD, (2.50) and (2.51). One can readily verify that, similarly to the 3D case, both energy

$$E = \int d^2x\left(\frac{1}{2}\rho v^2 + \frac{B^2}{8\pi}\right) = \int d^2x\left(\frac{1}{2}\rho(\nabla\phi)^2 + \frac{1}{8\pi}(\nabla\psi)^2\right) \qquad (2.77)$$

and cross-helicity

$$H^C = \int d^2x\,\boldsymbol{v}\cdot\boldsymbol{B} = -\int d^2x\,\omega\psi \qquad (2.78)$$

are conserved. Instead of the magnetic helicity, which vanishes in 2D, it follows from the scalar form of the magnetic-field equation (2.51) that any moment of ψ is conserved, in particular the quadratic one,

$$A = \int d^2x\,\psi^2 \qquad (2.79)$$

which is called the mean-square magnetic potential.

2.4 Equilibrium configurations

Though this book deals with turbulence, which is the most dynamic state of a fluid, static configurations have a certain relevance in this context, either as initial states from which, in the case of instability, turbulence develops similarly to the instability of a stationary shear flow in hydrodynamics, or as final states to which turbulence decays. For $\boldsymbol{v} = 0$ the momentum equation (2.7) reduces to

$$\nabla p = \frac{1}{c}\boldsymbol{j} \times \boldsymbol{B} + \rho \boldsymbol{g}. \tag{2.80}$$

We first consider the case of a strong magnetic field, for which gravity is negligible, $\nabla p = \boldsymbol{j} \times \boldsymbol{B}/c$. Hence ∇p is perpendicular both to the magnetic field and to the current. Since in general $\boldsymbol{j}$ and $\boldsymbol{B}$ are not parallel, they span a surface $\psi(x, y, z) =$ constant, which is called the magnetic, or flux, surface, where p is constant, $p = p(\psi)$. For a true equilibrium state this surface must extend all along any field line, continuing either to infinity as in most astrophysical systems or running on indefinitely in a finite volume, the typical case in a toroidal laboratory plasma. As one might expect, such a magnetic configuration must be sufficiently simple to satisfy the equilibrium condition in a strict mathematical sense. The existence of equilibria can be shown directly in the case of a continuous symmetry, for which all quantities, including the metric tensor, depend only on two coordinates, ξ and η, and the equilibrium relation reduces to an elliptic differential equation for the flux function $\psi(\xi, \eta)$. It turns out that the conditions on such a coordinate system are rather restrictive, the most general symmetric case being that of helical symmetry (Edenstrasser, 1980), of which plane symmetry and axisymmetry are special cases.

The equilibrium equation can also be written in terms of the total pressure P, (2.11),

$$\nabla P = \frac{1}{4\pi}\boldsymbol{B} \cdot \nabla \boldsymbol{B}. \tag{2.81}$$

One-dimensional (1D) equilibria can easily be discussed analytically. For the configuration in plane geometry, with $p(x)$ and $\boldsymbol{B} = \{0, B_y(x), B_z(x)\}$, we have

$$P = p + B^2/(8\pi) = \text{constant}, \tag{2.82}$$

since $\boldsymbol{B} \cdot \nabla \boldsymbol{B} = 0$. Such a configuration is called a sheet pinch[7]. In cylindrical geometry a magnetically confined plasma column is called a screw pinch, because of the way the magnetic-field lines wind around the axis, which is described by $p(r)$ and $\boldsymbol{B} = \{0, B_\theta(r), B_z(r)\}$ satisfying the equation

$$\frac{d}{dr}\left(p + \frac{B^2}{8\pi}\right) = -\frac{B_\theta^2}{4\pi r}. \tag{2.83}$$

[7] The term 'pinch' derives from the tendency of a current-carrying plasma to contract due to the attractive force between parallel current elements, which is balanced by the pressure gradient.

The field-line twist, defined as the rotation angle 'iota' of the field line around the axis, $\iota = LB_\theta/(rB_z)$, where L is the length of the plasma column, is a measure of the stability of a screw-pinch equilibrium, $\iota > 2\pi$ meaning instability. (In tokamak plasma physics one uses the inverse quantity, $q = 2\pi/\iota$, called the safety factor.)

Since we have only one equation for three unknown functions, two of them can be chosen arbitrarily, for instance the profiles of pressure and axial current density. This freedom exists also in more general 2D or 3D equilibria (for a detailed treatment of MHD equilibrium theory see, for instance, Lifshitz, 1989). The explicit computation of higher-than-1D equilibrium configurations requires sophisticated numerical techniques (see, e.g., Bauer *et al.*, 1978).

Magnetic configurations of particular interest in astrophysics arise in cases when the pressure force is negligible, $\boldsymbol{j} \times \boldsymbol{B} = 0$. These are called force-free equilibria or, more generally, force-free magnetic fields, since the Lorentz force vanishes, currents flowing only along the magnetic field. A special class is constituted by fields $\boldsymbol{j} = \lambda\boldsymbol{B}$, $\lambda =$ constant, called linear force-free fields, which play an important role in the relaxation of MHD turbulence, as we shall see in Section 4.2.1.

In the absence of a strong magnetic field, a pressure gradient can be supported by gravity, $\nabla p = \rho\boldsymbol{g}$, (2.36). The density profile depends on the equation of state. For polytropic pressure $p = p_0(\rho/\rho_0)^\gamma$ and, assuming that $\boldsymbol{g} = -g\boldsymbol{e}_z$, we have

$$\rho(z) = \rho_0\left(1 - \frac{\gamma - 1}{\gamma}\frac{z}{L_g}\right)^{1/(\gamma-1)}, \tag{2.84}$$

with the scale length $L_g = p_0/(g\rho_0)$. Hence the height of a stratified atmosphere is in general finite, ρ vanishing for $z > L_g\gamma/(\gamma - 1)$. Only in the special case $\gamma = 1$, corresponding to isothermal conditions, is the height not limited,

$$\rho(z) = \rho_0 \mathrm{e}^{-z/L_g}, \tag{2.85}$$

which is the well-known decrease in barometric density.

2.5 Linear waves

Quiescent plasmas are rarely found in nature. Usually one observes more or less strong fluctuations about an average state, which result either from a local instability of the system or, more often, have been excited elsewhere and propagated to their present location. Since, in principle, such oscillations form the basic elements of turbulence, it is useful to obtain an overview of the linear modes or waves in a MHD configuration.

2.5.1 Waves in a homogeneous magnetized system

In contrast to incompressible hydrodynamics, in which all perturbations are nonpropagating swirls or eddies, MHD supports several types of waves, even in the incompressible limit. We first consider the simplest case of a homogeneous plasma described by p_0 and ρ_0 embedded in a homogeneous magnetic field $\boldsymbol{B}_0$. For sufficiently small perturbations, $\widetilde{p} \ll p_0$, $\widetilde{\boldsymbol{b}} \ll \boldsymbol{B}_0$, we can linearize the MHD equations (2.7), (2.15), and (2.20). Fourier transformation in space and time, $\widetilde{\boldsymbol{v}}(\boldsymbol{x}, t) = \boldsymbol{v}_1 \exp(i\boldsymbol{k}\cdot\boldsymbol{x} - i\varpi t)$ etc.,[8] reduces the differential operators to products, which leaves us with a set of homogeneous algebraic equations:

$$-i\varpi\rho_0\boldsymbol{v}_1 = -i\boldsymbol{k}p_1 + \frac{1}{4\pi}(i\boldsymbol{k}\times\boldsymbol{b}_1)\times\boldsymbol{B}_0 - \mu k^2\boldsymbol{v}_1, \tag{2.86}$$

$$-i\varpi\boldsymbol{B}_1 = i\boldsymbol{k}\times(\boldsymbol{v}_1\times\boldsymbol{B}_0) - \eta k^2\boldsymbol{B}_1, \tag{2.87}$$

$$-i\varpi p_1 = -i\gamma p_0\boldsymbol{k}\cdot\boldsymbol{v}_1. \tag{2.88}$$

(The density perturbation $-i\varpi\rho_1 = -\rho_0 i\boldsymbol{k}\cdot\boldsymbol{v}_1$ decouples from the other equations.) These equations can be reduced to a single one for $\boldsymbol{v}_1$ by substituting for $\boldsymbol{B}_1$ and p_1:

$$\varpi^2\rho_0\boldsymbol{v}_1 = \left(\frac{\boldsymbol{B}_0\times(\boldsymbol{k}\times\boldsymbol{B}_0)}{4\pi} + \gamma p_0\boldsymbol{k}\right)\boldsymbol{k}\cdot\boldsymbol{v}_1 - \frac{1}{4\pi}\boldsymbol{k}\cdot\boldsymbol{B}_0(\boldsymbol{k}\times\boldsymbol{v}_1)\times\boldsymbol{B}_0, \tag{2.89}$$

neglecting dissipation for simplicity. We see that there are both longitudinal, or compressible, waves $\propto \boldsymbol{k}\cdot\boldsymbol{v}_1$ and transverse, or shear, waves $\propto \boldsymbol{k}\times\boldsymbol{v}_1$. Choosing the coordinate system such that $\boldsymbol{B}_0 = B_0\boldsymbol{e}_z$, $\boldsymbol{k} = k_\perp\boldsymbol{e}_y + k_\parallel\boldsymbol{e}_z$, (2.89) can be written in the matrix form

$$\begin{pmatrix} \varpi^2 - k_\parallel^2 v_A^2 & 0 & 0 \\ 0 & \varpi^2 - k_\perp^2 c_s^2 - k^2 v_A^2 & -k_\perp k_\parallel c_s^2 \\ 0 & -k_\perp k_\parallel c_s^2 & \varpi^2 - k_\parallel^2 c_s^2 \end{pmatrix}\begin{pmatrix} v_x \\ v_y \\ v_z \end{pmatrix} = 0. \tag{2.90}$$

Here $v_A = B_0/\sqrt{4\pi\rho_0}$ is the Alfvén velocity, $c_s = \sqrt{\gamma p_0/\rho_0}$ the sound speed, and $k^2 = k_\perp^2 + k_\parallel^2$. The condition of vanishing determinant gives the eigenvalue equation, usually called the dispersion relation,

$$\left(\varpi^2 - k_\parallel^2 v_A^2\right)\left[\varpi^4 - \varpi^2 k^2\left(c_s^2 + v_A^2\right) + k^2 c_s^2 k_\parallel^2 v_A^2\right] = 0. \tag{2.91}$$

The system (2.90) supports three types of eigenmodes.

(i) The shear Alfvén wave, or *Alfvén wave* in short,

$$\varpi^2 = \varpi_A^2 = k_\parallel^2 v_A^2. \tag{2.92}$$

[8] To avoid confusion with the vorticity, we denote the frequency by the somewhat exotic symbol ϖ, which is sometimes used in the literature to denote the radius variable in cylindrical coordinates.

In this mode the plasma motion $\boldsymbol{v}_1 = \{v_x, 0, 0\}$ is incompressible, or transverse, $\boldsymbol{k} \cdot \boldsymbol{v}_1 = 0$. The velocity is also perpendicular to $\boldsymbol{B}_0$, and so is the magnetic perturbation $\boldsymbol{b}_1 = \pm\sqrt{4\pi\rho_0}\,\boldsymbol{v}_1$, which originates from the $\boldsymbol{B} \cdot \nabla \boldsymbol{B}$ term in the Lorentz force (2.9) and corresponds to an elastic bending of the field lines.

(ii) The compressional Alfvén wave, usually called the *(fast) magnetosonic wave*,

$$\varpi^2 = \varpi_{\text{fast}}^2 = \tfrac{1}{2}k^2\left[v_{\text{A}}^2 + c_{\text{s}}^2 + \sqrt{(v_{\text{A}}^2 + c_{\text{s}}^2)^2 - 4v_{\text{A}}^2 c_{\text{s}}^2 k_\parallel^2/k^2}\,\right]. \quad (2.93)$$

This mode is, in general, compressible. The phase velocity is in the range $v_{\text{A}}^2 + c_{\text{s}}^2 \geq (\varpi/k)^2 \geq v_{\text{A}}^2$, being fastest for propagation perpendicular to $\boldsymbol{B}_0$, where the mode is longitudinal, $\boldsymbol{v}_1 \parallel \boldsymbol{k}$. Here both B^2 and p are compressed by the plasma motion, such that the restoring force is large and hence the frequency is high. For parallel propagation,

$$\varpi^2 = \varpi_{\text{fast}}^2 = \tfrac{1}{2}k^2\left(v_{\text{A}}^2 + c_{\text{s}}^2 + |v_{\text{A}}^2 - c_{\text{s}}^2|\right),$$

the mode is purely transverse and merges with the Alfvén wave (2.92) in the small-β case $v_{\text{A}} > c_{\text{s}}$, whereas for high β, with $c_{\text{s}} > v_{\text{A}}$, the mode becomes purely longitudinal, merging with the nonmagnetic sound wave $\varpi^2 = k^2 c_{\text{s}}^2$.

(iii) The slow magnetosonic wave or, simply, *slow mode*,

$$\varpi^2 = \varpi_{\text{slow}}^2 = \tfrac{1}{2}k^2\left[v_{\text{A}}^2 + c_{\text{s}}^2 - \sqrt{(v_{\text{A}}^2 + c_{\text{s}}^2)^2 - 4v_{\text{A}}^2 c_{\text{s}}^2 k_\parallel^2/k^2}\,\right]. \quad (2.94)$$

Also this mode is, in general, compressible with phase velocity in the range $0 \leq (\varpi/k)^2 \leq c_{\text{s}}^2$. For perpendicular propagation, the left-hand equality, it is longitudinal but, since the changes of B^2 and p are exactly opposite in phase, $\delta B^2/(8\pi) = -\delta p$, the restoring force vanishes, corresponding to a quasi-static equilibrium change. For parallel propagation the phase velocity reaches the upper limit,

$$\varpi_{\text{slow}}^2 = \tfrac{1}{2}k^2\left(v_{\text{A}}^2 + c_{\text{s}}^2 - |v_{\text{A}}^2 - c_{\text{s}}^2|\right),$$

the mode becomes the nonmagnetic sound wave for $v_{\text{A}} > c_{\text{s}}$ and the shear Alfvén wave for $c_{\text{s}} > v_{\text{A}}$. Clearly the phase velocity $v_{\text{ph}} = \varpi/k$ of these modes satisfies the relation

$$v_{\text{fast}} \geq v_{\text{A}} \geq v_{\text{slow}}.$$

The MHD modes (i)–(iii) are nondispersive, the phase velocity equalling the group velocity $v_{\text{g}} = \partial\varpi/\partial k$, since MHD theory contains no intrinsic spatial scale.

2.5.2 Waves in a stratified system

Let us now consider perturbations of a stratified equilibrium $\rho_0(z)$ under the influence of gravity. Since in the Boussinesq approximation we neglect the spatial variations of the equilibrium quantities, we can again Fourier transform the perturbations. Neglecting magnetic fields and viscosity, the Boussinesq equations (2.38) and (2.43), linearized about the equilibrium (2.36), read

$$-i\varpi\boldsymbol{\omega}_1 = \frac{g}{\rho_0}\boldsymbol{e}_z \times i\boldsymbol{k}\rho_1 + 2i\boldsymbol{k}\cdot\boldsymbol{\Omega}\boldsymbol{v}_1, \tag{2.95}$$

$$-i\varpi\rho_1 = -\rho_0' v_{1z}, \tag{2.96}$$

where we choose again $\boldsymbol{g} = -g\boldsymbol{e}_z$, following the sign convention. On applying the curl and using the relation $i\boldsymbol{k} \times \boldsymbol{\omega}_1 = k^2\boldsymbol{v}_1$, (2.95) becomes

$$i\varpi k^2\boldsymbol{v}_1 = -2i\boldsymbol{k}\cdot\boldsymbol{\Omega}\boldsymbol{\omega}_1 - \frac{1}{\rho_0}(gk_z\boldsymbol{k} - g\boldsymbol{e}_z k^2)\rho_1. \tag{2.97}$$

On substituting for ρ_1 and taking the z-components of (2.95) and (2.97) we obtain two homogeneous equations for v_{1z} and ω_{1z}, which yield the dispersion relation

$$\varpi^2k^2 + k_\perp^2 g\rho_0'/\rho_0 - 4(\boldsymbol{k}\cdot\boldsymbol{\Omega})^2 = 0. \tag{2.98}$$

Here the subscript $\perp$ means perpendicular to $\boldsymbol{g}$. The expression $-g\rho_0'/\rho_0$ is the square of a frequency called the *Brunt–Vaisala* frequency N,

$$N^2 = -g\rho_0'/\rho_0. \tag{2.99}$$

Thus we find the frequency of perturbations in a stratified system:

$$\varpi^2 = \frac{N^2k_\perp^2 + 4(\boldsymbol{k}\cdot\boldsymbol{\Omega})^2}{k^2}. \tag{2.100}$$

In the case of a nonrotating plasma, $\Omega = 0$, (2.100) describes *(internal) gravity waves* with real frequency

$$\varpi = \pm Nk_\perp/k, \quad \text{if } N^2 > 0, \text{ i.e., } g\rho_0' < 0, \tag{2.101}$$

which is called *stable stratification* (light fluid on top). We see that the frequency is much smaller than the sound frequency, $\varpi \leq N \ll NkL_g = c_s k$; hence gravity waves are incompressible. In the opposite case of *unstable stratification*, $g\rho_0' > 0$ (heavy fluid on top), the perturbation does not propagate but grows exponentially, which is called the *Rayleigh–Taylor* instability; this is considered in more detail in Section 3.2.2. In the other limit in (2.100), $N = 0$, one finds the *inertal waves*,

$$\varpi = \pm 2\boldsymbol{k} \cdot \boldsymbol{\Omega}/k. \tag{2.102}$$

We see from (2.100) that sufficiently fast rotation stabilizes the Rayleigh–Taylor instability.

2.6 Elsässer fields and Alfvén time normalization

Since interest in MHD turbulence theory is focussed on incompressible plasma motions, the Alfvén wave (2.92) is the most important linear mode. As we saw in the previous section, for an Alfvén wave velocity and magnetic-field perturbations are parallel, $\boldsymbol{v}_1 = \pm \boldsymbol{b}_1/\sqrt{4\pi\rho_0}$. The fundamental effect of Alfvén waves in MHD becomes evident on writing the (nonlinear) MHD equations in terms of the *Elsässer fields* (Elsässer, 1950):

$$\boldsymbol{z}^{\pm} = \boldsymbol{v} \pm \frac{1}{\sqrt{4\pi\rho_0}}\boldsymbol{b}. \tag{2.103}$$

To simplify the notation, we normalize the MHD equations with respect to the Alfvén time $\tau_A = L/v_A$:

$$t/\tau_A := t, \quad x/L := x, \quad b/B_0 := b, \quad p/(\rho_0 v_A^2) := p, \tag{2.104}$$

where L is a convenient scale length, B_0 a typical magnetic field, and $v_A = B_0/\sqrt{4\pi\rho_0}$ the corresponding Alfvén speed. In these units the magnetic diffusivity, or resistivity as we call it, is the inverse of the *Lundquist number* $S = v_A L/\eta$. The Elsässer fields (2.103) now are simply

$$\boldsymbol{z}^{\pm} = \boldsymbol{v} \pm \boldsymbol{b}. \tag{2.105}$$

By adding equations (2.7) and (2.15), or subtracting (2.15) from (2.7), assuming that we have incompressibility, we obtain

$$\partial_t \boldsymbol{z}^{\pm} + \boldsymbol{z}^{\mp} \cdot \nabla \boldsymbol{z}^{\pm} = -\nabla P + \tfrac{1}{2}(\nu + \eta)\nabla^2 \boldsymbol{z}^{\pm} + \tfrac{1}{2}(\nu - \eta)\nabla^2 \boldsymbol{z}^{\mp}, \tag{2.106}$$

$$\nabla \cdot \boldsymbol{z}^{\pm} = 0,$$

where P is again the total pressure, $P = p + \frac{1}{2}B^2$ in the units adopted. On linearizing these equations about a uniform magnetic field $\boldsymbol{B}_0$ and neglecting dissipation, one has

$$\partial_t \boldsymbol{z}^{\pm} \mp \boldsymbol{B}_0 \cdot \nabla \boldsymbol{z}^{\pm} = 0. \tag{2.107}$$

Equation (2.107) shows that z^- describes Alfvén waves propagating in the direction of $\boldsymbol{B}_0$, $z^-(\boldsymbol{x} - \boldsymbol{B}_0 t)$, and z^+ describes Alfvén waves propagating opposite to $\boldsymbol{B}_0$, $z^+(\boldsymbol{x} + \boldsymbol{B}_0 t)$. The interesting property of the Elsässer fields is that there is no self-coupling in the nonlinear term in (2.106), but only cross-coupling of z^+ and z^-. This is the basis of the Alfvén effect, which describes a fundamental interaction process, as will be discussed in detail in Section 5.3.

Since the Elsässer fields are, properly speaking, the more fundamental variables in incompressible MHD theory, we also express the ideal invariants in terms of these fields. The energy E, (2.62), becomes

$$E = \tfrac{1}{4} \int dV \left[(z^+)^2 + (z^-)^2 \right], \tag{2.108}$$

the cross-helicity H^C, (2.65), becomes

$$H^C = \tfrac{1}{4} \int dV \left[(z^+)^2 - (z^-)^2 \right], \tag{2.109}$$

while the magnetic helicity H^M, (2.72), is unrelated to the Alfvén-wave property and hence not conveniently expressed in the Elsässer-field formalism. There is, however, a further quantity that, even though it is not invariant, plays an important role in MHD turbulence, namely the difference between the kinetic energy and the magnetic energy, called the residual energy, which becomes simply

$$E^R = \tfrac{1}{2} \int dV \, (\boldsymbol{v}^2 - \boldsymbol{b}^2) = \tfrac{1}{2} \int dV \, z^+ \cdot z^-. \tag{2.110}$$

3

Transition to turbulence

One important aspect of turbulence theory is the need to understand how obviously random motions are generated from a smooth flow. There are essentially three approaches to this problem: the dynamic systems approach; the development of singular solutions of the ideal fluid equations, in particular the question of finite-time singularities; and the excitation of instabilities and their effects. The dynamic systems approach, i.e., the transition to a chaotic temporal behavior in some low-order nonlinear dynamic model such as the Lorentz model of thermal convection, had once been considered a very promising way to describe also the transition to turbulence in a fluid. However, these expectations have largely been frustrated, mainly because the low-order approximations of the fluid equations ignore the most important aspect of turbulence, namely the excitation and interactions of a broad range of different *spatial* scales. We will therefore not discuss dynamic systems theory in this treatise.

The problem of finite-time singularities has evoked considerable discussion. This is primarily a mathematical problem concerning the nature of the solution of the ideal fluid equations, whose relevance for the generation of turbulence in dissipative systems might be debatable. However, similarly to the theory of absolute equilibrium states of the ideal system considered in Section 5.2, which provides valuable information about the cascade dynamics in dissipative turbulence, the way in which the ideal solution becomes singular gives some indication of the spatial structure of eddies encountered in the dissipative system. We therefore discuss the problem of singularity in some detail, especially since remarkable progress has been made in recent years (Section 3.1).

The process directly connected with the generation of small-scale turbulence is the generation of instabilities, which arise during the evolution of the system from a smooth initial state. Here the first candidate is the Kelvin–Helmholtz instability excited by the sufficiently steep velocity gradients. In a stratified system, on the other hand, the Rayleigh–Taylor instablity driven by buoyancy is the most important process, which nonlinearly may suffer a secondary, Kelvin–Helmholtz-type instability. Both instabilities are strongly affected, and possibly even completely stabilized, by the presence of a magnetic field, due to their being frozen-in, which gives field lines a certain stiffness. Finite resistivity loosens this coupling of the field to the fluid and allows the growth of perturbations, notably the tearing mode, which are forbidden in an ideal MHD system. We give an overview of the linear properties of these instabilities and discuss briefly their nonlinear evolution (Section 3.2).

3.1 Singularities of the ideal equations

Turbulence is characterized by a broad range of spatial scales, extending down to very small eddies if the Reynolds number is large. It can therefore be expected that, in the limit of infinite Reynolds number, or vanishing dissipation, the fluid equations develop singularities and that their character is, somehow, related to the smallest structures in real, i.e., dissipative, turbulence. Two aspects are important, the time scale of the formation of a singularity and its spatial structure, these two aspects being tightly connected, as we shall see. Concerning the time scale, essentially two different behaviors are encountered, either some dynamic variable, for instance the vorticity, blows up, i.e., becomes infinite in a finite time, which is called a *finite-time singularity* (FTS), or the growth is only exponential, such that a singularity is reached only after an infinite period, i.e., the solution remains regular, formally speaking.

The simplest fluid model is the ideal Burgers equation describing a 1D compressible flow $u(x,t)$,

$$\partial_t u + u\,\partial_x u = 0. \tag{3.1}$$

Here the existence of a FTS can easily be shown. By taking the x-derivative of (3.1) we obtain

$$\partial_t \omega + u\,\partial_x \omega = -\omega^2, \qquad \omega = \partial_x u. \tag{3.2}$$

Since the l.h.s. is the Lagrangian time derivative $\dot{\omega}$, the equation can be integrated along the orbit of a fluid element, yielding the solution

$$\omega(t) = \omega[x(t)] = \frac{1}{t - t_0}. \tag{3.3}$$

Hence the solution becomes singular at time $t_0 = -\omega_0^{-1}$ for the fluid element with the most negative initial value of the velocity gradient.

3.1.1 FTS in the Euler equations

Incompressible flows, the most interesting case in turbulence theory, require at least two spatial dimensions. Here the simplest case is the 2D Euler equation

$$\partial_t \boldsymbol{v} + \boldsymbol{v} \cdot \nabla \boldsymbol{v} = -\nabla p, \quad \nabla \cdot \boldsymbol{v} = 0, \tag{3.4}$$

or, written in terms of the vorticity $\omega = (\nabla \times \boldsymbol{v})_z$,

$$\partial_t \omega + \boldsymbol{v} \cdot \nabla \omega = 0, \quad \boldsymbol{v} = \boldsymbol{e}_z \times \nabla \phi, \quad \omega = \nabla^2 \phi. \tag{3.5}$$

This equation means that the vorticity remains constant along the orbit of a fluid element; hence there is no FTS of ω. This is, however, no longer true for more general, axisymmetric flows v_r, v_z, v_θ, which are independent of the azimuthal coordinate θ.[1] In this case the Euler equations (2.71) can most suitably be written in terms of the angular momentum $L_\theta = r v_\theta$ and the azimuthal component of the vorticity $\omega_\theta = \partial_r v_z - \partial_z v_r$,

$$(\partial_t + \boldsymbol{v} \cdot \nabla) L_\theta = 0, \tag{3.6}$$

$$(\partial_t + \boldsymbol{v} \cdot \nabla) \frac{\omega_\theta}{r} = \frac{1}{r^4} \, \partial_z L_\theta^2, \tag{3.7}$$

where $\boldsymbol{v} = \{v_r, v_z\} = \{-\partial_z \phi, r^{-1} \, \partial_r (r\phi)\}$, and the streamfunction ϕ obeys Poisson's equation

$$\partial_r \left(\frac{1}{r} \, \partial_r r \, \phi \right) + \partial_z^2 \phi = \omega_\theta .$$

While (3.6) gives the conservation of angular momentum of a fluid element, (3.7) shows that the vorticity is no longer conserved, but rather is driven by the gradient of the angular momentum. This term has a certain similarity to the r.h.s. of Burgers' equation (3.2), and hence, for finite initial v_θ, called a flow with swirl, the existence of a FTS is conceivable. Numerical studies of the coupled equations (3.6) and (3.7) have been performed by Grauer and Sideris (1991) and Pumir and Siggia (1992a, 1992b). It was noticed by Pumir and Siggia that the equations can be made equivalent to the 2D Boussinesq equations of thermal convection with the centrifugal force corresponding to an effective gravity by introducing a local coordinate frame about some finite radius r_0 and restricting consideration to its vicinity, L_θ^2 playing the role of the temperature, which simplifies the numerical analysis considerably. Nonetheless, no definite

[1] At this point we denote the azimuthal coordinate by θ instead of the more conventional ϕ in order to avoid confusion with the streamfunction, (3.5).

answer about the character of the singularity has yet been reached. Because of the velocity shear, the temperature following (3.6) leads naturally to the generation of temperature-gradient sheets (cliffs in the temperature profile), which, in turn, serving as a source term in the vorticity equation, generate vorticity sheets. In the initial phase of sheet formation the vorticity grows rapidly, seemingly indicating a FTS, but subsequently growth slows down, becoming only exponential, a behavior expected for a sheet-like singularity, as we shall discuss in Section 3.1.2. Eventually the sheet becomes unstable, a mixture of Kelvin–Helmholtz and Rayleigh–Taylor instability (for a discussion of these instabilities, see Section 3.2.1), and thus disintegrates into smaller pieces, where ω starts to grow again in an explosive-like manner. Since at this point structures have already become very small, numerical computations performed to date could not treat them reliably. It is, however, rather likely that also these structures are flattened into sheets, which would again slow down the growth to exponential. A plausible asymptotic scenario is a sequence of sheet formation and subsequent instability of these sheets, where by successive generations of smaller and smaller sheets lead to increasingly faster exponential growth of the vorticity and hence effectively a FTS, but it is obviously extremely difficult to demonstrate this behavior numerically.

In the general 3D case the formation of singular structures in the solution of the Euler equation

$$\partial_t \boldsymbol{\omega} + \boldsymbol{v} \cdot \nabla \boldsymbol{\omega} = \boldsymbol{\omega} \cdot \nabla \boldsymbol{v}, \qquad \nabla \cdot \boldsymbol{v} = 0, \tag{3.8}$$

is driven by the vortex-stretching term $\boldsymbol{\omega} \cdot \nabla \boldsymbol{v}$, which is proportional to the component of the strain tensor (2.6) along the vorticity and hence leads to vortex-line stretching. Since, owing to incompressibility, this process entails a shrinking of the diameter of a vortex tube and, since the vorticity flux or circulation $\oint \boldsymbol{\omega} \cdot d\boldsymbol{S}$ is conserved, as discussed in Section 2.4, the vorticity increases. How fast this growth proceeds depends on the geometry of the flow in the neighborhood of the singular structure. It is well possible that the flow resists singularity formation by arranging the directions of $\boldsymbol{\omega}$ and $\nabla \boldsymbol{v}$ such that this term remains smaller than $O(\omega^2)$, a process called *nonlinearity depletion*. Because of the dependence on the flow geometry, it is even more difficult than in the 2D axisymmetric case to obtain analytical estimates of the temporal behavior of the solution, which makes a numerical treatment unavoidable. There are essentially two numerical approaches, either to follow the ideal system from some initial state as far as the available spatial resolution permits, and extrapolate to the asymptotic behavior, which is the usual approach also in the studies of axisymmetric flows discussed before, or, more indirectly, to consider the viscous system, reducing the viscosity as far as possible. If the maximum energy-dissipation rate $\epsilon = \mu \int \omega^2 \, d^3x$ becomes independent of the viscosity, this would indicate a FTS of ω in the ideal equations.

There are numerous articles on this topic in the literature, of which we here discuss only some studies of the ideal Euler equations (3.8). Brachet *et al.* (1992) chose for the initial condition the Taylor–Green vortex (Taylor and Green, 1937)

$$(v_x, v_y, v_z) = (\sin x \cos y \cos z, -\cos x \sin y \cos z, 0), \tag{3.9}$$

an incompressible 3D flow involving only the lowest Fourier components in the system. The Euler equations are integrated numerically in a periodic box using a Fourier representation in space, the method of choice in simulations of homogeneous turbulence. The symmetry of the initial state (3.9) reduces the number of Fourier modes considerably, so that Brachet *et al.* could reach a maximum resolution corresponding to 864^3 colocation points. The main result of this study was the formation of small-scale sheet-like structures of the vorticity, whose thickness shrinks exponentially in time, so that there does not seem to be a FTS. Also on choosing general random initial conditions (with only 256^3 colocation points) a similar behavior is found; hence the authors concluded that vorticity-sheet formation is a process generic to ideal incompressible hydrodynamics. They conjectured, however, that, at spatial resolutions higher than those achievable in their studies, the sheets may become unstable, breaking up into filaments, which may then blow up in a finite time.

The main drawback of the Fourier representation for studying isolated singularities is the equidistant distribution of the colocation points, since increasing the spatial resolution is very expensive in 3D (an increase in resolution by a factor of 2 means roughly a factor of 16 in computer time, i.e., a factor of 2^3 in space and a further factor of 2 because of the smaller timestep). Instead one should apply a grid refinement at the position of the singularity. Kerr (1993), while staying within the general framework of spectral methods, adopted a Chebyshev representation in one direction, which corresponds to a condensation of mesh points at the boundary, chosen as the plane of symmetry, where the maximum value of the vorticity is located. Kerr studied the interaction of two antiparallel vortex tubes and founds evidence of a FTS of the vorticity.

The most suitable and versatile numerical approach, however, is an adaptive local mesh refinement. Starting from a technique introduced by Berger and Collela (1989), Grauer *et al.* (1998) developed a sophisticated numerical method introducing mesh refinements at those grid points where the numerical resolution becomes insufficient, which results in a hierarchy of increasingly

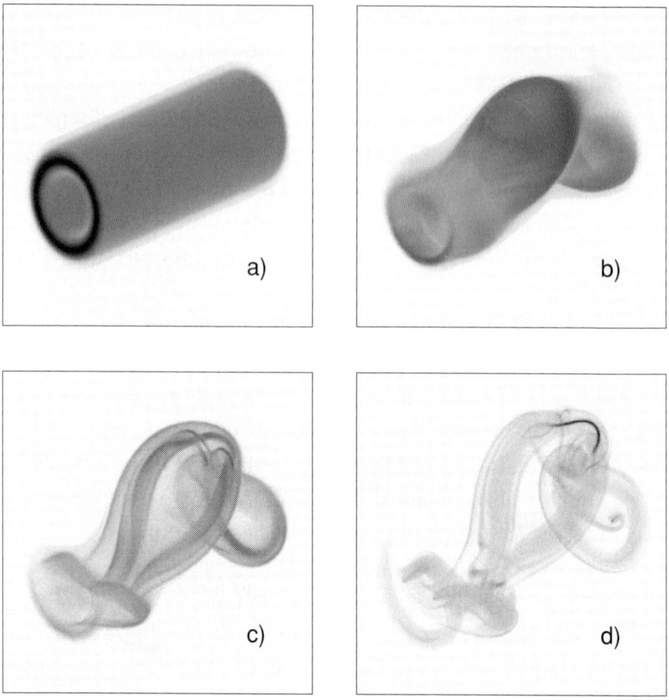

Figure 3.1: Vorticity $|\omega|$ in a 3D numerical simulation run by Grauer *et al.* (1998). (a) Initial state, and intermediate states at (b) $t = 0.99$, (c) $t = 1.18$, and (d) $t = 1.32$ close to the time of the FTS at $t \simeq 1.355$.

finer local grids capable of following the moving structures. The authors start from the cylindrical jet shown in Fig. 3.1(a). As the kink instability of the jet grows from the slight transverse perturbation imposed on the initial state, a sheet-like vorticity structure forms, Fig. 3.1(b), which subsequently breaks up into several filaments, Fig. 3.1(c). One in particular of these filaments, Fig. 3.1(d), continues to be stretched such that the vorticity seems to become singular at $t \simeq 1.355$. Figure 3.2 shows the temporal evolution of the maximum norm $\|\omega\|_{L^\infty}$. The computation starts from a uniform grid of 64^3 points, which is then locally refined in several steps up to a maximum resolution corresponding to a uniform grid of 2048^3 points. A similar behavior has been found by Pelz (1997) in a semi-analytical model, where in vortex filaments in a certain arrangment boost each other, leading, in fact, to a FTS.

Though it is still premature to draw a definite conclusion about the generic behavior, it appears from these studies that, at least for not-too-symmetric initial conditions, the 3D Euler equations do, in general, lead to a FTS in a filametary vorticity structure. The main difference, compared with the axisymmetric

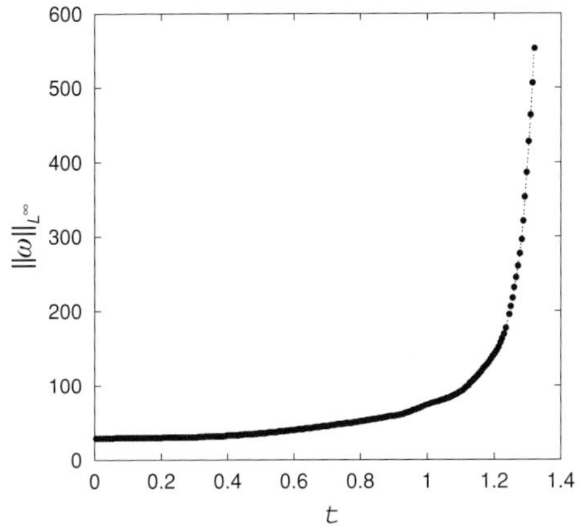

Figure 3.2: Evolution of the maximum norm of the vorticity $\|\omega\|_{L^\infty}$ and the fit $14.9 + 19.7/(1.355 - t)$ for the simulation run displayed in Fig. 3.1.

(or, equivalently, 2D Boussinesq) case, is that, in 3D, vortex filaments are mainly stretched instead of being flattened into sheets, and filaments are inherently more stable than sheets. These results from the solutions of the Euler equations are also consistent with vorticity filaments being the dominant small-scale structures of dissipative hydrodynamic turbulence (see Fig. 7.7). As a corollary, the energy-dissipation rate becomes independent of the viscosity at high Reynolds number.

3.1.2 Formation of current sheets in ideal MHD

Let us now discuss how the presence of magnetic fields modifies the results of ideal hydrodynamics. In terms of the Elsässer fields $\boldsymbol{z}^{\pm} = \boldsymbol{v} \pm \boldsymbol{b}$ (2.105), the ideal MHD equations (2.106) assume a form very similar to the Euler equations (3.4),

$$\partial_t \boldsymbol{z}^{\pm} + \boldsymbol{z}^{\mp} \cdot \nabla \boldsymbol{z}^{\pm} = -\nabla P, \quad \nabla \cdot \boldsymbol{z}^{\pm} = 0. \tag{3.10}$$

On taking the curl of (3.10) we find, after a short algebraic manipulation that is left to the reader,

$$\partial_t \boldsymbol{\omega}^{\pm} + \boldsymbol{z}^{\mp} \cdot \nabla \boldsymbol{\omega}^{\pm} = \boldsymbol{\omega}^{\pm} \cdot \nabla \boldsymbol{z}^{\mp} + \sum_{i=1}^{3} \nabla z_i^{\pm} \times \nabla z_i^{\mp}, \tag{3.11}$$

where

$$\boldsymbol{\omega}^{\pm} = \nabla \times \boldsymbol{z}^{\pm} = \boldsymbol{\omega} \pm \boldsymbol{j}. \tag{3.12}$$

Here the difference from the Euler equations (3.8) is the second term on the r.h.s. of (3.11), which becomes particularly clear in the 2D case, $\partial_z = 0$, in which only the z-components of the generalized vorticities $\omega^\pm = \omega \pm j$ appear,

$$\partial_t \omega^\pm + \boldsymbol{z}^\mp \cdot \nabla \omega^\pm = \sum_{i=1}^{2} \nabla z_i^\pm \times \nabla z_i^\mp, \tag{3.13}$$

with

$$\boldsymbol{z}^\pm = \boldsymbol{e}_z \times \nabla \phi^\pm, \quad \nabla^2 \phi^\pm = \omega^\pm, \quad \phi^\pm = \phi \pm \psi. \tag{3.14}$$

Let us first consider this case. Numerical simulations clearly show the formation of current (and vorticity) sheets at an X-type neutral point of the magnetic field. While for a quantitative description of this process a numerical treatment appears to be indispensible, the tendency toward sheet-like structures can be understood by considering a simple qualitative model. It is suitable to use the 2D MHD equations in the forms (2.50) and (2.51),

$$\partial_t \psi + \boldsymbol{v} \cdot \nabla \psi = 0, \tag{3.15}$$

$$\partial_t \omega + \boldsymbol{v} \cdot \nabla \omega = \boldsymbol{b} \cdot \nabla j. \tag{3.16}$$

Restricting consideration to the immediate vicinity of the X-point, the self-similar solution is

$$\psi = \frac{1}{2}\left(\frac{x^2}{\xi^2(t)} - \frac{y^2}{\eta^2(t)}\right), \tag{3.17}$$

$$\phi = \Lambda(t) x y, \tag{3.18}$$

corresponding to uniform current density $j = j(t)$ and vanishing vorticity $\omega = 0$, the local behaviors of a symmetric and an antisymmetric function, respectively. From (3.15) it follows that

$$\dot{\xi} = -\Lambda \xi, \quad \dot{\eta} = \Lambda \eta. \tag{3.19}$$

Since (3.16) is satisfied identically, the solution is not completely determined, $\Lambda(t)$ being a free function. We shall, however, see that $\Lambda =$ constant is the appropriate choice; hence

$$\xi = \xi_0 \mathrm{e}^{-\Lambda t}, \quad \eta = \eta_0 \mathrm{e}^{\Lambda t}, \tag{3.20}$$

an exponential flattening of the configuration and growth of the current density $j \propto \mathrm{e}^{\Lambda t}$. Returning to the Elsässer fields, the symmetry of (3.13) indicates that z^+ and z^- should develop in similar ways. In fact, both fields develop sheet-like structures in $\omega^\pm$ at an X-point. These sheets are slightly tilted with respect to each other, becoming more and more aligned as the sheet thickness

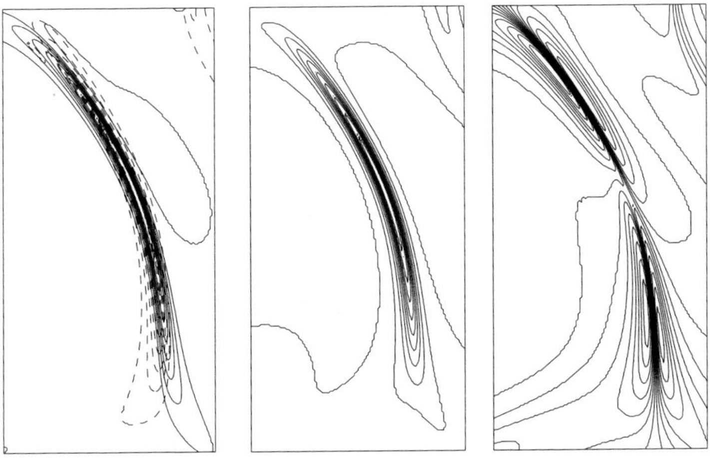

Figure 3.3: The structure of a current sheet. Contour plots of (a) ω^+ (continuous) and ω^- (dashed); (b) j; and (c) ω, of the current sheet in the lower part of the right-hand quadrant in Fig. 3.4.

shrinks. Vorticity and current density being the sum and the difference of $\omega^\pm$, respectively, break the symmetry of the Elsässer fields; j has the structure of an stretched monopole, ω that of quadrupole, as illustrated in Fig. 3.3. Following convention, we call these structures current sheets, though they are also the location of intense vorticity. Since the r.h.s. of (3.11), the source of the generalized vorticity $\omega^\pm$, is the cross product of the gradients of the z-fields, this term becomes effectively linear, as only one factor contains the large cross-sheet gradient, which is hence a case of nonlinearity depletion. Thus, in the vicinity of the sheets, (3.13) is essentially linear, $\dot{\omega}^\pm \sim \omega^\pm$, from which follows the exponential growth of j and ω.

The first numerical study of the problem solving (3.13) was performed by Sulem *et al.* (1985), who found clear evidence of exponential shrinking of the sheet width, implying exponential growth of the current density. These results were subsequently confirmed by Friedel *et al.* (1997). Here we present some numerical results starting from a slightly different initial state from that chosen by Friedel *et al.*,

$$\phi(x, y) = \cos(x + 1.4) + \cos(y + 2),$$
$$\psi(x, y) = \cos(2x + 2.3) + \cos(y + 6.2), \tag{3.21}$$

a modified Orszag–Tang vortex (Orszag and Tang, 1979) called A_1 by Biskamp and Welter (1989), which is not symmetric and hence is more generic of the

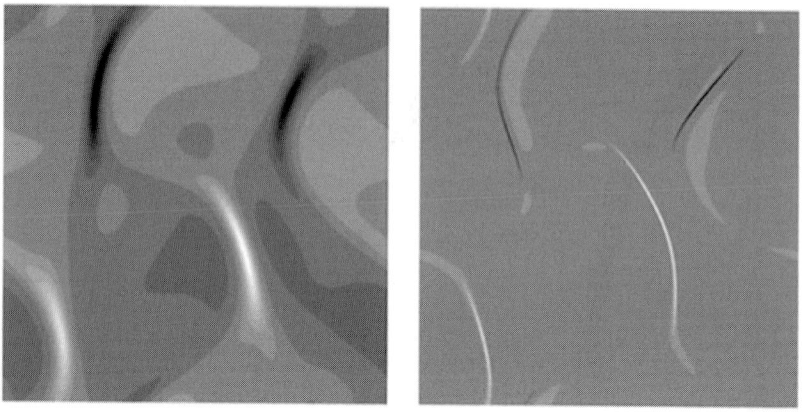

Figure 3.4: Thinning of current sheets in an ideal 2D MHD system, starting from the state (3.21). The current density is shown at two times, $t = 1.0$ (left) and $t = 1.2$ (right).

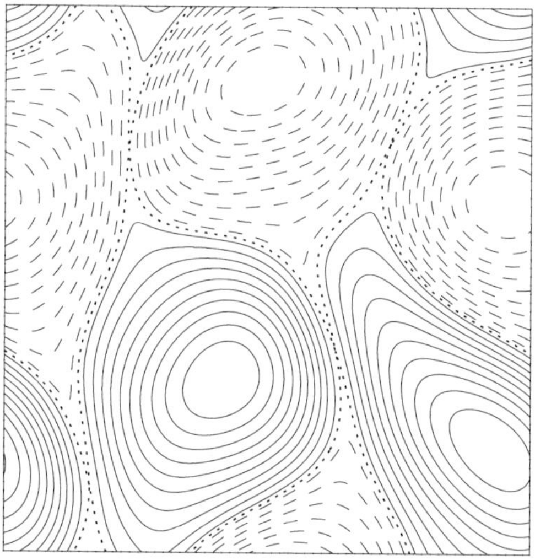

Figure 3.5: The magnetic-flux distribution ψ of the same state as the current distribution at $t = 1.0$ in Fig. 3.4; negative values are dashed. There are four (shallow) X-points between magnetic eddies ψ of equal sign, where current sheets form.

general behavior of MHD systems. Figure 3.4 shows the formation and thinning of the four major sheets in the system developing at the four X-points of the magnetic configuration shown in Fig. 3.5. Figure 3.6 gives the temporal development of the maximum current density $j_{\max}$ in the sheet in the lower part of the right-hand quadrant in Fig. 3.4, plotted on a log-linear scale, which

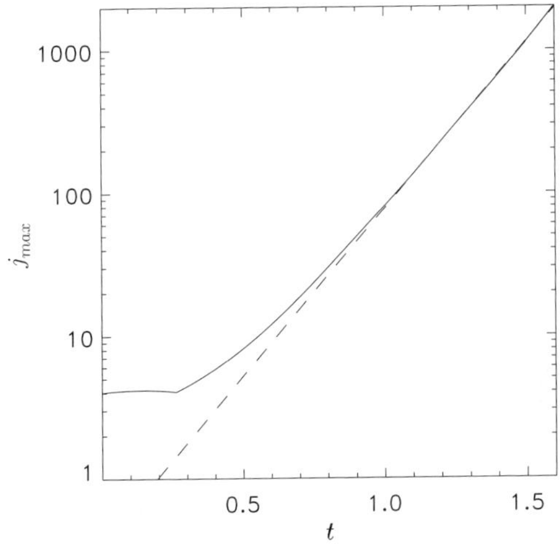

Figure 3.6: Evolution of the current density j_{max} in the lower part of the right-hand current sheet in Fig. 3.4. The dashed line, $e^{2.352(t-1.95)}$, fits the curve for $t > 1$.

demonstrates the exponential increase. Here we used a uniform grid, which was refined during the computation up to 8096^2 grid points at the final time $t = 1.6$ shown in Fig. 3.6. There is no indication of unstable breakup of the sheets, which could possibly accelerate the growth. Hence the solution remains regular, consisting of macroscopic sheets, whose global shape depends on the evolution of the system.

One thus finds that in 2D MHD no FTS seems to exist, in contrast to the dynamics in the Euler equations in the axisymmetric case and the general 3D case discussed in Section 3.1.1. The physical reason for this behavior is the stability of the current sheets, since the Kelvin–Helmholtz mode is stabilized by the magnetic field along the sheet, as we shall see in Section 3.2.1. Only if finite resistivity is included in Ohm's law is the sheet disrupted by the tearing mode (see Section 3.2.3).

In the case of 3D MHD flows, the vortex-stretching term, the first term on the r.h.s. of (3.11), in general does not vanish, which complicates the understanding of singularity formation. Depending on which of the two terms on the r.h.s. dominates, one could expect either a FTS in a filamentary structure as in the 3D Euler case, or an exponential temporal behavior in a sheet-like structure as in 2D MHD. High-resolution numerical studies have been performed by Grauer and Marliani (2000) using a method of adaptive mesh refinement similar to the one developed by these authors for the Euler equations discussed before. Several different initial flows have been chosen, a three-dimensional generalization of the Orszag–Tang vortex, two slightly perturbed parallel magnetic flux

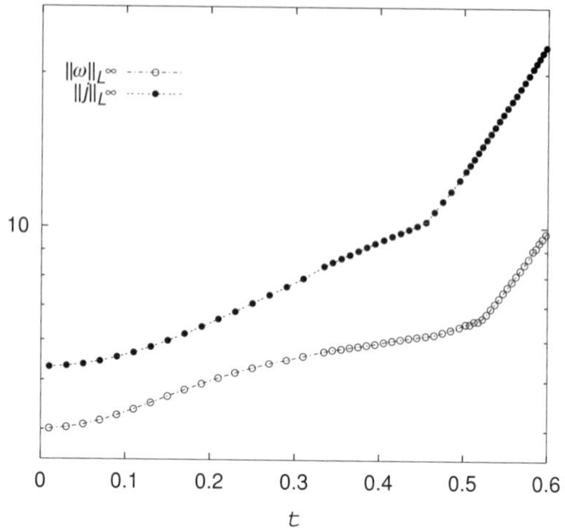

Figure 3.7: Logarithms of j_{max} (full dots) and ω_{max} (open dots) from the evolution of an ideal 3D MHD flow (from Grauer and Marliani, 2000).

tubes, and two interlinked flux tubes. The maximum local resolution reached corresponds to a uniform grid of 4096^3 points. In the early stages the temporal evolution varies considerably among these cases; there is in general a fast, almost explosive growth of current density and vorticity, which differs from the nearly purely exponential growth observed in 2D. However, as fine structures develop, a clear exponential behavior appears in all three cases, an example being shown in Fig. 3.7, and the corresponding spatial structures are pancake-like. These sheets, which seem to be very stable, similarly to the 2D case, constitute a fundamental property of MHD flows. Here magnetic-field lines of different orientations are pushed together, while the plasma is squeezed out of this region along the field. Since there is no normal component of the field across the sheet, this is also called a *tangential discontinuity*. Precisely because of the different orientations of the adjacent field lines, filamentary breakup of the sheet is inhibited. This will be possible only if reconnection is permitted by a finite resistivity.

3.2 Instabilities

We have seen in the preceding section that the solution of the ideal fluid equations develops isolated singular structures. Though this property appears to be related to the generation of turbulence, it does not explain how the *quasi-uniform* distribution of small scales, which is characteristic of turbulence, comes about.

A slowly evolving flow may suddenly give rise to rapid dynamics, which may eventually destroy the flow pattern. Such processes are usually attributed to the effect of an instability, whereby a disturbance imposed on a stationary state grows exponentially. Though real flows are usually not stationary, the concept of instability is nevertheless often very useful for understanding why fast changes of the flow occur.

A fluid, in particular a plasma, becomes unstable when the gradient of velocity, pressure, or magnetic field exceeds a certain threshold, which occurs, roughly speaking, when the convective transport of momentum, heat, or magnetic flux is more efficient than the correponding diffusive transport by viscosity, thermal conduction, or resistivity. There are hence three types of instabilities, which play a fundamental role in macroscopic plasma dynamics: the Kelvin–Helmholtz instability driven by a velocity shear; the Rayleigh–Taylor instability caused by the buoyancy force in a stratified system; and current-driven MHD instabilities in a magnetized plasma, in particular the tearing instability. On the other hand, the presence of magnetic fields, rotation, or stable stratification may exert a stabilizing influence on an otherwise unstable fluid. In this section we give a brief overview of the basic instabilities and the various stabilizing effects.

3.2.1 Kelvin–Helmholtz instability

We consider first the instability of a sheared flow in a neutral fluid neglecting viscosity. Assume that we have a plane equilibrium flow $\boldsymbol{v}_0 = \boldsymbol{e}_y V(x)$ with the perturbation $\boldsymbol{v}_1(x, y) = \boldsymbol{e}_z \times \nabla\phi_1$, $\nabla \cdot \boldsymbol{v}_1 = 0$, since the most unstable modes are incompressible. The perturbation follows the linearized Euler equation

$$\partial_t \omega_1 + \boldsymbol{v}_0 \cdot \nabla\omega_1 + \boldsymbol{v}_1 \cdot \nabla\omega_0 = 0, \tag{3.22}$$

$$\omega_1 = \nabla^2\phi_1, \qquad \omega_0 = V',$$

where the prime indicates the derivative. On making a Fourier *Ansatz* in y and t, $\phi_1 = \phi(x)\mathrm{e}^{i(ky-\varpi t)}$, (3.22) becomes

$$(c - V)(\phi'' - k^2\phi) + V''\phi = 0. \tag{3.23}$$

Here $c = \varpi/k$ is the phase velocity (not the velocity of light in this case) which, because of the advective term $\boldsymbol{v}_0 \cdot \nabla\omega_1$, is in general complex. The simplest unstable flow is a vortex sheet corresponding to the velocity profile

$$V(x) = \begin{cases} V_1 & x < 0 \\ V_2 & x > 0. \end{cases} \tag{3.24}$$

The perturbation is conveniently described by the perpendicular displacement $\xi_x = \xi$, which is related to the velocity perturbation,

$$v_x = -ik\phi = \dot{\xi} = \partial_t \xi + ikV\xi, \tag{3.25}$$

hence

$$\phi = \begin{cases} (c - V_1)\xi & x < 0 \\ (c - V_2)\xi & x > 0. \end{cases} \tag{3.26}$$

Integration of (3.23) across the sheet gives the condition

$$\int_{-\epsilon}^{\epsilon} \left[(c - V)(\phi'' - k^2\phi) + V''\phi\right] dx = c\phi'\Big|_{-\epsilon}^{\epsilon} + \int_{-\epsilon}^{\epsilon} (V''\phi - V\phi'')\, dx$$

$$= (c - V)\phi'\Big|_{-\epsilon}^{\epsilon} = 0. \tag{3.27}$$

Outside the sheet ϕ obeys the equation

$$\phi'' - k^2\phi = 0. \tag{3.28}$$

Since $\phi \to 0$ for $|x| \to \infty$, the solution is

$$\phi = \begin{cases} A_- \mathrm{e}^{kx} & x < 0 \\ A_+ \mathrm{e}^{-kx} & x > 0. \end{cases} \tag{3.29}$$

Use of (3.26) determines the coefficients $A_\pm$, such that we have

$$\phi' = \begin{cases} k(c - V_1)\xi & x < 0 \\ -k(c - V_2)\xi & x > 0. \end{cases} \tag{3.30}$$

Insertion into (3.27) gives the dispersion relation

$$(c - V_1)^2 + (c - V_2)^2 = 0, \tag{3.31}$$

with the solution

$$c = \tfrac{1}{2}(V_1 + V_2) \pm i\tfrac{1}{2}|V_1 - V_2|. \tag{3.32}$$

Hence a vortex sheet is unstable, which is the original Kelvin–Helmholtz instability (Kelvin, 1871; Helmholtz, 1868). Nowadays, the term is also used in a broader sense for general shear-flow instabilities. The growth rate $\gamma = \mathrm{Im}\,\varpi = k\frac{1}{2}|V_1 - V_2|$ is larger the larger k, limited only by the internal structure of the sheet, the maximum being of the order

$$\gamma_{\max} \sim |V_1 - V_2|/\delta, \quad \text{for } k\delta \sim 1, \tag{3.33}$$

where δ is the sheet thickness, in other words, $\gamma_{\max}$ is of the order of the velocity gradient in the sheet,

$$\gamma_{\max} \sim |V'|. \tag{3.34}$$

Instability requires that the flow has at least one inflection point $V'' = 0$ (Rayleigh's criterion; see, e.g., Drazin and Reid, 1981), which is practically always satisfied for free flows without close boundaries. However, if the velocity gradients are too shallow, finite viscosity suppresses the instability. On adding the viscous term in (3.22), the eigenmode equation (3.23) becomes the well-known Orr–Sommerfeld equation

$$(c - V)(\phi'' - k^2\phi) + V''\phi = \frac{i}{k\mathrm{Re}}(\nabla^2)^2\phi, \tag{3.35}$$

where $\nabla^2 \equiv d^2/dx^2 - k^2$. The equation is written in nondimensional form with x normalized with respect to a characteristic length L and t with respect to L/v, where v is a typical velocity. The parameter

$$\mathrm{Re} = vL/\nu \tag{3.36}$$

is the Reynolds number, the normalized inverse kinematic viscosity.[2] One can show that, under rather general conditions, viscosity has a damping effect and hence $\mathrm{Re} > \mathrm{Re}_{\mathrm{crit}}$ is required for instability, where the critical Reynolds number depends on the geometry of the flow, typically $\mathrm{Re}_{\mathrm{crit}} \sim 10^2$.

The Reynolds number is the control parameter in incompressible hydrodynamics, the only dimensionless quantity entering the Navier–Stokes equations when it is normalized in the way given above,

$$\partial_t \boldsymbol{\omega} + \boldsymbol{v} \cdot \nabla \boldsymbol{\omega} - \boldsymbol{\omega} \cdot \nabla \boldsymbol{v} = \frac{1}{\mathrm{Re}} \nabla^2 \boldsymbol{\omega}. \tag{3.38}$$

It determines not only the transition to turbulence at low Re but, even more importantly in the context of this book, the properties of the turbulence at high Re, as we shall see. For fast flows, for which finite compressibility is important, there is a second control parameter, the Mach number $\mathrm{M_s} = v/c_s$, which was introduced in (2.32).

In the nonlinear development of the Kelvin–Helmholtz instability a vortex sheet disrupts (rolls up) into a chain of vortices, which subsequently coalesce

[2] A similar dimensionless parameter arises for thermal convection, for which (2.44) can be written as

$$\partial_t T + \boldsymbol{v} \cdot \nabla T = \frac{1}{\mathrm{Pe}} \nabla^2 T$$

with the Péclet number

$$\mathrm{Pe} = vL/\kappa. \tag{3.37}$$

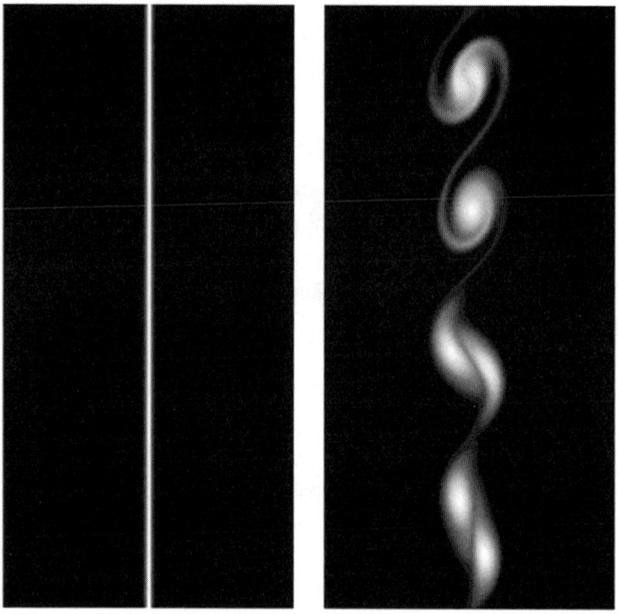

(a) (b)

Figure 3.8: Nonlinear development of the Kelvin–Helmholtz instability of a vortex sheet corresponding to the velocity profile $V(x) = \tanh(x/d)$: (a) the vorticity distribution in the initial state; (b) breakup of the vortex sheet into a series of vortices.

into eddies of ever larger size; see Fig. 3.8. The nonlinear process is qualitatively described by the quasi-linear broadening of the sheet, which stabilizes modes of increasingly long wavelength, since a mode is practically stable if $kd > 1$, where d is the momentary sheet width. The nonlinear evolution of the instability of a jet, a vorticity double sheet, is shown in the left-hand column of Fig. 3.9, the jet being finally completely disrupted.

Let us now add a magnetic field to equilibrium flow. If the field is perpendicular to the flow, it cannot stabilize the Kelvin–Helmholtz mode, since the Lorentz force $\boldsymbol{j}_1 \times \boldsymbol{B}_0$ is parallel to the flow and hence does not affect the transverse motion of the kink mode. By contrast, a parallel field is expected to exert a damping effect, since the mode involves field-line bending. Consider the incompressible MHD equations (2.28), again in the ideal limit, linearized about the stationary flow $\boldsymbol{v}_0 = V(x)\boldsymbol{e}_y$ as used above and with a parallel field $\boldsymbol{B}_0 = B(x)\boldsymbol{e}_y$. Assuming that the field perturbation is again in the x, y-plane, $\boldsymbol{b}_1 = \boldsymbol{e}_z \times \nabla\psi_1$, and making the Fourier *Ansatz* $\psi_1 = \psi(x)e^{i(ky-\varpi t)}$, the

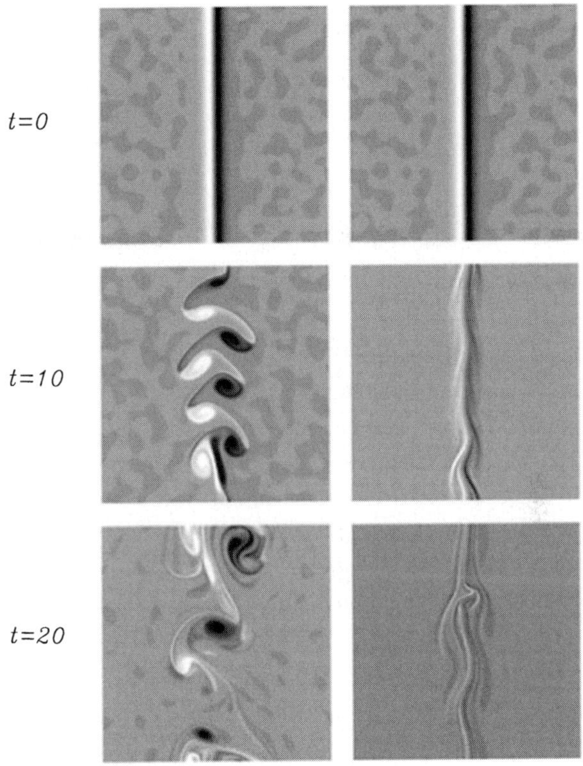

Figure 3.9: The effect of a parallel magnetic field on the evolution of instability of a plasma jet, $v_y(x) = \mathrm{sech}^2(x/d)$, corresponding to a vorticity double sheet. The left-hand column gives the evolution of the unmagnetized jet, the jet being completely disrupted by the instability. In the right-hand column there is a homogeneous parallel field with $v_A = 0.5$. Here the jet is only broadened. The striations at the jet boundaries represent Alfvén waves radiated into the unperturbed plasma (note that the grayscales in this plot are determined by the intense secondary current sheets).

equations take the form

$$(c - V)(\phi'' - k^2\phi) + V''\phi + B(\psi'' - k^2\psi) - B''\psi = 0, \tag{3.39}$$

$$(c - V)\psi + B\phi = 0. \tag{3.40}$$

These equations can be combined into one for $f = \psi/(kB)$:

$$\frac{d}{dx}\left(\left[(c - V)^2 - v_A^2\right]\frac{df}{dx}\right) - k^2\left[(c - V)^2 - v_A^2\right]f = 0, \tag{3.41}$$

where $v_A = B(x)$ is the local Alfvén velocity in the units adopted. We consider again a vortex sheet (3.24), now embedded in a constant parallel field $B = B_0$.

Integration of (3.41) across the sheet gives

$$\left[(c - V)^2 - v_{\rm A}^2\right] \frac{df}{dx}\bigg|_{-\epsilon}^{\epsilon} = 0,$$

and with $df/dx = -\phi'/[k(c - V)]$ and use of (3.30) we obtain the dispersion relation

$$(c - V_1)^2 + (c - V_2)^2 - 2v_{\rm A}^2 = 0, \tag{3.42}$$

which has the solution

$$c = \tfrac{1}{2}(V_1 + V_2) \pm \sqrt{v_{\rm A}^2 - \tfrac{1}{4}(V_1 - V_2)^2}. \tag{3.43}$$

Hence a parallel magnetic field with $v_{\rm A} > \frac{1}{2}|V_1 - V_2|$ stabilizes the Kelvin–Helmholtz mode. The physical interpretation of this effect is that field-line bending consumes more energy than is set free from the sheared velocity, the perturbation being just carried away as an Alfvén wave. In general the magnetic field is not constant across a vortex sheet, i.e., a vortex sheet is also a current sheet, as discussed in Section 3.1.2. Though in this case a quantitative analysis of the eigenvalue problem requires a numerical solution of (3.41), the qualitative result is not changed; the Kelvin–Helmholtz mode is stabilized if the Alfvén velocity of the average parallel field in the sheet exceeds the change in velocity across the sheet. Since in current sheets the Alfvén velocty of the adjacent magnetic field is always larger than the flow velocity along the sheet (see Section 3.2.3), current sheets are stable in ideal MHD, in agreement with the observations in the numerical simulations, which is the reason for the absence of a FTS. However, in the presence of finite resistivity, current sheets may break up due to the tearing mode considered in Section 3.2.3.

Nonlinearly, a parallel magnetic field is very powerful in stabilizing the Kelvin–Helmhotz dynamics, even if the flow is linearly unstable. Consider a plasma jet along a magnetic field. If $v_{\max} < v_{\rm A}$, the jet is stable. In the opposite case $v_{\max} > v_{\rm A}$ initially, the instability starts growing but is switched off as soon as the jet has broadened and hence the velocity has decreased such that $v_{\max} \simeq v_{\rm A}$ (note that $\int v_y\, dx$ = constant). The evolution in this case is shown in the right-hand column of Fig. 3.9. In contrast to the nonmagnetized jet in the left-hand column, for which the vorticity is conserved along the fluid orbit, in the MHD case vorticity (and current) sheets are generated. Viscosity and resistivity determine the width of these sheets, but do not affect the dynamics.

Finally, we discuss briefly the stability of a rotating fluid, since most astrophysical plasmas are in a state of rotation, which is in general differential, $\Omega(r)$, and thus corresponds to a shear flow. The stability of such flows follows from a simple criterion first derived by Rayleigh (1916). Restricting the analysis to axisymmetric perturbations of the rotating fluid, we know that the angular momentum of a fluid element is conserved, $rv_\theta = L_\theta$ =constant. Consider two fluid elements (or rings, because of the assumed axisymmetry) at radii r_1 and $r_2, r_2 > r_1$, with the angular momenta $L_{\theta 1}$ and $L_{\theta 2}$. We calculate the change of energy on exchanging their positions to be

$$\delta W = \left(\frac{L_{\theta 2}}{r_1}\right)^2 + \left(\frac{L_{\theta 1}}{r_2}\right)^2 - \left(\frac{L_{\theta 2}}{r_1}\right)^2 - \left(\frac{L_{\theta 1}}{r_2}\right)^2 = (L_{\theta 2}^2 - L_{\theta 1}^2)\left(\frac{1}{r_1^2} - \frac{1}{r_2^2}\right). \tag{3.44}$$

assuming, without loss of generality, that we have equal masses, $\int_{V_1} \rho\, dV = \int_{V_2} \rho\, dV$. If $\delta W < 0$, energy is set free, corresponding to instability, which occurs if anywhere in the system $L_{\theta 2}^2 < L_{\theta 1}^2$, or $dL_\theta/dr < 0$. Hence the system is stable if the angular momentum increases with radius throughout the entire system,

$$\frac{dL_\theta^2}{dr} = \frac{d(r^2\Omega)^2}{dr} > 0. \tag{3.45}$$

This is Rayleigh's stability criterion for a rotating flow[3]. It follows that a fluid rotating in Keplerian orbits, $\Omega \propto r^{-3/2}$, is stable. Since this rotation pattern is expected for dilute systems such as accretion disks, no turbulence should be excited, which causes a major difficulty for understanding the transport of angular momentum in these systems. The problem has, however, been solved by discovering the crucial role of the magnetic field. The magnetorotational instability will be treated in detail in Section 11.4.

3.2.2 Rayleigh–Taylor instability

In Section 2.6 we found that gravity waves in a stratified fluid are unstable, (2.100), if the density gradient is opposite to the gravity, which is called Rayleigh–Taylor instability (Rayleigh, 1883; Taylor, 1950). This instability is the basic mechanism of thermal convection. Usually it is not the density gradient which is controlled, but rather the temperature gradient. A fluid heated from below becomes unstable if the temperature gradient exceeds a critical value. The basic physics of the instability can be understood by considering the following argument. Consider a small fluid element, originally located at height z, that is displaced upward to $z + \delta z$, where it is decompressed to the ambient pressure at the new position, $p(z + \delta z) = p(z) + (dp/dz)\,\delta z$. Neglecting conduction of

[3] Strictly speaking, Rayleigh's criterion applies only to axisymmetric perturbations. Nonaxisymmetric modes may be unstable even if Rayleigh's criterion is satisfied, if the velocity profile is quasi-discontinuous, resembling a vortex sheet (see Drazin and Reid, 1981). For smooth profiles Rayleigh's criterion applies to general perturbations.

heat, the change of state is adiabatic, i.e., δp and $\delta\rho$ satisfy the relation

$$\frac{\delta\rho}{\rho} = \frac{1}{\gamma}\frac{\delta p}{p} = \frac{1}{\gamma p}\frac{dp}{dz}\,\delta z. \tag{3.46}$$

If the density of the displaced fluid element is smaller than the surrounding density,

$$\rho + \delta\rho < \rho(z + \delta z) = \rho + (d\rho/dz)\,\delta z, \tag{3.47}$$

the fluid element is buoyant and hence pushed further up; in other words, the fluid is convectively unstable. Inserting (3.46) gives the condition for instability

$$\frac{1}{\gamma p}\frac{dp}{dz} < \frac{1}{\rho}\frac{d\rho}{dz}, \tag{3.48}$$

which, using the ideal-gas law

$$\frac{1}{p}\frac{dp}{dz} = \frac{1}{\rho}\frac{d\rho}{dz} + \frac{1}{T}\frac{dT}{dz}, \tag{3.49}$$

is usually written in the following form:

$$-\frac{dT}{dz} > -(\gamma - 1)\frac{T}{\rho}\frac{d\rho}{dz} = -\frac{\gamma - 1}{\gamma}\frac{T}{p}\frac{dp}{dz} = -\left.\frac{dT}{dz}\right|_{\mathrm{adiab}}, \tag{3.50}$$

i.e., instability occurs if the temperature gradient is superadiabatic. Using (3.49), relation (3.50) can also be written as

$$\frac{d}{dz}\ln(p\rho^{-\gamma}) < 0, \quad \text{or } N^2 < 0, \tag{3.51}$$

where N is the Brunt–Vaisala frequency (2.99) for a general compressible gas,

$$N^2 = \frac{g}{\gamma}\frac{d}{dz}\ln(p\rho^{-\gamma}). \tag{3.52}$$

Using the equilibrium relation (2.36) and the equation of state (2.17) with $\boldsymbol{g} = -g\boldsymbol{e}_z$, (3.50) reads

$$-\frac{dT}{dz} > \frac{(\gamma - 1)}{\gamma}\frac{m_i g}{2k_{\mathrm{B}}}, \tag{3.53}$$

which is known as the *Schwarzschild condition.* In the incompressible limit, i.e., the Boussineq approximation, which formally corresponds to $\gamma \to \infty$, one recovers $N^2 = -g\,d\rho'/\rho$, (2.99), and the condition for convective instability becomes $d\rho/dz > 0$, or $dT/dz < 0$, since $\delta p = 0$.

These considerations give, however, only an approximate stability limit, since the actual threshold depends also on other effects exerting a damping on the buoyancy drive, namely the rotation Ω, as seen in (2.100), the magnitude and direction of the magnetic field, and the diffusion coefficients, i.e., the kinematic viscosity ν, the thermal diffusivity κ, and the magnetic diffusivity η. Thermal convection is often considered in the Boussinesq approximation, (2.38), with $\widetilde{\rho}$ replaced by $\widetilde{T}$ using (2.41) or, more generally, (2.42), and the magnetic-field equation (2.15):

$$\partial_t \boldsymbol{\omega} + \boldsymbol{v} \cdot \nabla \boldsymbol{\omega} - \boldsymbol{\omega} \cdot \nabla \boldsymbol{v} = [1/(c\rho_0)](\boldsymbol{B} \cdot \nabla \boldsymbol{j} - \boldsymbol{j} \cdot \nabla \boldsymbol{B}) - \alpha_\rho \nabla T \times \boldsymbol{g} + 2\boldsymbol{\Omega} \cdot \nabla \boldsymbol{v} + \nu \nabla^2 \boldsymbol{\omega}, \quad (3.54)$$

$$\partial_t T + \boldsymbol{v} \cdot \nabla (T_0 + T) = \kappa \nabla^2 T, \quad (3.55)$$

$$\partial_t \boldsymbol{B} + \boldsymbol{v} \cdot \nabla \boldsymbol{B} - \boldsymbol{B} \cdot \nabla \boldsymbol{v} = \eta \nabla^2 \boldsymbol{B}, \quad (3.56)$$

omitting the tilde to simplify the notation.

Since in thermal convection the buoyancy force is the dominant effect, we normalize the Boussinesq equations with respect to the time scale characteristic of buoyant motions $\tau_g = (\alpha_\rho g \, \Delta T/L)^{-1/2} = N^{-1}$ instead of the Alfvén time τ_A in the usual MHD normalization (2.104), introducing the nondimensional quantities[4]

$$t/\tau_g := t, \quad x/L := x, \quad T/\Delta T := T, \quad B/B_0 := B, \quad (3.57)$$

where L is a length of the order of the size of the system, ΔT a typical temperature difference, $T_0' = \Delta T/L$, and B_0 a typical magnetic field. Note that, in the Boussinesq approximation, the average gradient scale is much larger than the size of the system considered, $L_g \gg L$, and the variation in temperature is much smaller than the average temperature, $\Delta T \ll T_0$. Thus normalized the Boussinesq equations read

$$\partial_t \boldsymbol{\omega} + \boldsymbol{v} \cdot \nabla \boldsymbol{\omega} - \boldsymbol{\omega} \cdot \nabla \boldsymbol{v} = S_B(\boldsymbol{B} \cdot \nabla \boldsymbol{j} - \boldsymbol{j} \cdot \nabla \boldsymbol{B}) - \nabla T \times \boldsymbol{e}_g + S_\Omega \boldsymbol{e}_\Omega \cdot \nabla \boldsymbol{v} + \widehat{\nu} \nabla^2 \boldsymbol{\omega}, \quad (3.58)$$

$$\partial_t T + \boldsymbol{v} \cdot \nabla (T_0 + T) = \widehat{\kappa} \nabla^2 T, \quad (3.59)$$

$$\partial_t \boldsymbol{B} + \boldsymbol{v} \cdot \nabla \boldsymbol{B} - \boldsymbol{B} \cdot \nabla \boldsymbol{v} = \widehat{\eta} \nabla^2 \boldsymbol{B}, \quad (3.60)$$

$$\boldsymbol{\omega} = \nabla \times \boldsymbol{v}, \quad \boldsymbol{j} = \nabla \times \boldsymbol{B},$$

[4] This normalization differs from the traditional one used in thermal convection, for which the diffusion time L^2/κ is taken as the unit. Here we want to stay close to the units used in hydrodynamics and MHD, i.e., the coefficients in the dynamic terms should be of order unity and those in the dissipative terms small.

where $\boldsymbol{e}_g$ and $\boldsymbol{e}_\Omega$ are unit vectors in the directions of $\boldsymbol{g}$ and $\boldsymbol{\Omega}$. The coefficients

$$S_B = \frac{\tau_g^2}{\tau_A^2} = \frac{S^2}{\mathsf{Ra}}\frac{\mathsf{Pr}}{\mathsf{Pr}_m^2}, \quad S_\Omega = 2\Omega\tau_g = \left(\frac{\mathsf{TaPr}}{\mathsf{Ra}}\right)^{1/2}, \tag{3.61}$$

$$\widehat{\nu} = \left(\frac{\mathsf{Pr}}{\mathsf{Ra}}\right)^{1/2}, \quad \widehat{\eta} = \left(\frac{\mathsf{Pr}}{\mathsf{Ra}}\right)^{1/2}\frac{1}{\mathsf{Pr}_m}, \quad \widehat{\kappa} = \frac{1}{(\mathsf{RaPr})^{1/2}} \tag{3.62}$$

are written in terms of the Lundquist number S, the Rayleigh number Ra, the Taylor number Ta, the Prandtl number Pr, and the magnetic Prandtl number Pr_m,

$$S = \frac{v_A L}{\eta} = \frac{\tau_\eta}{\tau_A}, \quad \mathsf{Ra} = \frac{\alpha_\rho g\, \Delta T\, L^3}{\nu\kappa} = \frac{\tau_\nu \tau_\kappa}{\tau_g^2}, \quad \mathsf{Ta} = \left(\frac{2\Omega L^2}{\nu}\right)^2 = (2\tau_\nu\Omega)^2, \tag{3.63}$$

$$\mathsf{Pr} = \nu/\kappa, \quad \mathsf{Pr}_m = \nu/\eta. \tag{3.64}$$

The Lundquist number is the ratio of the Alfvén time and the magnetic diffusion time, the Rayleigh number measures the buoyancy effect compared with the effects of viscosity and thermal diffusion, and the Taylor number compares the rate of rotation with the viscous dissipation rate.[5] S, Ra, and Ta are very large numbers in most applications. The Prandtl numbers are formally of order unity, though in practice they may vary considerably depending on the medium, for instance $\mathsf{Pr}_m \simeq 7$ for water, $\simeq 0.7$ for air, and $\sim 10^{-5}$ in the liquid core of the Earth. In a hot plasma it is difficult to assign definite numbers to Pr and Pr_m because of the extreme differences between parallel and perpendicular viscosities and thermal conductivities. Here these quantities should be considered as phenomenological parameters, which we can assume to be of order unity. With this in mind, we see that the constants S_B and S_Ω in the Boussinesq equations are formally of order unity, while $\widehat{\nu}$, $\widehat{\eta}$, and $\widehat{\kappa}$ are small.

We linearize (3.54)–(3.56) about a stratified equilibrium. Because of the Boussinesq approximation, which we can also call the quasi-local approximation when the size of the system is small compared with the equilibrium gradient scales, T_0, T_0', and B_0 are considered constant and the variation of the magnetic field is sufficiently weak that the equilibrium current density can be neglected, $B_0' = j_0 = 0$. After Fourier transformation the linearized equations read

$$-i\varpi\omega = S_B i\boldsymbol{k}\cdot\boldsymbol{e}_B j - i\boldsymbol{k}\times\boldsymbol{e}_g T + S_\Omega i\boldsymbol{k}\cdot\boldsymbol{e}_\Omega \boldsymbol{v} - \widehat{\nu}k^2\boldsymbol{\omega}, \tag{3.65}$$

$$-i\varpi T = \boldsymbol{e}_g\cdot\boldsymbol{v}T_0' - \widehat{\kappa}k^2 T, \tag{3.66}$$

[5] One often uses the Ekman number $\mathsf{E} = \mathsf{Ta}^{-1/2}$ instead of the Taylor number. The Taylor number is also related to the Rossby number $\mathsf{Ro} = vL/\nu = 2\mathsf{Re}/\mathsf{Ta}^{1/2}$.

$$-i\varpi \boldsymbol{b} = -i\boldsymbol{k}\cdot\boldsymbol{e}_B\boldsymbol{v} - \widehat{\eta}k^2\boldsymbol{b}. \tag{3.67}$$

To solve these equations it is useful to take the curl of (3.65) by multiplying it by $i\boldsymbol{k}\times$:

$$-i\varpi k^2\boldsymbol{v} = iS_B\boldsymbol{k}\cdot\boldsymbol{e}_Bk^2\boldsymbol{b} - (k^2\boldsymbol{e}_g - \boldsymbol{k}\cdot\boldsymbol{e}_g\boldsymbol{k})T + iS_\Omega\boldsymbol{k}\cdot\boldsymbol{e}_\Omega\boldsymbol{\omega} - \widehat{\nu}k^4\boldsymbol{v}. \tag{3.68}$$

To simplify the notation, we choose $\boldsymbol{e}_g = -\boldsymbol{e}_z$. On substituting for T and $\boldsymbol{b}$ and taking the z-components of (3.65) and (3.68), we obtain two equations for ω_z and v_z,

$$\left(-i\varpi + \widehat{\nu}k^2 + S_B\frac{\boldsymbol{k}\cdot\boldsymbol{e}_B}{-i\varpi + \widehat{\eta}k^2}\right)\omega_z - iS_\Omega\boldsymbol{e}_\Omega\cdot\boldsymbol{k}v_z = 0, \tag{3.69}$$

$$\left((-i\varpi + \widehat{\nu}k^2)k^2 + S_B\frac{(\boldsymbol{k}\cdot\boldsymbol{e}_B)^2}{-i\varpi + \widehat{\eta}k^2} + \frac{k_\perp^2T_0'}{-i\varpi + \widehat{\kappa}k^2}\right)v_z - iS_\Omega\boldsymbol{k}\cdot\boldsymbol{e}_\Omega\omega_z = 0, \tag{3.70}$$

where $k_\perp^2 = (\boldsymbol{k}\times\boldsymbol{e}_g)^2$. The general dispersion relation derived from these equations is rather complicated algebraically. We therefore consider only two special cases.

(a) Stabilizing effects of $\widehat{\nu}$ and $\widehat{\kappa}$, assuming that $B_0 = \Omega = 0$. From (3.70) one obtains

$$(-i\varpi + \widehat{\nu}k^2)(-i\varpi + \widehat{\kappa}k^2) = -(k_\perp/k)^2T_0'. \tag{3.71}$$

In the nondissipative limit instability arises for any negative temperature gradient $T_0' < 0$. Since viscosity and heat conductivity exert a damping effect, the marginally stable temperature gradient is finite,

$$T_{\mathrm{crit}}' = -\widehat{\nu}\widehat{\kappa}k^4(k/k_\perp)^2, \tag{3.72}$$

or, in dimensional form,

$$T_{\mathrm{crit}}' = -\frac{1}{\mathsf{Ra}}\frac{\Delta T}{L}\frac{(kL)^6}{(k_\perp L)^2}. \tag{3.73}$$

Hence the control parameter for the instability is the Rayleigh number. The most unstable mode requiring the lowest Rayleigh number depends on the geometry of the system and the boundary conditions. For a cubic box and periodic boundary conditions, where $\{k_x, k_y, k_z\} = (2\pi/L)\{n_x, n_y, n_z\}$, this mode has $n_x = 1, n_y = n_z = 0$, hence $\mathsf{Ra}_{\mathrm{crit}} = 16\pi^4 \simeq 1559$. For a fluid between two infinite plates separated by a distance L, and non-slip boundary conditions for the velocity, the value is somewhat higher,

$\mathsf{Ra}_{\mathrm{crit}} \simeq 1708$. If Ra is smaller than the threshold value, the transport of heat is dominated by diffusion.

(b) In the second special case we demonstrate the stabilizing effect of the magnetic field, neglecting for simplicity dissipation. The stabilization of the Rayleigh–Taylor instability by rotation has already been discussed in Section 2.6, see (2.100); hence we also set $\Omega = 0$. The dispersion relation now reads

$$[S_B(\boldsymbol{k}\cdot\boldsymbol{e}_B)^2 - \varpi^2][S_B(\boldsymbol{k}\cdot\boldsymbol{e}_B)^2 + T_0' k_\perp^2/k^2 - \varpi^2] = 0 \tag{3.74}$$

While the first factor gives the Alfvén wave (2.92), the second describes the gravity wave in the presence of a magnetic field,

$$\varpi^2 = \alpha_\rho g T_0' k_\perp^2/k^2 + k_\parallel^2 v_{\mathrm{A}}^2 = N^2 k_\perp^2/k^2 + k_\parallel^2 v_{\mathrm{A}}^2, \tag{3.75}$$

written again in dimensional form, $k_\parallel^2 = (\boldsymbol{k}\cdot\boldsymbol{e}_B)^2$. Equation (3.75) shows that field-line bending has a stabilizing effect on the Rayleigh–Taylor instability, which is similar to the stabilization of the Kelvin–Helmholtz instability, (3.43). The dispersion relation (3.75) can also be interpretated in the sense that, for a sufficiently large negative temperature gradient, buoyancy destabilizes the Alfvén wave. To minimize the stabilizing effect of the magnetic field, convective motions tend to be interchange- or flute-like, i.e., perpendicular to $\boldsymbol{B}_0$.

3.2.3 Kelvin–Helmholtz instability in a stratified medium

Equation (3.75) is also interpreted in the sense that Alfvén waves may be destabilized by unstable stratification. A stable stratification, on the other hand, can stabilize certain unstable modes. As an example we return to the Kelvin–Helmholtz mode of a sheared flow $v_x = V(z)$, now embedded in a stably stratified medium, $d\rho(z)/dz < 0$. The basic mechanism can be understood by a simple energetic consideration (Chandrasekhar, 1961). Take two fluid elements of equal volume at heights z_1 and $z_2 = z_1 + \delta z$ and velocities $v_1 = v$, $v_2 = v + \delta v$, $\delta v = (dV/dz)\delta z$, and densities $\rho_1, \rho_2 = \rho_1 + \delta\rho$, $\delta\rho = (d\rho/dz)\delta z$, and interchange their vertical positions, which requires a net amount of gravitational energy,

$$\delta W^G = -g\,\delta\rho\,\delta z \tag{3.76}$$

Along with the displacement the horizontal velocities of the elements change, but not necessarily in such a way that the original velocities are simply interchanged, the only constraint being the conservation of total momentum. We therefore write the final velocities,

$$v_1 \to v + a\,\delta v, \quad v_2 \to v + (1-a)\,\delta v,$$

in terms of the free parameter a, $0 < a < 1$. One can now easily calculate the change of kinetic energy

$$\begin{aligned}\delta W^V &= \tfrac{1}{2}\rho(v + a\,\delta v)^2 + \tfrac{1}{2}(\rho + \delta\rho)[v + (1-a)\,\delta v]^2 \\ &\quad - \tfrac{1}{2}v^2 - \tfrac{1}{2}(\rho + \delta\rho)(v + \delta v)^2 \\ &= -a(1-a)\rho(\delta v)^2 - v\,\delta v\,\delta\rho \\ &\geq -\tfrac{1}{4}\rho(\delta v)^2 - v\,\delta v\,\delta\rho,\end{aligned}$$

the minimum of the expression being reached for $a = \frac{1}{2}$. The Kelvin–Helmholtz mode is stable if the net change in energy is positive, $\delta W^G + \delta W^V > 0$, i.e., if throughout the system

$$-g\,\frac{d\rho}{dz} > \frac{1}{4}\rho\left(\frac{dV}{dz}\right)^2 + \frac{1}{2}\,\frac{dV^2}{dz}\,\frac{d\rho}{dz}, \tag{3.77}$$

or, neglecting the variation of the density in the kinetic energy,

$$\mathsf{Ri} \equiv -\frac{g\,d\rho/dz}{\rho(dV/dz)^2} > \frac{1}{4}. \tag{3.78}$$

Here Ri is called the Richardson number.

3.2.4 The tearing instability

In the preceding sections we have discussed the stabilizing effect of a magnetic field, which forces buoyancy-driven modes in a stratified equilibrium to be flute-like and may completely stabilize a shear flow. The effect is due to the coupling between the fluid and the magnetic field, which gives field lines stiffness and elasticity, turning unstable perturbations into propagating Alfvén waves. The coupling is, however, not perfect, since finite resistivity allows some slippage of field across the fluid. Field lines lose their identity; they may be cut, pictorially speaking, and reconnected in a different way, a process commonly called magnetic reconnection; for a review of reconnection theory and important applications see, e.g., Biskamp (2000). Reconnection allows one to tap the reservoir of free magnetic energy more efficiently by exciting a new class of modes called tearing modes. The simplest case is the tearing mode of a plane current sheet. The driving mechanism is readily understood by considering the attractive forces on parallel current elements. If the homogeneous current distribution along the sheet is slightly modulated, or clumped, the modulation tends

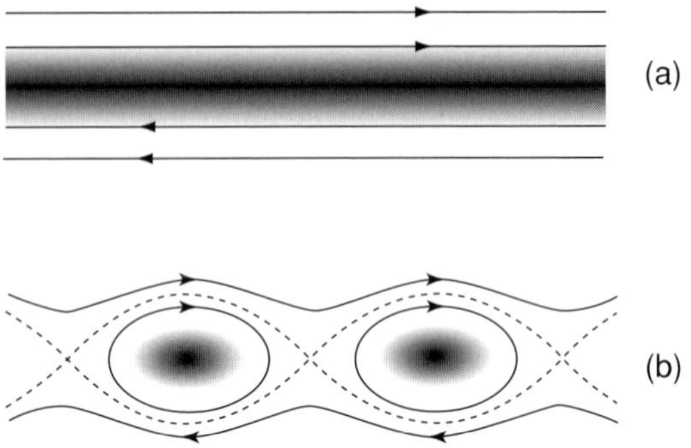

Figure 3.10: Tearing instability of a current sheet. (a) The initial current distribution; (b) local condensation of current density by the tearing mode.

to grow, since the elements of each current clump are more strongly attracted by each other than by the neighboring clumps. The change of the magnetic configuration is illustrated in Fig. 3.10.

For a quantitative treatment we start from the 2D incompressible MHD equations (2.50) and (2.51), written in the usual Alfvén time units (2.104),

$$\partial_t \omega + \boldsymbol{v} \cdot \nabla \omega = \boldsymbol{B} \cdot \nabla j + \nu \nabla^2 \omega, \tag{3.79}$$

$$\partial_t \psi + \boldsymbol{v} \cdot \nabla \psi = \eta \nabla^2 \psi, \tag{3.80}$$

with $\boldsymbol{v} = \boldsymbol{e}_z \times \nabla \phi$, $\omega = \nabla^2 \phi$, and the poloidal, or in-plane, magnetic field $\boldsymbol{B} = \boldsymbol{e}_z \times \nabla \psi$, $j = \nabla^2 \psi$. The equilibrium is a sheet pinch $B_{0y}(x) = \psi_0'(x)$ reversing sign at $x = 0$, $B_{0y}(0) = 0$ corresponding to the current density $j(x) = B_{0y}' = \psi_0''$. The classical paradigm is the Harris sheet, $B_{0y} = \tanh(x/a)$, $j = a^{-1} \operatorname{sech}^2(x/a)$, which is an exact equilibrium solution in a collisionless plasma. Note that there may also be an axial, or out-of-plane, magnetic-field component B_{0z}, which for incompressibility does not, however, affect the dynamics. On imposing a small sinusoidal perturbation, $\boldsymbol{B} = \boldsymbol{B}_0 + \boldsymbol{b}_1$, the flux function in the vicinity of the neutral surface, where the field reverses sign, has the form

$$\psi(x, y) = \psi_0(x) + \delta\psi(x, y) \simeq \tfrac{1}{2} \psi_0'' x^2 + \psi_1 \cos(ky). \tag{3.81}$$

In the presence of an axial field component B_z, the term neutral surface is not quite appropriate. Instead one generally uses the term *resonant surface* because of the vanishing of the operator $\boldsymbol{B}_0 \cdot \nabla$ for the perturbation considered. To be definite, we choose ψ_0'' and ψ_1 positive. The flux perturbation corresponds to a vertical field at the resonant surface $b_{1x} = k\psi_1 \sin(ky)$. There is a sequence of neutral lines, or simply neutral points in projection, where the poloidal field

$|\nabla\psi|$ vanishes, alternately X-type and O-type neutral points or simply X-points and O-points. The line $\psi = \psi_1$, the separatrix between reconnected and merely deformed field lines, forms a chain of magnetic islands (Fig. 3.10(b)), whose width, defined as the distance between the two branches at an O-point, is given approximately by

$$w_I \simeq 4\sqrt{\psi_1/\psi_0''}. \tag{3.82}$$

The perturbation follows the linearized equations (3.79) and (3.80),

$$-i\varpi\, \nabla^2\phi_1 = \boldsymbol{B}_0 \cdot \nabla j_1 + \boldsymbol{b}_1 \cdot \nabla j_0, \tag{3.83}$$

$$-i\varpi\, \psi_1 = \boldsymbol{B}_0 \cdot \nabla\phi_1 + \eta\, \nabla^2\psi_1, \tag{3.84}$$

neglecting viscosity, which is not essential for the tearing mode. It can be shown that, in the case of instability, the frequency is imaginary (Furth *et al.*, 1963), i.e., the mode is purely growing; hence we shall replace $-i\varpi$ by the growth rate γ. Since γ is found to be small, $\gamma \sim \eta^\alpha$, $0 < \alpha < 1$, inertia and diffusion are important only in a narrow layer δ_η around the resonant surface, where $B_{0y} = 0$. Thus the stability analysis leads to a boundary-layer problem, for which one calculates separately the solutions in the diffusion layer using simplified geometry and in the external region using simplified physics and matches the two solutions asymptotically.

First consider the diffusion layer. Writing $\psi_1(x, y) = \psi_1(x)\mathrm{e}^{iky}$, (3.83) and (3.84) become

$$\gamma\phi_1'' = ikxB_0'\psi_1'' - ikj_0'\psi_1, \tag{3.85}$$

$$\gamma\psi_1 = ikxB_0'\phi_1 + \eta\psi_1'', \tag{3.86}$$

where we make the approximations $\nabla^2\phi_1 \simeq \phi_1''$, $\nabla^2\psi_1 \simeq \psi_1''$, and $B_{y0}(x) \simeq xB_0'$. Since the instability is relatively slow, the perturbation ψ_1 can diffuse across the layer during a growth time, such that its variation in the layer is weak, $\psi_1 \simeq \psi_1(0)$, which is called the constant-ψ property. Only the derivative varies across the layer, which variation is described by the quantity

$$\Delta_i' = (\psi_{1+}' - \psi_{1-}')/\psi_1(0), \tag{3.87}$$

the change of the logarithmic derivative; the subscripts $+$ and $-$ refer to positions outside the layer, where ψ_1' is smooth, varying only on the global scale a. Even though it is formally of the same order as the other terms in the equation, the j_0' term in (3.85) can be neglected, since it practically does not change the stability threshold and the growth rate (Bertin, 1982). The functions ψ_1 and ϕ_1 may then be chosen with definite parity, ψ_1 even (to have $\psi_1(0) \neq 0$) and ϕ_1 odd, which makes the following order-of-magnitude analysis semi-quantitative.

Writing $\phi_1'' \sim -\phi_1/\delta_\eta^2$ and $\psi_1'' \sim \Delta_i'\psi_1/\delta_\eta$, and equating individual terms in (3.85) and (3.86) gives γ and the sheet width δ_η as functions of Δ_i',

$$\gamma \sim \eta^{3/5}(\Delta_i')^{4/5}(kB_0')^{2/5}, \tag{3.88}$$

$$\delta_\eta \sim \eta\Delta_i'/\gamma \sim \eta^{2/5}(\Delta_i')^{1/5}(kB_0')^{-2/5}. \tag{3.89}$$

Let us also write γ in dimensional form:

$$\gamma \sim \tau_\eta^{-3/5}\tau_A^{-2/5}(\Delta' a)^{4/5}(ka)^{2/5}(B_0'a/B_0)^{2/5}, \tag{3.90}$$

where $\tau_A = a/v_A$ is the equilibrium Alfvén time and $\tau_\eta = a^2/\eta$ the resistive diffusion time. Relations (3.88) and (3.89) are exact apart from numerical factors of order unity (for details, see, e.g., Biskamp, 1993a, Chapter 4). We have tacitly assumed that $\Delta_i' > 0$. In fact, the tearing mode is unstable if, and only if, this quantity is positive (Furth *et al.*, 1963).

The configuration in the external region outside the diffusion layer relaxes on the Alfvén time scale, i.e., quasi-instantaneously, to a perturbed equilibrium state, since the instability is weak, $\gamma \ll \tau_A^{-1}$, such that inertia and diffusion terms can be neglected and (3.83) reduces to

$$\boldsymbol{B}_0 \cdot \nabla j_1 + \boldsymbol{b}_1 \cdot \nabla j_0 = 0. \tag{3.91}$$

This equation is solved for ψ_1 by integrating from the boundary values on both sides up to the resonant surface and equating the values of $\psi_1(0)$. The derivative of ψ_1 is, however, in general not continuous, an example being shown in Fig. 3.12. One defines the quantity Δ_e', the jump of the derivatives of the external solution,

$$\Delta_e' = [\psi_1'(0_+) - \psi_1'(0_-)]/\psi_1(0). \tag{3.92}$$

The solutions in the resistive layer and the external region are now matched asymptotically by equating the logarithmic deriviatives of ψ_1, i.e., the jump parameters $\Delta_{i,e}'$,

$$\Delta_i' = \Delta_e' = \Delta'. \tag{3.93}$$

This gives the dispersion relation determining γ in (3.88) in terms of the quantity Δ', which depends only on the equilibrium and on the boundary conditions for the perturbation.

In the case of a sheet pinch, (3.91) becomes

$$\psi_1'' - \left(k^2 + \frac{j_0'}{B_0(x)}\right)\psi_1 = 0, \tag{3.94}$$

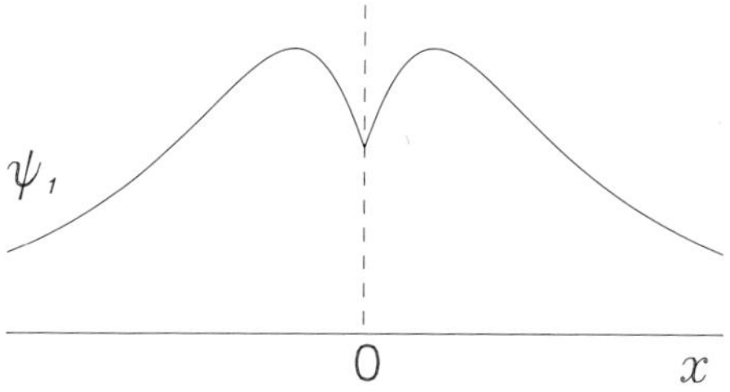

Figure 3.11: The external solution $\psi_1(x)$ of the tearing mode in a sheet pinch. The derivative exhibits a discontinuity at the resonant surface $x = 0$.

which can be solved analytically for the Harris sheet with the boundary conditions $\psi_1 = 0$ for $x \to \pm\infty$,

$$\psi_1 = e^{-k|x|}\left[1 + \frac{1}{ka}\tanh\left(\frac{|x|}{a}\right)\right], \tag{3.95}$$

whence

$$\Delta' = \frac{2}{a}\left(\frac{1}{ka} - ka\right). \tag{3.96}$$

We thus find that the tearing mode is unstable, $\Delta' > 0$, for long wavelengths $ka < 1$ and stable, $\Delta' < 0$, for short wavelengths $ka > 1$. These properties remain qualitatively valid for rather general current profiles $j_0(x)$, in particular the long-wavelength behavior $\Delta' \sim k^{-1}$. Hence the constant-ψ property breaks down for small k, $k \sim \delta_\eta/a^2$, where $\Delta' \sim \delta_\eta^{-1}$. The analysis without the constant-ψ assumption shows that the growth rate, which increases with decreasing k, $\gamma \sim (\Delta')^{4/5}k^{2/5} \sim k^{-2/5}$, reaches a maximum at $k \simeq \eta^{1/4}$,

$$\gamma_{\max} \simeq 0.6\eta^{1/2} = 0.6(\tau_A\tau_\eta)^{-1/2}. \tag{3.97}$$

The quantity Δ' determines not only the linear growth rate but also the nonlinear dynamics. As shown by Rutherford (1973), the growth rate is reduced already at very low amplitude, much below the final saturation level. The dominant nonlinear effect is the diffusive broadening of the current perturbation j_1. While in the linear phase of the instability $j_1(x)$ is concentrated inside the layer δ_η, $j_1 \simeq \psi_1'' \sim \psi_1\Delta'/\delta_\eta$, the layer width is replaced by the island width w_I when the latter exceeds the former. Using the definitions (3.82) and (3.89), this cross-over occurs at amplitude $\psi_1 \sim \eta^{4/5}$. Further growth is a purely diffusive process, independent of inertia,

$$\dot{\psi}_1 = \eta j_1 \sim \eta\frac{\Delta'}{w_I}\psi_1, \tag{3.98}$$

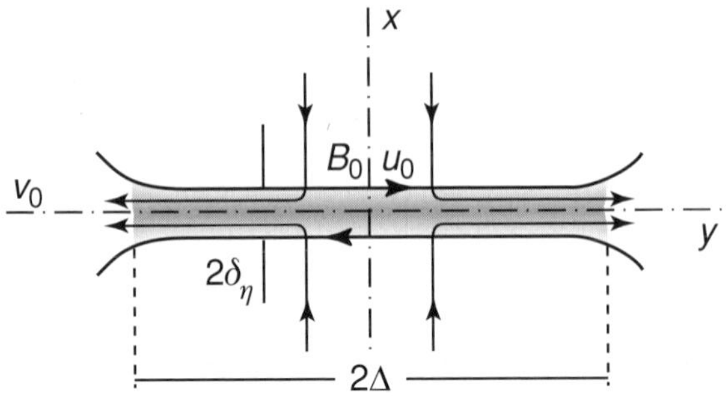

Figure 3.12: A schematic drawing of a Sweet–Parker sheet. The heavy lines indicate the magnetic field, the lighter lines the flow.

or, in terms of the island width (3.82),

$$\dot{w}_I \sim \eta \Delta'. \tag{3.99}$$

Hence the island no longer grows exponentially but rather grows linearly in time, $w_I \propto t$.

The saturation of the tearing mode is given by the nonlinear reduction of Δ' defined by

$$\Delta'(w_I) \simeq [\psi_1'(\tfrac{1}{2}w_I) - \psi_1'(-\tfrac{1}{2}w_I)]/\psi_1(0), \tag{3.100}$$

where $\psi_1(x)$ is the solution of the linear equation (3.91). As is intuitively clear from Fig. 3.10, $\Delta'(w_I)$ decreases to zero with increasing island width, whereupon growth stops, which gives a saturation island width of the order of the current-profile width a, the actual width depending on the wavenumber. Since the nonlinear evolution proceeds on the diffusion time scale, the final state depends also on the spatial and temporal behavior of the resistivity η. In addition, neighboring islands tend to coalesce to give islands of larger wavelength, which may then grow to widths greatly exceeding the equilibrium width a.

Current sheets that form dynamically, as discussed in Section 3.1.2, are not simply static configurations but also carry strong flows. Adding finite resistivity results in a stationary state, in which the transport of magnetic flux into the sheet is balanced by resistive diffusion. Such a sheet is called a Sweet–Parker sheet (Sweet, 1958; Parker, 1963); see Fig. 3.12. It is characterized by six quantities, the inflow and outflow velocities u_0 and v_0, the upstream field pushed against the sheet B_0, the length Δ, the width, or thickness, δ_η, and the resistivity η. These obey three relations, conservation of mass (assuming incompressibility)

$$v_0 \delta_\eta \simeq u_0 \Delta, \tag{3.101}$$

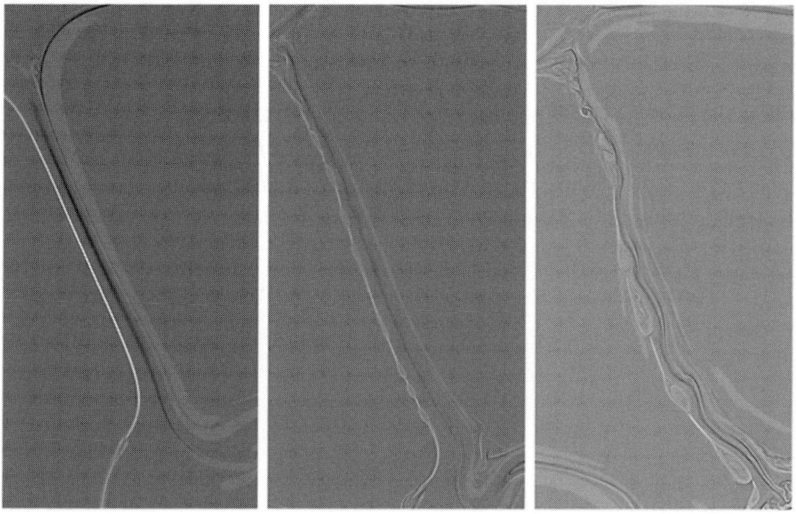

Figure 3.13: The development of the tearing instability of a current sheet (the lower right-hand sheet in Fig. 3.4) developing from the initial state (3.21), shown on grayscale plots of j taken at three different times.

Ohm's law, for which stationarity requires that E_z = constant,

$$u_0 B_0 = \eta j_{\max} \simeq \eta B_0/\delta_\eta, \tag{3.102}$$

and the force balance along the sheet $v_y\,\partial_y v_y = -\partial_y p$, where the pressure is determined by the cross-sheet force balance $\partial_x(p + \frac{1}{2}B^2) = 0$, which gives

$$v_0 \simeq B_0 = v_A \tag{3.103}$$

in our Alfvén time normalization. As a result of the last relation, the Kelvin–Helmholtz mode is stable, as discussed in Section 3.2.1.

However, the sheet is also stabilized against the tearing mode by the inhomogeneity of the flow resulting from the acceleration along the sheet, $v_y(y) = v_A y/\Delta$. The basic mechanism can readily be understood. The local current-density bunching caused by the tearing mode is counteracted by the differential stretching due to the inhomogeneous flow, such that mode growth is essentially suppressed if the stretching of a current-density bunch during one growth time exceeds a substantial fraction of the wavelength, say one quarter, $v_A/(\gamma_{\max}\Delta) > 0.25$. Here $\gamma_{\max}$ is the maximum tearing-mode growth rate in the absence of a flow (3.97). Using (3.101) and (3.102), we have

$$v_A/\Delta \simeq \eta/\delta_\eta^2 = \tau_\eta^{-1}, \quad v_A/\delta_\eta = \tau_A^{-1},$$

and hence the stability condition

$$\delta_\eta/\Delta \gtrsim 0.02, \tag{3.104}$$

where the numerical value on the r.h.s. should not be taken too literally. We thus find that a Sweet–Parker sheet remains stable against tearing up to a rather large aspect ratio Δ/δ_η. Since, however, the sheet width shrinks with decreasing η, the tearing mode does eventually appear, disrupting the macro-sheets, as shown in Fig. 3.13.

4

Macroscopic turbulence theory

In the preceding chapters we considered the dynamics of an individual system. Starting from a smooth state, fine structures develop, which, in general, become unstable at some point. After the onset of instability the structure of the flow is very complex and irregular and, most importantly, the further behavior is unpredictable in the sense that minimal changes would soon lead to a completely different state. Such a behavior is commonly called turbulent. Though a direct view of the continuously changing patterns is certainly most eyecatching and fascinating, a pictorial description of these structures is not very suitable for a quantitative analysis. On the other hand, it is just this chaotic behavior which makes turbulence accessible to a theoretical treatment involving statistical methods. While individual shapes and motions are intricate and volatile, the average properties of the turbulence described by the various correlation functions are, in general, smooth and follow rather simple laws. A well-known paradigm is the turbulent behavior in our atmosphere. We try to predict the short-term changes, called weather, in a deterministic way for as long as is feasible, which, as daily experience shows, is not very long, while predictions of the long-term behavior, called climate, can be made only on a statistical basis.

Dividing the fields into mean and fluctuating parts, we derive equations for the average quantities, the generalized Reynolds equations, which contain second-order moments of the fluctuating parts, the turbulent stresses. A rigorous treatment would require one to consider also the equations for these stresses, which introduces third-order moments and, when it is continued, leads to the well-known infinite hierarchy of correlation functions, the approximate treatment of which, the problem of closure, is the central task in turbulence theory. In this chapter we consider only the Reynolds equations, using physical arguments to obtain phenomenological models of the turbulent stresses, eddy viscosity, and eddy resistivity. This level of approximation, called one-point closure theory, is the basis for describing the global properties of a turbulent system, i.e., the actual way in which turbulence is excited by boundary effects and the turbulent transport properties, such as the drag on a fast-moving car, which depends on the shape of its body, or, to give a magnetic example, the transport of angular momentum in an accretion disk.

The second part of the chapter deals with macroscopic self-organization, the formation of large-scale coherent structures, which has a fundamental influence on the development of MHD turbulence, in contrast to Navier–Stokes turbulence, for which no such processes seem to exist. Self-organization is intimately connected with the existence of several ideal invariants and the difference in their dissipation rates, which gives rise to selective decay processes, either the tendency toward a force-free magnetic state or the tendency toward an aligned, or Alfvénic, state. The latter process is based on the Alfvén effect in MHD turbulence. The decay of the turbulence energy is essentially determined by these selective decay processes.

4.1 One-point closure

Most applications are primarily concerned with the properties of the mean quantities, such as the mean flow averaged over the turbulent fluctuations, for which the turbulence enters only through enhancement of transport effects. We split the velocity and magnetic field into mean and fluctuating parts,

$$\boldsymbol{v} = \langle \boldsymbol{v} \rangle + \widetilde{\boldsymbol{v}}, \quad \boldsymbol{B} = \langle \boldsymbol{B} \rangle + \widetilde{\boldsymbol{b}}. \tag{4.1}$$

Since, however, the mean quantities are in general neither homogeneous nor stationary – we are, for instance, interested in the evolution of the mean velocity profile of a turbulent shear flow –, the distinction regarding which scales should be averaged and which should be regarded as mean profiles is to a certain extent arbitrary. Such a two-scale approach, which is frequently applied in turbulence theory, relies, strictly speaking, on there being a spectral gap between the small scales, over which the average is taken, and the large scales, which are treated explicitly. Though such a gap rarely exists, one may nonetheless in many cases rather clearly distinguish between the smaller inertial-range scales and the macroscales of the system. However, in so-called large-eddy simulations, which are very convenient in practical turbulence computations, the break is assumed to occur within the inertial range at some wavenumber k_c, at which only the larger scales $k_c l > 1$ are computed directly, while the effect of the smaller scales $k_c l < 1$ is accounted for by phenomenological expressions, which depend explicitly on the cutoff k_c.

4.1.1 Reynolds equations for MHD

We restrict our consideration to incompressible motions (effects of compressibility will be discussed in Chapter 9), but treat the MHD equations in the original velocity form (2.7) or in terms of the Elsässer fields (2.106), instead of the vorticity form (2.28). The equations for the mean velocity and the magnetic field, the Reynolds equations for MHD, are

$$\partial_t \langle \boldsymbol{v} \rangle + \langle \boldsymbol{v} \rangle \cdot \nabla \langle \boldsymbol{v} \rangle = -\nabla \langle P \rangle + \langle \boldsymbol{B} \rangle \cdot \nabla \langle \boldsymbol{B} \rangle - \nabla \cdot \langle \widetilde{\boldsymbol{v}} \widetilde{\boldsymbol{v}} - \widetilde{\boldsymbol{b}} \widetilde{\boldsymbol{b}} \rangle + \nu \nabla^2 \langle \boldsymbol{v} \rangle, \quad (4.2)$$

$$\partial_t \langle \boldsymbol{B} \rangle = \nabla \times \big(\langle \boldsymbol{v} \rangle \times \langle \boldsymbol{B} \rangle \big) + \nabla \times \langle \widetilde{\boldsymbol{v}} \times \widetilde{\boldsymbol{b}} \rangle + \eta \nabla^2 \langle \boldsymbol{B} \rangle, \quad (4.3)$$

with $\nabla \cdot \langle \boldsymbol{v} \rangle = \nabla \cdot \widetilde{\boldsymbol{v}} = 0$. The equations are written in Alfvén time units assuming that we have a homogeneous density. The effect of the turbulence is contained in the turbulent-stress tensor

$$-\langle \widetilde{v}_i \widetilde{v}_j - \widetilde{b}_i \widetilde{b}_j \rangle = R_{ij}, \quad (4.4)$$

which consists of the Reynolds tensor $R^V_{ij} = -\langle \widetilde{v}_i \widetilde{v}_j \rangle$ and the Maxwell tensor $R^M_{ij} = \langle \widetilde{b}_i \widetilde{b}_j \rangle$, and the turbulent electromotive force

$$\boldsymbol{\epsilon} = \langle \widetilde{\boldsymbol{v}} \times \widetilde{\boldsymbol{b}} \rangle, \quad (4.5)$$

the negative of the electric field induced by the turbulence, $\boldsymbol{\epsilon} = -\boldsymbol{E}$ apart from resistive effects. The electromotive force can also be written in tensorial form:

$$\epsilon_i = -\varepsilon^{ijk} S_{jk}, \qquad S_{ij} = -\langle \widetilde{v}_i \widetilde{b}_j - \widetilde{v}_j \widetilde{b}_i \rangle. \quad (4.6)$$

Here ε^{ijk} is the completely antisymmetric unit tensor. Expressed in terms of the Elsässer fields, the mean-field equations (4.2) and (4.3) become

$$\begin{aligned} \partial_t \langle z^\pm \rangle + \langle z^\mp \rangle \cdot \nabla \langle z^\pm \rangle = -\nabla \langle P \rangle - \nabla \cdot \langle \widetilde{z}^\pm \widetilde{z}^\mp \rangle + \tfrac{1}{2}(\nu + \eta) \nabla^2 \langle z^\pm \rangle \\ + \tfrac{1}{2}(\nu - \eta) \nabla^2 \langle z^\mp \rangle, \quad (4.7) \end{aligned}$$

where the effect of the turbulence appears in the tensor,

$$-\langle \widetilde{z}_i^\pm \widetilde{z}_j^\mp \rangle = R^\pm_{ij}, \quad (4.8)$$

from which R_{ij} and S_{ij} can readily be obtained as the symmetric and antisymmetric parts. The stress tensors R_{ij} and S_{ij} are special cases of the two-point correlation functions $\langle \widetilde{\boldsymbol{v}}(\boldsymbol{x}, t) \widetilde{\boldsymbol{v}}(\boldsymbol{x}', t') \rangle$ etc., the equations for which are readily derived by multiplying the equations for the fluctuations $\widetilde{\boldsymbol{v}}(\boldsymbol{x}, t)$ by $\widetilde{\boldsymbol{v}}(\boldsymbol{x}', t')$ and averaging, and similarly for the other correlation functions. Since at this point we are not concerned with these equations, we only point out their structure,

$$\partial_t \langle uu' \rangle = \mathcal{L}_2 \langle uu' \rangle + \mathcal{L}_3 \langle uu'u'' \rangle, \tag{4.9}$$

where u represents any component $\widetilde{v}_i$, $\widetilde{b}_i$, and $\mathcal{L}_{2,3}$ are differential operators depending on the mean fields $\langle \boldsymbol{v} \rangle$ and $\langle \boldsymbol{B} \rangle$. Hence, to calculate the second-order correlation functions, we need to know the third-order functions and so forth, which leads to the problem of closure in turbulence theory, which is discussed in more detail in Chapter 6.

4.1.2 Turbulent transport coefficients

In the one-point-closure approximation expressions for the turbulent stresses, the one-point correlations, are obtained by phenomenological modeling. Let us first discuss the case of hydrodynamic turbulence. One introduces the concept of the eddy viscosity ν_t (index t for 'turbulent') by writing the turbulent Reynolds stresses in a form analogous to the viscous-stress tensor (2.5),

$$R_{ij}^V = -\langle \widetilde{v}_i \widetilde{v}_j \rangle = -\tfrac{1}{3} \langle \widetilde{\boldsymbol{v}}^2 \rangle \delta_{ij} + \nu_t \big(\partial_i \langle v_j \rangle + \partial_j \langle v_i \rangle \big). \tag{4.10}$$

The first term on the r.h.s. is the isotropic part, which can simply be added to the pressure $\langle p \rangle$, while the second is the deviatoric part (for the definition, see the footnote preceding equation (2.5)). Equation (4.10) implies isotropy of the small-scale turbulence, since one expresses the five independent components of R_{ij}^V essentially in terms of two quantities, the mean turbulence energy $E = \frac{1}{2} \langle \widetilde{\boldsymbol{v}}^2 \rangle$ and the eddy viscosity ν_t, in a way similar to what we assumed for the effects of collisions in molecular viscosity. In the simplest model one assumes that ν_t depends only on E and the energy-dissipation rate $\epsilon = \nu \sum_{ij} \langle (\partial_i \widetilde{v}_j)^2 \rangle$. From dimensional arguments – E and ϵ have the dimensions $[E] = L^2/T^2$ and $[\epsilon] = L^2/T^3$ – we obtain the form

$$\nu_t = C E^2 / \epsilon, \tag{4.11}$$

where C is a free numerical parameter. The quantities E and ϵ, which are functions of the large-scale coordinates, are obtained from additional coupled advection–diffusion equations. The *Ansatz* (4.11) is usually called the K–ϵ model, since K is the traditional notation for the kinetic energy in hydrodynamic turbulence.

A more physically motivated model of the eddy viscosity is based on the concept of the *mixing length* l_m, a turbulent 'mean free path',

$$\nu_t = \widetilde{v} l_m, \tag{4.12}$$

where $\widetilde{v}$ represents a typical velocity-fluctuation amplitude. The mixing length, which was introduced by Prandtl (1925), is, roughly speaking, the scale over which momentum is transported during an eddy turnover time and is hence of the order of the size of the dominant eddies. In applications l_m is usually considered as a free parameter to be adjusted to fit the observations. (For turbulence close to a wall l_m has to be a function of the distance x from the wall, $l_m \propto x$.) The fluctuation amplitude can be estimated by assuming that the gradient of the velocity fluctuation compensates the mean velocity gradient driving the fluctuation, $|\nabla \widetilde{\boldsymbol{v}}| \sim d\langle v\rangle/dx$; hence

$$\widetilde{v} \sim l_m \frac{d\langle v\rangle}{dx}, \tag{4.13}$$

which is called the mixing-length estimate of the turbulence level. From (4.12) and (4.13) we now obtain

$$\nu_t = l_m^2 \left| \frac{d\langle v\rangle}{dx} \right|, \tag{4.14}$$

where the vertical bars indicate that the eddy viscosity is to be taken positive, since the effect of the turbulence should be dissipative. For practical applications these concepts are used in large-eddy simulations, which will be discussed in Section 4.1.3.

We now return to the MHD case and give expressions for R_{ij}, (4.4), and S_{ij}, (4.6), generalizing the hydrodynamic *Ansatz* (4.10) (Yoshizawa, 1990),[1]

$$R_{ij} = -\tfrac{1}{3}\langle \widetilde{\boldsymbol{v}}^2 - \widetilde{\boldsymbol{b}}^2\rangle \delta_{ij} + \nu_t^V\Big(\partial_i\langle v_j\rangle + \partial_j\langle v_i\rangle\Big) - \nu_t^M\Big(\partial_i\langle B_j\rangle + \partial_j\langle B_i\rangle\Big), \tag{4.15}$$

$$S_{ij} = -\alpha_t\, \varepsilon^{ijk}\langle B_k\rangle + \beta_t^M\Big(\partial_i\langle B_j\rangle - \partial_j\langle B_i\rangle\Big) - \beta_t^V\Big(\partial_i\langle v_j\rangle - \partial_j\langle v_i\rangle\Big). \tag{4.16}$$

The last equation can also be written in the more familiar vector form,

$$\boldsymbol{\epsilon} = \alpha_t\langle \boldsymbol{B}\rangle - \beta_t^M\langle \boldsymbol{j}\rangle + \beta_t^V\langle \boldsymbol{\omega}\rangle, \tag{4.17}$$

with $\boldsymbol{j} = \nabla \times \boldsymbol{B}$, the current density, and $\boldsymbol{\omega} = \nabla \times \boldsymbol{v}$, the vorticity. (Similar expressions have been derived by Zhou and Vahala (1991.) Let us first discuss some general properties of the various eddy-viscosity-type coefficients appearing in these expressions. Since $\boldsymbol{\epsilon}$ and $\boldsymbol{j}$ are vectors, while $\boldsymbol{B}$ and $\boldsymbol{\omega}$ are axial vectors, ν_t^V and β_t^M are scalars, while ν_t^M, β_t^V, and α_t are pseudoscalars (a pseudoscalar reverses sign under coordinate reflection). The physical meaning of the coefficients ν_t^V and β_t^M can easily be understood. ν_t^V is an eddy viscosity similar

[1] α_t is conventionally denoted simply by α, but we prefer to add the subscript to distinguish this quantity from the α characterizing the eddy viscosity in an accretion disk, Section 11.2.2.

to ν_t in the hydrodynamic relation (4.10), while comparison of (4.17) with the simple version of Ohm's law $\boldsymbol{E} = \eta \boldsymbol{j}$ shows that β_t^M is a turbulent resistivity (remember that $\epsilon = -\boldsymbol{E}$); hence both ν_t^V and β_t^M are expected to be positive. The ν_t^M-term in (4.15) results from the magnetic tension force exerted on $\langle \boldsymbol{v} \rangle$ by the fluctuation $\widetilde{\boldsymbol{b}}$, which explains the minus sign of this term. Its counterpart is the β_t^V-term in (4.16) representing the effect of the velocity fluctuation $\widetilde{\boldsymbol{v}}$ on the magnetic field $\langle \boldsymbol{B} \rangle$. The coefficients ν_t^V and β_t^M are primarily connected with the turbulence energy $E = \frac{1}{2}\langle \widetilde{v}^2 + \widetilde{b}^2 \rangle$, while ν_t^M and β_t^V are connected with the cross-helicity (2.65), $H^C = \langle \widetilde{\boldsymbol{v}} \cdot \widetilde{\boldsymbol{b}} \rangle$. Yoshizawa found the members of both pairs of analogous quantities to be of the same order, $\nu_t^V \simeq \beta_t^M$ and $\nu_t^M \simeq \beta_t^V$; however, as numerical simulations (Section 4.1.3) indicate, the latter two coefficients are significantly smaller than the former and can be neglected to a good approximation.

4.1.3 Large-eddy simulations of MHD turbulence

Numerical simulations have become an indispensible tool for studying turbulent phenomena. In basic turbulence theory, in which one deals with scaling properties and small-scale structures, one usually needs the exact information down to the dissipative scales. Hence, in order to cope with the limitations imposed by the available computer power, approximations are made at the large scales, such that the computational system is restricted to a small section of the entire turbulence field, where conditions are approximately homogeneous. Direct numerical simulations of homogeneous turbulence will be discussed in the following chapters.

For practical, or even technical, applications, in which the primary interest is in the large energy-containing turbulent eddies, it is natural to make approximations at the small scales. In such computations (called large-eddy simulations (LESs)) one introduces a cutoff wavenumber K, such that a function $v(x)$ is split into two parts, a low-pass-filtered part $v_K^<(x)$ and a high-pass-filtered part $v_K^>(x)$,

$$v_K^<(x) = \sum_{k \le K} v_k e^{ikx}, \quad v_K^>(x) = \sum_{k > K} v_k e^{ikx}, \tag{4.18}$$

$v = v_K^< + v_K^>$, of which only $v_K^<$ representing the large scales $l > K^{-1}$ is computed directly, while the effect of the smaller scales $l < K^{-1}$ contained in $v_K^>$ is treated in some phenomenological way called subgrid-scale modeling. The average flow $\langle \boldsymbol{v} \rangle$ in (4.1) is now the filtered velocity field $v^<$, which is actually resolved numerically, and the role of the mixing length is played by the grid spacing Δ related to the cutoff wavenumber, $K = \pi/\Delta$. In the most common form, which as introduced by Smagorinsky (1963), the eddy viscosity becomes

$$\nu_t = (C_S\Delta)^2 \left(\sum_{ij} \tfrac{1}{2}(\partial_i v_j^< + \partial_j v_i^<)^2 \right)^{1/2}, \tag{4.19}$$

which has the same basic form as the simple mixing-length *Ansatz* (4.14). Comparison of LESs with corresponding experiments determines the parameter C_S; one finds $C_S \simeq 0.1$ for turbulence far away from boundaries, whereas in boundary-layer turbulence C_S should decrease on approaching the wall (for a more detailed discussion see Yakhot *et al.*, 1989). A review of the physical aspects of LES in hydrodynamic turbulence can be found in an article by Canuto (1994). Chapter 12 of Lesieur's book (Lesieur, 1997), and a recent monograph by Sagaut (2001) present comprehensive introductions to the field.

LES is not as well established in MHD turbulence as it is in hydrodynamics. On the one hand the physics is more complex, since the MHD equations contain two fields, which introduces considerably more freedom into the dynamics, for instance the presence of both direct and inverse spectral cascades (Section 5.2). On the other hand experimental verification of the validity of a subgrid-scale model and the calibration of free parameters are very difficult, since MHD turbulence is mainly observed in astrophysical systems – *in situ* measurements are possible only in the solar wind – not in controllable laboratory experiments. At present, the most reliable way to check the accuracy of LESs of MHD turbulence is by comparison with high-resolution direct numerical simulations.

Yoshizawa (1991) presented a systematic derivation of a subgrid-scale theory, but the corresponding expressions are too unwieldy for actual computations and their physical meaning remains unclear. Let us therefore briefly describe several models based more on physical reasoning and numerical practicality. One possibility is a direct generalization of Smagorinsky's *Ansatz* (4.19), considering only the dissipative effects ν_t^V and β_t^M in (4.15) and (4.16) (e.g., Theobald *et al.*, 1994),

$$\nu_t^V = (C^V\Delta)^2 \left(\sum_{ij} \tfrac{1}{2}(\partial_i v_j^< + \partial_j v_i^<)^2 \right)^{1/2}, \quad \beta_t^M = (C^M\Delta)^2 |\boldsymbol{j}^<|. \tag{4.20}$$

It appears, however, that this *Ansatz* is too dissipative (Agullo *et al.*, 2001). One indication of such behavior is that, while the dynamics of MHD turbulence is suppressed when either $\boldsymbol{b}$ or $\boldsymbol{v}$ vanishes, the turbulent diffusivities (4.20) are not directly affected. A possible remedy is a combination of the effects of ν_t^V and ν_t^M, and β_t^M and β_t^V, respectively. In addition, to account for the inverse cascade of magnetic helicity, and along with it also of magnetic energy, one should allow negative values of the turbulent resistivity. Müller and Carati (2001) studied the following subgrid model:

$$\nu_t = (C\Delta)^2 \left| \sum_{ij} (\partial_i v_j^< + \partial_j v_i^<)(\partial_i b_j^< + \partial_j b_i^<) \right|^{1/2},$$

$$\beta_t = (C'\Delta)^2 s |\boldsymbol{j}^< \cdot \boldsymbol{\omega}^<|^{1/2}, \tag{4.21}$$

where s is a sign factor, which has been chosen to be $s = \mathrm{sgn}(\boldsymbol{j}^< \cdot \boldsymbol{\omega}^<)$. This model is found to give good agreement of the evolution of kinetic and magnetic energies E^K and E^M with previous direct numerical simulations. The indefinite sign of β_t in (4.21) substitutes in some way for the α_t-term in (4.16) and (4.17), the dynamo effect, which is neglected in most LES studies.

If no direct numerical simulations of sufficiently high Reynolds number are available, which is the case in most practical applications, in particular for inhomogneous systems, the free coefficients in the subgrid model can be determined by a dynamic procedure (Germano, 1992; see also Meneveau and Katz, 2000), which exploits the self-similarity of the turbulence. Here one introduces a second filtering process, the test filter, with the cutoff $K_t = \gamma K$, $\gamma < 1$, which is applied to the resolved fields. Since the fields are known explicitly in the range $K_t < k \leq K$, one knows also their contribution to the stress tensors. If the turbulence is self-similar, the same form of the subgrid model can be chosen both on the original grid and on the coarser test grid with the scale parameter Δ suitably adjusted. This gives a relation for the stress tensors on the test-grid level, which should hold if the subgrid model is exact. Optimal values of the free parameters of the subgrid model chosen, such as C and C' in (4.21), are then obtained by minimizing the actual error occurring in this relation.

4.1.4 Mean-field electrodynamics

On inserting the α_t-term into Faraday's law (4.3), we find that $\partial_t \boldsymbol{B}$ is proportional to $\boldsymbol{j}$. Since the most interesting applications of the dynamo effect are found in axisymmetric systems – in spherical shells such as in planetary and stellar dynamos and in the toroidal plasma column of a reversed-field pinch –, it is natural to divide $\boldsymbol{B}$ into a poloidal (or meridional) and a toroidal (or azimuthal) component. The relation $\partial_t \boldsymbol{B} = \alpha_t \boldsymbol{j}$ then shows that a poloidal field, corresponding to a toroidal current, amplifies the toroidal field (for the appropriate sign of α_t), while the latter, corresponding to a poloidal current, in turn amplifies the poloidal field, which is the basis of the dynamo mechanism. To see how the dynamo process is driven by turbulence, the simplest approach is to consider the kinematic phase, in which the magnetic field is still sufficiently weak to allow us to neglect the Lorentz force in the equation of motion, such that the velocity can be regarded as an independent given quantity in Faraday's law. In this case we treat only the latter, i.e., the equations for the mean magnetic field, (4.3), and for the field fluctuation,

$$\partial_t \widetilde{\boldsymbol{b}} - \nabla \times (\langle \boldsymbol{v} \rangle \times \widetilde{\boldsymbol{b}}) = \nabla \times (\widetilde{\boldsymbol{v}} \times \langle \boldsymbol{B} \rangle) + \nabla \times \left(\widetilde{\boldsymbol{v}} \times \widetilde{\boldsymbol{b}} - \langle \widetilde{\boldsymbol{v}} \times \widetilde{\boldsymbol{b}} \rangle \right) + \eta \nabla^2 \widetilde{\boldsymbol{b}}. \tag{4.22}$$

It is also assumed that the fluctuations are small compared with the mean field, $\widetilde{\boldsymbol{b}} \ll \langle \boldsymbol{B} \rangle$, such that the nonlinear term in (4.22) can be omitted. This is called the quasi-linear approximation, whereby only the nonlinear term in the mean-field equation (4.3) is retained. These equations are referred to as *mean-field electrodynamics* and were developed in a pioneering paper by Steenbeck *et al.* (1966) (for a review of the theory, see Krause and Rädler, 1980). To obtain a simple explicit expression for the magnetic fluctuation, one assumes, in addition, that the spatial variation of the mean velocity is weak, such that the second term on the l.h.s. of (4.22) reduces to a simple advective term. Equation (4.22) now reads

$$\partial_t \widetilde{\boldsymbol{b}} + \langle \boldsymbol{v} \rangle \cdot \nabla \widetilde{\boldsymbol{b}} = \nabla \times (\widetilde{\boldsymbol{v}} \times \langle \boldsymbol{B} \rangle) \tag{4.23}$$

neglecting resistive effects. By integrating along the orbit of a fluid element, (4.22) can formally be solved, giving

$$\widetilde{\boldsymbol{b}}(\boldsymbol{x}, t) = \int_{-\infty}^{t} dt' \, \nabla \times \left[\widetilde{\boldsymbol{v}}(\boldsymbol{x}', t') \times \langle \boldsymbol{B} \rangle \right]. \tag{4.24}$$

By inserting $\widetilde{\boldsymbol{b}}$ into the electromotive force (4.5) one obtains

$$\boldsymbol{\epsilon} = \int_{-\infty}^{t} dt' \left\langle \widetilde{\boldsymbol{v}} \times \left[\nabla \times (\widetilde{\boldsymbol{v}}' \times \langle \boldsymbol{B} \rangle) \right] \right\rangle. \tag{4.25}$$

Here the mean field can be regarded as constant, since its variation over a correlation time of the fluctuations is small. Expression (4.25) is most easily evaluated by considering vector components $\boldsymbol{\epsilon} = \{\epsilon_i\}$. The ∇-operator acts either on $\widetilde{\boldsymbol{v}}'$ or on the average field; hence ϵ_1 contains $\langle B_1 \rangle$ and the derivatives $\partial_2 \langle B_3 \rangle$ and $\partial_3 \langle B_2 \rangle$,

$$\begin{aligned} \epsilon_1 = & \int_{-\infty}^{t} dt' \left(\langle \widetilde{v}_2 \partial_1 \widetilde{v}_3' \rangle - \langle \widetilde{v}_3 \, \partial_1 \widetilde{v}_2' \rangle \right) \langle B_1 \rangle \\ & - \int_{-\infty}^{t} dt' \left(\langle \widetilde{v}_2 \widetilde{v}_2' \rangle \, \partial_2 \langle B_3 \rangle - \langle \widetilde{v}_3 \widetilde{v}_3' \rangle \, \partial_3 \langle B_2 \rangle \right). \end{aligned} \tag{4.26}$$

Since the effect of the mean magnetic field on the turbulence is neglected, the velocity fluctuations may be assumed isotropic, which implies invariance of (4.26) under cyclic permutations,

$$\langle\widetilde{v}_2\partial_1\widetilde{v}_3'\rangle - \langle\widetilde{v}_3\partial_1\widetilde{v}_2'\rangle = \langle\widetilde{v}_2(\partial_1\widetilde{v}_3' - \partial_3\widetilde{v}_1')\rangle = -\tfrac{1}{3}\langle\widetilde{\boldsymbol{v}}\cdot\nabla\times\widetilde{\boldsymbol{v}}'\rangle, \tag{4.27}$$

$$\langle\widetilde{v}_1\widetilde{v}_1'\rangle = \langle\widetilde{v}_2\widetilde{v}_2'\rangle = \langle\widetilde{v}_3\widetilde{v}_3'\rangle = \tfrac{1}{3}\langle\widetilde{\boldsymbol{v}}\cdot\widetilde{\boldsymbol{v}}'\rangle. \tag{4.28}$$

When these expressions are inserted into (4.26), comparison with (4.17) gives

$$\alpha_t = -\tfrac{1}{3}\int_{-\infty}^{t} dt'\,\langle\widetilde{\boldsymbol{v}}\cdot\nabla\times\widetilde{\boldsymbol{v}}'\rangle = -\tfrac{1}{3}\tau\langle\widetilde{\boldsymbol{v}}\cdot\nabla\times\widetilde{\boldsymbol{v}}\rangle = -\tfrac{1}{3}\tau H^K, \tag{4.29}$$

$$\beta_t^M = \tfrac{1}{3}\int_{-\infty}^{t} dt'\,\langle\widetilde{\boldsymbol{v}}\cdot\widetilde{\boldsymbol{v}}'\rangle = \tfrac{1}{3}\tau'\langle\widetilde{\boldsymbol{v}}^2\rangle = \tfrac{2}{3}\tau' E^K, \tag{4.30}$$

where τ and τ' are velocity-correlation times, $\tau \simeq \tau'$, H^K is the kinetic helicity and E^K is the kinetic energy of the turbulence. The third coefficient β_t^V in (4.17) is not included in this derivation, since derivatives of the mean velocity, in particular the vorticity $\langle\boldsymbol{\omega}\rangle$, are neglected in (4.23).

Expression (4.29) forms the basis of the theory of magnetic-field generation in most cosmic objects, especially in stars, where a natural cause of helical fluid motions is thermal convection in a rotating system. Here one estimates the magnitude of α_t by a mixing-length approach. The helicity H^K is, roughly, proportional to the two vectors characterizing convective turbulence in such systems, the angular frequency $\boldsymbol{\Omega}$ (an axial vector) and the density (or temperature) stratification $\nabla \ln \rho$. Hence by simple dimensional arguments one finds $H^K \sim \widetilde{v} l_m \boldsymbol{\Omega}\cdot\nabla\ln\rho$, where the mixing length is assumed to be of the order of the density scale height, $l_m \sim |\nabla \ln \rho|^{-1}$. Since the correlation time is $\tau \sim l_m/\widetilde{v}$, we obtain from (4.29)

$$\alpha_t \sim l_m^2 \boldsymbol{\Omega}\cdot\nabla\ln\rho. \tag{4.31}$$

One should, however, keep in mind that the expression (4.29) is valid only in the kinematic limit, i.e., for sufficiently weak magnetic fields. Estimates of the reaction on the fluid motions show that the nonlinear decrease of α_t sets in when the spectral intensity $|B_k|^2$ becomes of the order of the kinetic energy $|v_k|^2$, which is first reached at the small scales and is connected with the Alfvén effect considered in Section 4.2.2. Yoshizawa (1990) derived that $\alpha_t \propto \langle\widetilde{\boldsymbol{v}}\cdot\widetilde{\boldsymbol{\omega}} - \widetilde{\boldsymbol{b}}\cdot\widetilde{\boldsymbol{j}}\rangle$; hence α_t may be strongly reduced for $|B_k|^2 \sim |v_k|^2$. This nonlinear quenching justifies neglecting the α_t-term in the subgrid-scale models used for LESs of MHD turbulence discussed in Section 4.1.2.

4.2 Self-organization processes

Self-organization means the spontaneous generation of large-scale coherent structures. Since the term 'structure' implies already that its appearance is 'coherent', we will henceforth talk only of structures. The qualifier 'spontaneous' is, however, important, since most systems of fully developed turbulence exhibit large-scale structures, which reflect the properties of the turbulence drive, for instance the large-scale vortices in the wake behind some object or the primary vortices of the Kelvin–Helmholtz instability in a turbulent jet, which are continuously regenerated simply by the geometry of the system. By contrast, the structures generated by self-organization arise spontaneously out of a sea of homogeneous turbulence. Such processes are intimately connected with the presence of an inverse spectral cascade, which will be considered in more detail in Chapter 5, whereas this section is focussed on the macroscopic aspects of self-organization.

4.2.1 Selective decay

It is often said that self-organization in turbulence is due to the presence of several ideal invariants. The classical example is 2D hydrodynamic turbulence, which is characterized by the ideal conservation both of energy and of the mean-square vorticity, or enstrophy. Here self-organization leads to formation of large-scale vortices, such as the cyclones and anticyclones in the atmosphere, whose size is limited only by the variation of the normal component of the Earth's rotation vector caused by the curvature of the Earth. By contrast, fully 3D hydrodynamic turbulence does not seem to exhibit self-organization. In MHD theory there are three (quadratic) invariants, the energy E, the cross-helicity H^C, and the magnetic helicity H^M. Let us give again the conservation relations written in Alfvén time units, assuming that incompressibility holds and ignoring surface terms by assuming, for instance, periodic boundary conditions,

$$\frac{dE}{dt} \equiv \frac{d}{dt}\int \tfrac{1}{2}(v^2 + B^2)\,dV = -\eta \int j^2\,dV - \nu \int \omega^2\,dV, \tag{4.32}$$

$$\frac{dH^C}{dt} \equiv \frac{d}{dt}\int \boldsymbol{v}\cdot\boldsymbol{B}\,dV = -(\nu+\eta)\int \boldsymbol{j}\cdot\boldsymbol{\omega}\,dV, \tag{4.33}$$

$$\frac{dH^M}{dt} \equiv \frac{d}{dt}\int \boldsymbol{A}\cdot\boldsymbol{B}\,dV = -\eta\int \boldsymbol{j}\cdot\boldsymbol{B}\,dV. \tag{4.34}$$

Dissipation of turbulence occurs primarily at small scales. Since the dissipation terms contain different orders of spatial derivatives, the decay rates of the ideal invariants are, in general, different; in particular, H^M is expected to decay more slowly than E. In addition, the integral on the r.h.s. of (4.33) is not positive definite, such that also H^C will, in general, decay more slowly than the energy. (H^C and H^M can, in principle, even increase with time.) There are hence two selective decay processes, which govern the decay of MHD turbulence. On the one hand, turbulence may relax to a minimum-energy state under the constraint of constant helicity H^M (if $H^M \neq 0$; in the special case of a reflectionally symmetric state with $H^M = 0$ a different behavior occurs, as discussed in Section 4.2.3), which is described by the following variational principle (Woltjer, 1958; Taylor, 1974):

$$\delta\left(\int \tfrac{1}{2}(v^2 + B^2)\,dV - \tfrac{1}{2}\lambda \int \boldsymbol{A}\cdot\boldsymbol{B}\,dV\right) = 0. \tag{4.35}$$

Variation with respect to $\boldsymbol{A}$ gives the equation

$$\nabla \times \boldsymbol{B} - \lambda \boldsymbol{B} = 0, \tag{4.36}$$

while variation with respect to $\boldsymbol{v}$ gives $\boldsymbol{v} = 0$. Hence the minimum-energy state (properly speaking, the extremum-energy solution; the minimum property has to be proved separately) is a static constant-λ force-free field, called a linear force-free field,[2] where λ, the Lagrange multiplier in the variational equation, is determined by the value of H^M, $\lambda = \int B_{\rm min}^2\,dV/H^M$, and $B_{\rm min}$ itself depends on λ through (4.36). Such force-free states play an important role in reversed-field-pinch plasmas (see, for instance Biskamp, 1993a). They also serve as model fields in low-β astrophysical systems, for instance coronal loops.

Competing with this relaxation to a static force-free field, there is a further process of self-organization, which leads to an alignment of $\boldsymbol{v}$ and $\boldsymbol{B}$. This behavior can be associated with the slow decay of the cross-helicity compared with that of the energy, such that the relaxed state is the minimum-energy state for a given value of H^C, which follows from the variational principle

$$\delta\left(\int \tfrac{1}{2}(v^2 + B^2)\,dV - \lambda' \int \boldsymbol{v}\cdot\boldsymbol{B}\,dV\right) = 0. \tag{4.37}$$

Variation with respect to $\boldsymbol{v}$ or $\boldsymbol{B}$ yields the equations

$$\boldsymbol{v} - \lambda'\boldsymbol{B} = 0,$$
$$\boldsymbol{B} - \lambda'\boldsymbol{v} = 0.$$

[2] A magnetic field is called force-free if the Lorentz force vanishes, $\boldsymbol{j} \times \boldsymbol{B} = 0$, such that $\nabla \times \boldsymbol{B} = \lambda\boldsymbol{B}$. In general, λ varies in space, satisfying only the condition $\boldsymbol{B}\cdot\nabla\lambda = 0$, which follows from $\nabla\cdot\boldsymbol{B} = 0$.

Hence the Lagrange multiplier satifies $\lambda'^2 = 1$, such that

$$v = \pm \boldsymbol{B}, \tag{4.38}$$

which is valid locally at each point in space. The relaxed state (4.38) is a so-called pure Alfvénic state since it corresponds to a finite-amplitude Alfvén wave. Since in an aligned state (4.38) the dynamics is turned off, as can be seen directly in the Elsässer formulation of the MHD equations (2.106) (in an aligned state either z^+ or z^- vanishes), this is also the final state to which the turbulence decays, apart from the very slow collisional diffusion. Which of the two relaxation processes just discussed dominates depends on the initial values of H^M and H^C. In a strongly helical system, the final state is the force-free field, whereas for sufficiently large initial alignment the system ends up in an Alfvénic state. Numerical computations by Stribling and Matthaeus (1991) confirmed this picture.

4.2.2 Alfvén effect and dynamic alignment

We now show, following Dobrowolny *et al.* (1980), that the relaxation to an aligned state is indeed likely to occur. For this purpose it is convenient to use the Elsässer fields presented in Section 2.7, which describe Alfvén waves propagating along a guide field (2.107). This guide field need not be an external static field, but can also be the field in the large-scale energy-containing eddies. A remarkable property of the dynamic equations for the Elsässer fields, (2.106), is the absence of self-interactions in the nonlinear term, which just couples z^+ and z^-. Hence only Alfvén waves propagating in opposite directions along the guide field interact. This is the basis of the *Alfvén effect* introduced independently by Iroshnikov (1964) and Kraichnan (1965b), which may play a crucial role in MHD turbulence, assuming that the cascade dynamics is mainly due to scattering of Alfvén waves.

We distinguish between two dynamic time scales, the Alfvén time $\tau_A = l/v_A$ and the time for distortion of a wave packet, or eddy, δz_l^+ of scale l by a similar eddy δz_l^- and vice versa (a more precise definition of the amplitudes z_l is given in Section 5.3.2),

$$\tau_l^{\pm} = l/\delta z_l^{\mp}, \tag{4.39}$$

where in general $\tau_A \ll \tau_l^{\pm}$. Since the interaction time of two oppositely propagating wave packets is τ_A, the change of amplitude $\Delta\delta z_l$ during a single collision of two wave packets is small,

$$\frac{\Delta\delta z_l^\pm}{\delta z_l^\pm} = \frac{\tau_A}{\tau_l^\pm} \ll 1.$$

Because of the diffusive nature of the process, $N \sim (\delta z_l/\Delta\delta z_l)^2$ elementary interactions are needed in order to produce a relative change in amplitude of order unity. Hence the energy-transfer time is

$$T_l^\pm \sim N\tau_A \sim (\tau_l^\pm)^2/\tau_A. \tag{4.40}$$

Since both $E = E^+ + E^-$ and $H^C = \frac{1}{2}(E^+ - E^-)$ are ideal invariants, so are E^+ and E^-, where $E^\pm = \frac{1}{4}\int (z^\pm)^2\,dV$. The spectral densities $(\delta z_l^+)^2$ and $(\delta z_l^-)^2$ are therefore cascading quantities as discussed in Section 5.2, and the corresponding energy fluxes are

$$\epsilon_l^\pm \sim (\delta z_l^\pm)^2/T_l^\pm \sim (\delta z_l^+)^2(\delta z_l^-)^2\tau_A/l^2. \tag{4.41}$$

Since these fluxes are constant across the inertial range, we write $\epsilon_l^\pm = \epsilon^\pm$, where $\epsilon^\pm$ are the dissipation rates of $E^\pm$, the symmetry in (4.41) indicates that $\epsilon^+ = \epsilon^-$, and hence the cross-helicity is not dissipated in this picture,

$$dH^C/dt = \tfrac{1}{4}(\epsilon^+ - \epsilon^-) = 0. \tag{4.42}$$

(Strictly speaking, the relation $\epsilon^+ = \epsilon^-$ holds only in the local approximation in wavenumber space, and will be somewhat modified by considering interactions within a certain band of wavenumbers, as discussed in Section 5.3.)

Let us now consider the consequences of (4.42) on the dynamics of decaying turbulence, assuming that initially $\delta z_l^+ > \delta z_l^-$. In this case the energy fluxes are time-dependent $\epsilon^\pm(t)$ but still preserve the property $\epsilon^+ = \epsilon^-$, which follows from the symmetry of (4.41) and does not require global stationarity. The energy-transfer times are, however, different. From (4.40) one finds that

$$T_l^+/T_l^- = (\delta z_l^+/\delta z_l^-)^2 > 1, \tag{4.43}$$

which means that energy transfer, and hence damping, of the minority field δz_l^- is more rapid, leading to a continuous increase of the ratio E^+/E^-, until the dynamics is finally switched off in a pure Alfvénic state. Thus dynamic alignment is a direct consequence of the MHD equations, valid both in 3D and in 2D. The process has been studied quantitatively in the framework of closure theory (Grappin *et al.*, 1982 and 1983) as well as in direct numerical simulations in 2D (Pouquet *et al.*, 1988; Biskamp and Welter, 1989).

4.2.3 Energy-decay laws

In the preceding sections we used only the property that the turbulence energy decays more rapidly than does the cross-helicity or magnetic helicity, but we have not yet specified the energy-decay law. Whereas the rate of decay a of smooth laminar field or flow is proportional to the resistivity or the viscosity, $E_k \sim e^{-2\eta k^2 t}$ for a mode k, in fully developed (3D) hydrodynamic turbulence both experimental results and numerical simulations indicate that the decay of energy is essentially independent of the Reynolds number, and this property seems to be valid also in MHD turbulence even in 2D (in contrast to 2D hydrodynamic turbulence, which was discussed in Section 8.1).

The decay properties of turbulence not only constitute an intriguing academic problem but also have considerable practical consequences. Think of the turbulence in the wake of a large airplane, which can produce a violent disturbance on a smaller plane crossing this wake. It is therefore clearly important to know how fast the turbulence decays. Or consider the interstellar magnetic field in a galaxy, which is presumably highly turbulent. The rate of decay of this turbulence gives a measure of the lifetime of the galactic field and helps us to decide whether this field can be of primordial origin.

Decay of hydrodynamic turbulence

Self-similarity of the decay of turbulence, if this property is indeed satisfied, suggests that there should be a power-law behavior $E \sim t^{-\lambda}$, where the exponent λ is a characteristic property of the turbulent system.[3] A theory of the decay of hydrodynamic turbulence was first derived by Kolmogorov (1941b) on the basis of the assumed invariance of the Loitsiansky integral $\Lambda = \int_0^\infty \langle v_\parallel(\boldsymbol{x}) v_\parallel(\boldsymbol{x}+\boldsymbol{r})\rangle r^4\, dr$, where $v_\parallel = \boldsymbol{v}\cdot\boldsymbol{r}/r$. The basic idea is rather simple. Let us define the integral scale length L by the relation

$$\Lambda = EL^5 = \text{constant}. \tag{4.44}$$

On the other hand, the integral length should be independent of the Reynolds number and hence depend only on the integral quantities of the system, the turbulence energy E and the energy-decay rate ϵ; hence, by simple dimensional arguments,

$$L \sim E^{3/2}/\epsilon, \quad \text{with} \quad \epsilon = -dE/dt. \tag{4.45}$$

[3] In the following analysis, which is mainly built on dimensional arguments, the integral quantities such as E and H^M refer to a fixed volume, which we can assume to be the unit volume, such that for instance $E = V^{-1}\int \frac{1}{2}v^2\, dV$ has the same dimensions as the integrand, $[E] = L^2T^{-2}$, or the energy-dissipation rate $[\epsilon] = L^2T^{-3}$.

Combining (4.44) and (4.45) gives the differential equation

$$-dE/dt \sim E^{17/10}, \tag{4.46}$$

which has the asymptotic solution $E \sim t^{-10/7}$. Unfortunately, however, the Loitsiansky integral is not preserved, as noted by Batchelor and Proudman (1956), which invalidates Kolmogorov's approach. Lesieur and Schertzer (1978) considered the decay problem in the framework of closure theory. The authors relate the decay of energy to the 'principle of permanence of big eddies', the invariance of the energy spectrum at small wavenumbers,

$$E_k = Ck^s = \text{constant}, \quad \text{for} \quad k < k_{\text{in}}, \tag{4.47}$$

where $k_{\text{in}} \sim L^{-1}$ is the injection wavenumber and s is restricted to $1 \simeq s \leq 4$. The energy-decay law can be obtained using the self-similarity of the spectrum during the decay of turbulence. Thus we may write the energy spectrum E_k in terms of an invariant function $f(kL)$,

$$E_k(t) = E(t)L(t)f(kL), \tag{4.48}$$

which, by use of (4.47), yields

$$E_k \simeq EL^{s+1} = \text{constant} \quad \text{for} \quad kL < 1, \tag{4.49}$$

generalizing relation (4.44). Using (4.45) as in Kolmogorov's theory, one now finds the exponent of the energy-decay law

$$\lambda = 2(s+1)/(s+3), \tag{4.50}$$

while the integral length increases, $L \sim t^p$,

$$p = 2/(s+3). \tag{4.51}$$

Since, as has been assumed, the shape of the small-k spectrum is invariant for $s \leq 4$, the energy-decay law depends on the initial spectrum. For instance, one has $\lambda = 1$ for $s = 1$ or $\lambda = \frac{6}{5}$ for $s = 2$. If the spectrum is steeper, $s > 4$, initially, it is found to relax rapidly to $s = 4$, which hence plays a special role. For $s = 4$ (4.50) seems to indicate the Kolmogorov value $\lambda = 10/7 \simeq 1.43$, but in this case the parameter C in (4.47) is no longer constant but varies in time, which is related to the noninvariance of the Loitsiansky integral mentioned above and leads to a slightly lower decay exponent, $\lambda \simeq 1.38$.

Experimental observations of the decay of turbulence are mainly performed on grid-generated turbulence in a wind tunnel by measuring the turbulence level at various distances x using Taylor's hypothesis to interpret the variation with time at a fixed position in terms of the spatial variation $t = x/U$, where

U is the mean velocity. A decay exponent close to 1.38 has indeed been found by Warhaft and Lumney (1978), but in other studies also significantly lower values are observed, typically $\lambda \sim 1$. Since it is difficult to measure the low-k spectrum, the relevance of this theory cannot yet be established.

Decay of MHD turbulence

We now consider the decay of energy in MHD turbulence. Since for a high Reynolds number the magnetic helicity is well conserved, we can use similar arguments to those in Kolmogorov's theory given above. Let us therefore first discuss the case of finite H^M, which is characteristic of most astrophysical plasmas, since magnetic turbulence usually occurs in rotating systems, wherein the combined action of Coriolis and buoyancy forces naturally leads to twisted field lines. We define a magnetic integral scale L_M of the turbulence by

$$E^M L_M = H^M = \text{constant}, \tag{4.52}$$

where E^M is the magnetic energy, $E = E^M + E^K$. On the other hand, we have the dynamic scale length L, (4.45). Assuming that self-similarity of the decay holds, in particular $E \sim E^M \sim E^K$, and combining (4.52) and (4.45) identifying the two scales $L_M \sim L$, one obtains

$$-dE/dt \sim E^{5/2}, \tag{4.53}$$

which has the asymptotic solution $E \sim t^{-2/3}$. (This result was originally derived by Hatori (1984), who, unnecessarily, assumed the existence of a specific inertial-range spectrum.) Since experiments on MHD turbulence are difficult to perform, the only practical way, at present, to test this prediction is by direct numerical simulations. Several numerical studies have been reported, in particular by Biskamp and Müller (2000a, 2000b), who used a relatively high spatial resolution and hence high Reynolds number, which is important in order to insure that H^M remains constant. Figure 4.1 shows $E(t)$ from a typical simulation run. It is, however, difficult to test the theoretical prediction for the decay exponent λ directly from a log–log plot of $E(t)$ as given in this figure, since the exact solution of (4.53) is $E = E_0/(t - t_0)^{2/3}$, which contains the integration constant t_0 of the order of the initial eddy-turnover time and thus depends on the initial conditions, while the asymptotic behavior $t^{-\lambda}$ becomes visible only at sufficiently large times $t \gg t_0$. To check the validity of the theoretical assumptions, it is therefore simpler and more reliable to compare the simulation results with the differential decay law (4.53),

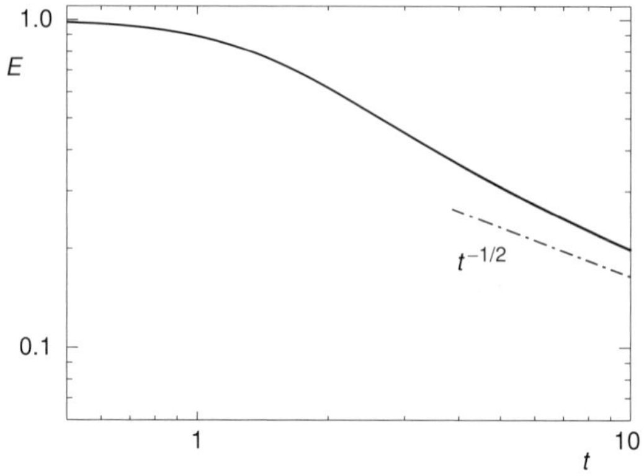

Figure 4.1: A log–log plot of the energy $E(t)$ from a simulation run with magnetic helicity $H^M = 0.68 H^M_{\mathrm{max}}$. The dash–dotted line indicates the asymptotic law $t^{-1/2}$. (From Biskamp and Müller, 2000b.)

which can be written in the form

$$\frac{E^{5/2}}{\epsilon H^M} = \text{constant}. \tag{4.54}$$

However, simulations show that expression (4.54) is not constant but increases with time, indicating that the decay is slower than $t^{-2/3}$. The origin of this discrepancy lies in the assumption of self-similarity, which is actually not satisfied; in particular, the energy ratio $\Gamma = E^K/E^M$ is not constant but decreases. Let us therefore generalize the analysis by including the variation with time of the energy ratio Γ. Assuming that the most important nonlinearities in the MHD equations arise from the advective terms, we write

$$-\frac{dE}{dt} = \epsilon \sim \boldsymbol{v} \cdot \nabla E \sim (E^K)^{1/2} \frac{E}{L} \tag{4.55}$$

instead of (4.45). Making the substitutions $L \sim L_M$ from (4.52) and $E^K = E\Gamma/(1+\Gamma)$ gives

$$\frac{E^{5/2}}{\epsilon H^M} \frac{\Gamma^{1/2}}{(1+\Gamma)^{3/2}} = \text{constant}. \tag{4.56}$$

Figure 4.2 shows that this ratio is indeed constant for $t > 2$, after turbulence has become fully developed, with little spread of the curves obtained from a number of runs with different values of H^M and Rm, which indicates that (4.55) is rather generally valid and also shows that the decay of turbulence is independent of the Reynolds number.

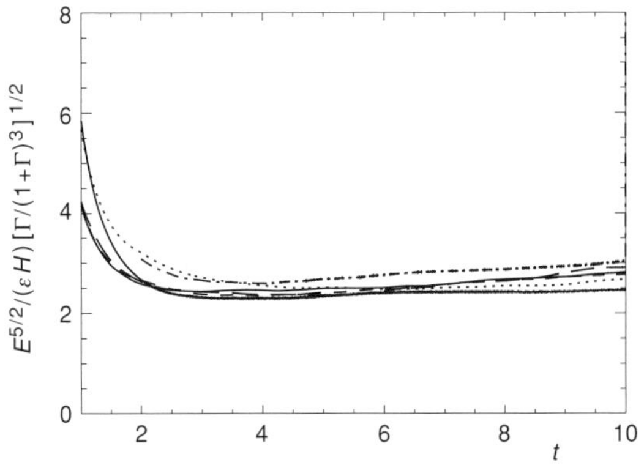

Figure 4.2: Expression (4.56) for a number of simulation runs with different values of the Reynolds number and of H^M (from Biskamp and Müller, 2000b).

Relation (4.55) does not yet give the energy-decay law, since it contains the energy ratio Γ, which varies with time. The simulations show that Γ is proportional to the total energy,

$$\Gamma \sim E/H^M. \tag{4.57}$$

By inserting this relation into (4.56) we obtain the differential equation for $E(t)$, which in the limit $\Gamma \ll 1$ reduces to

$$-\frac{dE}{dt} \sim \frac{E^3}{(H^M)^{3/2}}. \tag{4.58}$$

It has the similarity solution $t^{-1/2}$; hence energy decays more slowly than the $t^{-2/3}$ decay predicted for a self-similar turbulence decay. Note, however, that this behavior is reached only asymptotically, whereas for finite Γ the energy decays more rapidly, as can also be seen in Fig. 4.1. Using this result and the asymptotic behavior $\Gamma = E^K/E^M \simeq E^K/E$, (4.57) gives the decay law of the kinetic energy

$$E^K \sim t^{-1}. \tag{4.59}$$

The decay of the energy ratio is consistent with the relaxation of the turbulence to a static force-free state (4.36). In fact, relation (4.58) has a simple interpretation in terms of the inverse cascade of the magnetic helicity. As will be shown in the following chapter, for finite magnetic helicity turbulence continuously develops larger magnetic structures, whose size is limited only by external constraints. Physically this process occurs by a sequence of coalescence events of adjacent helical flux tubes, which preserve the total helicity, while there is no

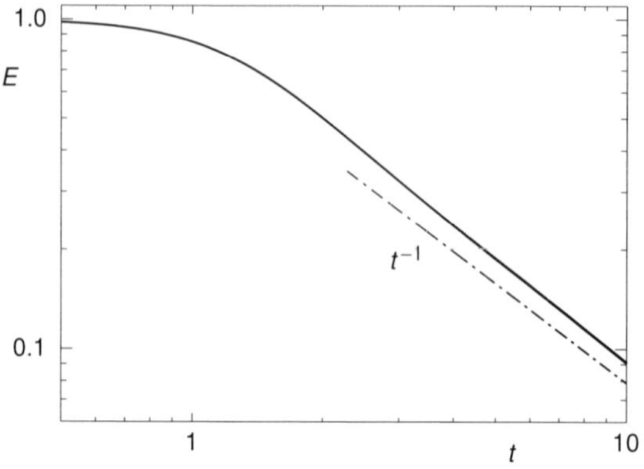

Figure 4.3: A log–log plot of the energy for a simulation run with $H^M = 0$. The dash–dotted line gives the power law t^{-1}.

correponding process for velocity structures. Hence the decrease of the energy ratio is a consequence of the inverse magnetic cascade. Using (4.52) relation (4.57) can be written as

$$\frac{E^K}{E^M} \sim \frac{L(t=0)}{L}, \tag{4.60}$$

where the size of the magnetic eddies increases as $L \sim t^{1/2}$.

We now consider the case of small helicity, $H^M \simeq 0$, for which the derivation of the decay law given above fails. Since a nonhelical field is less constrained, one expects a faster decay of the energy. This is indeed numerically observed; one finds the differential decay law $\epsilon/E^2 =$ constant and hence the asymptotic behavior $E \sim t^{-1}$, as illustrated in Fig. 4.3. In contrast to the finite-H^M case, the energy ratio is approximately constant. The behavior for $H^M = 0$ is reminiscent of the decay of energy in 2D MHD; see Section 8.2.2. Since, however, no macroscopic mechanism such as a selective decay process seems to exist for $H^M = 0$, the decay law need not be universal, but could depend on the initial conditions, similarly to the behavior in 3D hydrodynamic turbulence discussed before.

The preceding analysis of the decay of turbulence energy was restricted to weak velocity–magnetic-field alignment $|H^C| \ll E$. Effects of finite initial alignment and their competition with the selective decay due to finite magnetic helicity have been studied extensively for low-order 3D MHD systems by Stribling and Matthaeus (1991), though the relevance of the results for high-Reynolds-number turbulence remains unknown. The simulations by Biskamp and Müller (2000a) indicate that, for an initial correlation $|\rho_C| > 0.1 - 0.2$,

$|\rho_C(t)|$ increases, implying that H^C decays more slowly than the energy. The correlation is defined by[4]

$$\rho_C = H^C/E. \tag{4.61}$$

Since the alignment reduces the $\boldsymbol{v} \times \boldsymbol{B}$ nonlinearity in the magnetic-field equation, the turbulence dynamics is weakened and hence the decay of energy is slowed down. In contrast to magnetic-helicity-dominated turbulence, the energy ratio Γ does not decrease, which is to be expected, since the system tends toward an aligned state with $\Gamma \simeq 1$.

[4] Since, as we have seen, the energy ratio may become very small, $E^K \ll E^M$, this definition does not give a direct measure of the mean alignment. In this case one should define instead $\rho_C = \langle \boldsymbol{v} \cdot \boldsymbol{B} \rangle / (vB) = H^C/(4E^K E^M)^{1/2}$.

5

Spectral properties and phenomenology

Until now we have viewed the turbulent motions and fields mainly in space, or configuration space. Though spatial structures are important in MHD turbulence, for instance as final states of selective decay processes, the most characteristic property of fully developed turbulence is the presence of a *wide spectrum of different scales*. Turbulence scales have already been used in a very loose way in Section 4.2.2 in discussing the mechanism of dynamic alignment. In this chapter these ideas will be given a more precise meaning. In Section 5.1 we introduce the concept of homogeneous turbulence, a very useful idealization of a turbulence field far away from boundary layers. Here the Fourier components of the field play the role of the amplitudes at a certain scale $l \sim k^{-1}$. Section 5.2 considers an approximation that, at first sight, has little resemblance to real turbulence, namely a nondissipative system of Fourier modes truncated at a finite wavenumber and its relaxed states, which are called absolute equilibrium distributions. In spite of their artificial character such states can provide valuable information about the tendencies of the spectral evolution in dissipative systems, in particular the direction of the spectral fluxes called cascades. In Section 5.3 we then switch on dissipation in order to study the spectral properties of MHD turbulence. After defining appropriate Reynolds numbers, we derive two different power laws for the inertial-range spectrum proposed in the literature, namely the Kolmogorov spectrum $k^{-5/3}$ residing on a local spectral-transfer process, and the Iroshnikov–Kraichnan spectrum $k^{-3/2}$ based on the Alfvén effect, the nonlocal influence of the large-scale magnetic field on the small-scale turbulent eddies. Using the dissipation scales to normalize the wavenumber, we introduce the time-invariant spectra, which are particularly useful for decaying turbulence. Finally, recent direct numerical simulations of 3D MHD turbulence at relatively high Reynolds numbers, which seem to favor a local transfer mechanism, are discussed. Though 3D and 2D MHD turbulence have many features in common, there are also significant differences, as will be seen later in Chapter 8, when we treat the 2D case.

5.1 Homogeneous isotropic turbulence

On a global scale, turbulence is inhomogeneous, being driven by the gradient of some quantity such as the velocity in turbulent shear flow, the temperature in thermal convection, or the current density in a magnetized plasma. In fact, the gradient scale length is often taken as the integral scale L in the definition of the Reynolds number (3.36). However, away from strongly inhomogeneous boundary layers, variations of the average quantities are usually rather weak, for instance in the central part of a wind tunnel behind a grid or of a turbulent jet, such that a limited region thereof can indeed be considered homogeneous. For homogeneous turbulence, one-point quantities, for instance the kinetic-energy density

$$E = \tfrac{1}{2}\langle \boldsymbol{v}(\boldsymbol{x}) \cdot \boldsymbol{v}(\boldsymbol{x})\rangle, \tag{5.1}$$

are independent of $\boldsymbol{x}$, while two-point correlation functions

$$C_{ij} = \langle v_i(\boldsymbol{x}) v_j(\boldsymbol{x}')\rangle \tag{5.2}$$

depend only on the difference vector $\boldsymbol{x} - \boldsymbol{x}'$. Homogeneous turbulence constitutes an open system, for which periodic boundary conditions are usually the most adequate choice, since they are least intrusive at the boundaries in the sense that eddies close to a boundary do not feel its presence, and the effect of tying opposite boundaries together is weak if the turbulence correlation length is smaller than the size of the system. For periodic boundary conditions a fluctuating field quantity f can be Fourier analyzed:

$$f(\boldsymbol{x}) = \sum_{\boldsymbol{k}} f_{\boldsymbol{k}} \mathrm{e}^{i\boldsymbol{k}\cdot\boldsymbol{x}}, \tag{5.3}$$

$$k_j = 2\pi n_j / L, \quad n_j = \pm 1, \pm 2, \ldots,$$

where L is the box size. Reality of $f(\boldsymbol{x})$ requires $f_{-\boldsymbol{k}} = f_{\boldsymbol{k}}^*$. The Fourier components are called modes, meaning spatial modes, which are not necessarily oscillation eigenmodes of the system. Configuration-space and Fourier-space representations complement each other. The closure theory considered in Chapter 6 is usually performed in Fourier space, whereas deviations from self-similarity of the turbulence scales, especially the small-scale intermittency and structures discussed in Chapter 7, are more conveniently treated in configuration space.

While diffusion terms become very simple in k-space,

$$\nabla^2 f \to -k^2 f_k,$$

nonlinear terms become more complicated,

$$f(\boldsymbol{x})g(\boldsymbol{x}) \to \sum_{k'} f_{k-k'} g_{k'}.$$

Since in fully developed turbulence the nonlinear terms are dominant, being formally larger than the dissipation terms by a factor of the order of the Reynolds number, the value of the Fourier representation might appear somewhat doubtful. Its main advantage, however, is a conceptual one, namely the direct representation of spatial scales. Since turbulent dynamics is characterized by processes involving the transfer of certain quantities, such as energy and helicity, between different scales, the Fourier-space representation, which directly yields the corresponding spectral densities, is very suitable.

In addition to homogeneity, one usually assumes the turbulence to be isotropic, which simplifies the formalism of the theory as well as the interpretation of experiments and numerical simulations considerably, for instance the correlation function (5.2) is given in terms of only two scalar functions,

$$C_{ij} = f(r)\delta_{ij} + g(r)\frac{r_i r_j}{r^2}, \quad \boldsymbol{r} = \boldsymbol{x} - \boldsymbol{x}' \tag{5.4}$$

(for more details of the geometrical properties of isotropic turbulence see Monin and Yaglom, 1975). The justification for the assumption of isotropy is that, because of random mixing, the fluid rapidly forgets the often anisotropic way in which turbulence is generated. Results of experiments on hydrodynamic turbulence show that isotropy is usually (though not always) well satisfied in a coordinate frame moving with the average flow velocity. By contrast, a mean magnetic field in a plasma cannot be eliminated by a Galilean transformation, but has a crucial influence on the turbulence dynamics. Since in nature most turbulent plasmas do indeed carry a more or less strong mean field, about which turbulent fluctuations occur, the conditions for isotropy are rarely satisfied. Isotropic 3D MHD turbulence should therefore be considered as a theoretical concept generalizing isotropic hydrodynamic turbulence. A particular trick has been used in closure theory, namely the presence of a large homogeneous field of random orientation is assumed, such that the ensemble average is again isotropic; see Section 6.2.2. Only in the limiting case of a strongly magnetized plasma, where turbulence is restricted to the plane perpendicular to the field, can isotropic 2D turbulence be expected, which we will consider in Chapter 8.

5.2 Ideal systems and turbulent cascades

In fully developed turbulence the energy-dissipation rate is finite, even for very small values of the coefficients ν and η, such that conditions differ fundamentally from the ideal case, in which ν vanishes exactly, i.e., the limit $\nu \to 0$ differs from the behavior for $\nu = 0$. Nevertheless, important information about the nonlinear transfer processes can be obtained by considering the statitistical equilibrium properties of the ideal system, which is characterized by certain integral quantities, namely the ideal invariants.

In order to apply the formalism of equilibrium statistical mechanics to continuum fluid turbulence, it is convenient to introduce some discretization. This can be achieved in configuration space by applying a finite grid or in Fourier space by limiting the number of modes included. In homogeneous turbulence theory the latter is more practical, i.e., we truncate the Fourier series

$$\sum_k \to \sum_k{}',$$

where the prime indicates restriction to modes $\boldsymbol{k}$ within the band $k_{\min} \le k \le k_{\max}$. In general, ideal invariants of the continuum system are not strictly conserved in the truncated system. Quadratic invariants, however, and presumably only these, are sufficiently robust, or 'rugged', to survive truncation. This property is based on the validity of a detailed conservation relation for the elementary interaction between any triad of wave vectors $\boldsymbol{k}$, $\boldsymbol{p}$, $\boldsymbol{q}$ forming a triangle, i.e., satisfying $\boldsymbol{k} + \boldsymbol{p} + \boldsymbol{q} = 0$. In the presence of only three such modes the spectral mode energy $E_{\boldsymbol{k}} = \frac{1}{2}(|\boldsymbol{v}_k|^2 + |\boldsymbol{b}_k|^2)$, for instance, obeys the equation

$$\dot{E}_{\boldsymbol{k}} = T(\boldsymbol{k}, \boldsymbol{p}, \boldsymbol{q}), \tag{5.5}$$

where T is called the nonlinear transfer function, which follows from the dynamic equation. The detailed conservation relation is

$$\dot{E}_{\boldsymbol{k}} + \dot{E}_{\boldsymbol{p}} + \dot{E}_{\boldsymbol{q}} = T(\boldsymbol{k}, \boldsymbol{p}, \boldsymbol{q}) + T(\boldsymbol{p}, \boldsymbol{q}, \boldsymbol{k}) + T(\boldsymbol{q}, \boldsymbol{k}, \boldsymbol{p}) = 0. \tag{5.6}$$

The validity of this relation, and similar ones for the cross-helicity H_k^C and the magnetic helicity H_k^M, can be verified by direct calculation from the MHD equations written in Fourier space. For incompressibility these are

$$\dot{\boldsymbol{v}}_k = -i\left(\boldsymbol{I} - \frac{\boldsymbol{k}\boldsymbol{k}}{k^2}\right) \cdot \sum_p \left[\boldsymbol{v}_p(\boldsymbol{k} \cdot \boldsymbol{v}_{k-p}) - \boldsymbol{b}_p(\boldsymbol{k} \cdot \boldsymbol{b}_{k-p})\right], \tag{5.7}$$

$$\dot{\boldsymbol{b}}_k = -i\left(\boldsymbol{I} - \frac{\boldsymbol{k}\boldsymbol{k}}{k^2}\right) \cdot \sum_p \left[\boldsymbol{b}_p(\boldsymbol{k} \cdot \boldsymbol{v}_{k-p}) - \boldsymbol{v}_p(\boldsymbol{k} \cdot \boldsymbol{b}_{k-p})\right]. \tag{5.8}$$

Multiplying (5.7) by $\boldsymbol{v}_{-k}$ and (5.8) by $\boldsymbol{b}_{-k}$ and adding the results gives the explicit form of the energy-transfer function (5.5), from which the balance relation (5.6) is easily derived, assuming that only three modes are present.

5.2.1 Absolute equilibrium states

Let us assume that we have an ensemble of equivalent ideal truncated turbulent systems. Statistical theory shows that the equilibrium probability distribution in phase space is given by Gibbs' functional

$$\rho_{\rm G} = Z^{-1} \exp(-W), \quad W = \alpha E + \beta H^M + \gamma H^C, \tag{5.9}$$

where E, H^C, and H^M are quadratic forms of the phase-space variables, the real and imaginary parts of the Fourier components $\boldsymbol{v}_{\boldsymbol{k}}$ and $\boldsymbol{b}_{\boldsymbol{k}}$,

$$E = \tfrac{1}{2} \sum_{\boldsymbol{k}} (\boldsymbol{v}_{\boldsymbol{k}} \cdot \boldsymbol{v}_{-\boldsymbol{k}} + \boldsymbol{b}_{\boldsymbol{k}} \cdot \boldsymbol{b}_{-\boldsymbol{k}}), \tag{5.10}$$

$$H^M = \int \boldsymbol{A} \cdot \boldsymbol{B} \, dV = \sum_{\boldsymbol{k}} i(\boldsymbol{k} \times \boldsymbol{b}_{\boldsymbol{k}}) \cdot \boldsymbol{b}_{-\boldsymbol{k}} / k^2, \tag{5.11}$$

$$H^C = \int \boldsymbol{v} \cdot \boldsymbol{B} \, dV = \sum_{\boldsymbol{k}} \boldsymbol{v}_{\boldsymbol{k}} \cdot \boldsymbol{b}_{-\boldsymbol{k}}. \tag{5.12}$$

α, β, and γ are Lagrange multipliers to be determined from the values of the invariants and Z is a normalizing factor. The mathematical basis of (5.9) is the validity of Liouville's theorem as discussed, for instance, by Kraichnan and Montgomery (1980). If initially the ensemble is not in equilibrium, it will relax to the distribution (5.9), which is called the absolute equilibrium. The most interesting quantities are the Fourier spectra of the fields, such as $E_{\boldsymbol{k}} = \frac{1}{2}\langle |v_{\boldsymbol{k}}|^2 \rangle$ or, more conveniently, the angle-integrated spectra $E_k = \int E_{\boldsymbol{k}} \, d\Omega_{\boldsymbol{k}}$. To derive these one makes use of the lemma that a multivariate Gaussian distribution for the variables u_i,

$$Z^{-1} \exp\left(-\tfrac{1}{2} \sum_{i,j} A_{ij} \, u_i u_j \right), \tag{5.13}$$

has the second-order moments

$$\langle u_i u_j \rangle = A_{ij}^{-1}, \tag{5.14}$$

where A_{ij}^{-1} is the inverse of the matrix A_{ij} (see, e.g., Lumley, 1970, Chapter 2). By inserting the expressions for E, H^C, and H^M, Frisch *et al.* (1975), after some straightforward, though somewhat tedious, algebraic manipulations, obtained the following ideal spectral densities:

$$E_k^K = \frac{4\pi k^2}{\alpha}\left(1 + \frac{k^2 \tan^2\phi}{k^2 - k_0^2}\right), \tag{5.15}$$

$$E_k^M = \frac{4\pi k^2}{\alpha}\frac{k^2 \sec^2\phi}{k^2 - k_0^2}. \tag{5.16}$$

Hence

$$E_k = \frac{4\pi k^2}{\alpha}\left(1 + \frac{k^2(\tan^2\phi + \sec^2\phi)}{k^2 - k_0^2}\right); \tag{5.17}$$

and

$$H_k^M = -\frac{8\pi k^2}{\alpha\cos^2\phi}\frac{k_0}{k^2 - k_0^2} = -\frac{2k_0}{k^2}E_k^M, \tag{5.18}$$

$$H_k^C = -\frac{8\pi k^2\gamma}{\alpha^2\cos^2\phi}\frac{k^2}{k^2 - k_0^2} = -\frac{2\gamma}{\alpha}E_k^M, \tag{5.19}$$

with $\sin\phi = \gamma/(2\alpha)$, $k_0 = \beta/(\alpha\cos^2\phi)$, and $E = \int_{k_{\min}}^{k_{\max}} E_k\, dk$ and similarly for H^M and H^C. The values of the multipliers α, β, and γ are restricted by the requirement that ρ_G must be integrable, which means that W must be a positive-definite form; in particular, E_k^K and E_k^M must be positive for any k in the range $k_{\min} \le k \le k_{\max}$, from which follow the conditions

$$\alpha > 0, \quad |\gamma| < 2\alpha, \quad k_0^2 < k_{\min}^2. \tag{5.20}$$

The ratio β/α is a measure of the magnetic helicity, while γ/α is a measure of the cross-helicity. Note that the residual energy spectrum, the difference of kinetic and magnetic energies,

$$E_k^R = E_k^K - E_k^M = -\frac{4\pi k^2}{\alpha}\frac{k_0^2}{k^2 - k_0^2} \tag{5.21}$$

is always negative, being extremal for the smallest $k = k_{\min}$. Hence the ideal energy spectrum is dominated by the magnetic contribution, especially at long wavelengths.

For comparison, we also give the absolute equilibrium distributions for (3D) hydrodynamic turbulence, for which one has two invariants, the kinetic energy

E^K and the kinetic helicity $H^K = \sum_k i(\boldsymbol{k} \times \boldsymbol{v}_k) \cdot \boldsymbol{v}_{-k}$. Hence the Gibbs distribution is

$$\rho_G = Z^{-1} \exp(-\alpha E^K - \beta H^K).$$

By a similar analysis to that described above, Kraichnan (1973) derived the expressions

$$E_k^K = \frac{4\pi\alpha k^2}{\alpha^2 - \beta^2 k^2}, \quad H_k^K = \frac{8\pi\beta k^4}{\alpha^2 - \beta^2 k^2}. \tag{5.22}$$

Here $\alpha > 0$ and $|\beta| k_{max} < \alpha$ are required. In the case of vanishing helicity, $\beta = 0$, the equilibrium distribution is simply $E_k^K \propto k^2$, which corresponds to equipartition, $E_{\boldsymbol{k}}^K = \text{constant}$.

5.2.2 Cascade directions

The spectral densities of the ideal invariants are conserved in the nonlinear interactions, following detailed conservation relations such as (5.6). Hence, if such a quantity is excited, or injected to use the terminology in turbulence theory, within a certain spectral range $k \sim k_{in}$, it can be scattered only into other regions of k-space. If triad interactions are local, i.e., dominated by wavenumbers $\boldsymbol{k}, \boldsymbol{p}, \boldsymbol{q}$ of similar magnitude (which can be shown for hydrodynamic turbulence), transfer in k-space occurs in relatively small steps, such that many steps are required from the injection range to the dissipation range, the highest wavenumbers excited. The transfer process of a conserved quantity is therefore called a *cascade*.

In genuine dissipative turbulence two cases may occur. A *normal*, or *direct*, cascade describes the spectral transfer from k_{in} to larger wavenumbers; normal, since this is the well-known transfer in ordinary hydrodynamic turbulence (the 'Richardson cascade', Richardson, 1922). In an *inverse* cascade, transfer proceeds from k_{in} to smaller wavenumbers, i.e., to larger eddies corresponding to self-organization of the turbulence.

Cascade directions can essentially be determined from the absolute equilibrium distributions, which depend only on the structure of the ideal invariants of the system. Though the absolute equilibrium states are far from dissipative turbulence, they nonetheless indicate the direction of the spectral evolution in a real system, since the nonlinear dynamics is identical. One expects a direct cascade for the spectral density of some ideal invariant if the ideal spectrum is peaked at high k since, when it is injected in some intermediate spectral range, the quantity relaxes toward larger wavenumbers. In the opposite case of an ideal spectrum peaked at small k, the quantity should exhibit an inverse cascade. On inspecting the ideal spectra for the MHD invariants E_k, H_k^M, and H_k^C, (5.17)–(5.19), we find that E_k and H_k^C should have direct cascades, while H_k^M should have an inverse cascade. The existence of an inverse cascade of the magnetic helicity was first discussed by Frisch *et al.* (1975). Hence, when the selective decay is dominated by finite magnetic helicity leading to a force-free field (4.36), the average wavenumber should decrease and the final state should occupy the largest possible scales, whereas the pure Alfvénic states (4.38) resulting from the process of dynamic alignment dominated by finite cross-helicity may exhibit structures of arbitrarily small scales. In Section 8.1.1 we compare the cascade directions of 2D and 3D turbulence in hydrodynamic and MHD turbulence.

It is often said that an inverse cascade process requires the existence of several ideal invariants. The converse is, however, not true; the presence of several invariants does not necessarily give rise to inverse cascades. In (3D) hydrodynamic turbulence we have two invariants, energy E^K and helicity H^K, but both ideal spectra, (5.22), are peaked at high k; hence they exhibit direct cascades and no large-scale structures are formed spontaneously. A condition for an inverse cascade seems to be that, in the presence of dissipation, the decay of turbulence is selective, i.e., one of the ideal invariants decays much more slowly than the other.

5.3 Spectra in dissipative MHD turbulence

5.3.1 Magnetic Reynolds numbers

Until now dissipation was either assumed to be sufficiently weak without further specification, or, in the preceding section, was neglected altogether. For the following, in which we treat the spectral properties of dissipative turbulence, it is useful to have a more quantitative measure of the dissipation effects, which is provided by the Reynolds number. The classical definition of the Reynolds number has already been given in (3.36), $\mathsf{Re} = vL/\nu$, where L is a typical mean gradient scale depending on the geometry of the system, such as the diameter of the pipe in Poiseuille flow. In MHD turbulence we have, in addition, the *magnetic Reynolds number*

$$\mathsf{Rm} = vL/\eta, \tag{5.23}$$

which is related to Re by $\mathsf{Re} = \mathsf{Rm}\,\mathsf{Pr}_m$ with $\mathsf{Pr}_m = \nu/\eta$, the *magnetic Prandtl number*.

Since in turbulence theory special interest lies in the intrinsic turbulence properties, it is preferable to choose a scale length in the definition of the Reynolds number which is independent of the geometry of the system. In the literature one often finds the following definition, depending on the energy-spectrum properties at small k,

$$L = \frac{1}{E}\int_0^\infty \frac{E_k}{k}\,dk \sim k_0^{-1}.$$

k_0 is the wavenumber at which the spectrum is peaked, which in hydrodynamic turbulence is of the order of the injection wavenumber $k_{\rm in}$; see (5.28) in Section 5.3.2. Here we choose a definition of the integral scale that depends only on the macroscopic properties of the turbulence. The simplest such expression is (4.45), which contains only the turbulence energy E and the energy-injection rate ϵ,

$$L = E^{3/2}/\epsilon, \tag{5.24}$$

whence we have the Reynolds numbers[1]

$$\mathsf{Re} = \frac{E^2}{\epsilon\nu}, \qquad \mathsf{Rm} = \frac{E^2}{\epsilon\eta}. \tag{5.25}$$

Instead of the integral scale L, it is customary in hydrodynamic turbulence to use a microscale λ, loosely defined by $\nabla^2 u \sim u/\lambda^2$, which is intermediate between the integral scale and the dissipation scale l_d (to be defined in Section 5.3.3), and therefore depends also on the dissipation coefficients.[2] The simplest expression is

$$\lambda^2 = E\nu/\epsilon, \tag{5.26}$$

[1] MHD turbulence does not always behave in a (macroscopically) self-similar way, as we have seen in Section 4.2.3; in particular, kinetic and magnetic energies may develop differently. Since the Reynolds number is a measure of the turbulence dynamics, the definition (5.25) is not appropriate when $E^K \ll E^M \simeq E$, i.e., for a quasi-static magnetic field, where Re in this form would be large, even though the system is practically not turbulent. In such a case we choose a different integral length L, which was defined in (4.55), namely

$$\Lambda = (E^K)^{1/2}E/\epsilon,$$

which results in the macroscale Reynolds numbers

$$\mathsf{Re} = \frac{E^K E}{\epsilon\nu}, \quad \mathsf{Rm} = \frac{E^K E}{\epsilon\eta}.$$

[2] The microscale was introduced by Taylor (1935), who defined λ as the radius of curvature of the longitudinal velocity correlation function $C_l(r) = \langle v_l(\boldsymbol{x})v_l(\boldsymbol{x}+\boldsymbol{r})\rangle$, $v_l = \boldsymbol{v}\cdot\boldsymbol{r}/r$, at the origin:

$$\lambda^2 = -\left.\frac{C_l}{C_l''}\right|_{r=0} = \frac{15\nu\langle v_l^2\rangle}{\epsilon} = \frac{10\nu E}{\epsilon},$$

where the numerical coefficient is obtained from the von Kármán–Howarth equation for $C_l(r,t)$, (7.26). Hence the microscale Reynolds number is

$$\mathsf{Re}_\lambda = \left(\frac{10E^2}{\epsilon\nu}\right)^{1/2}.$$

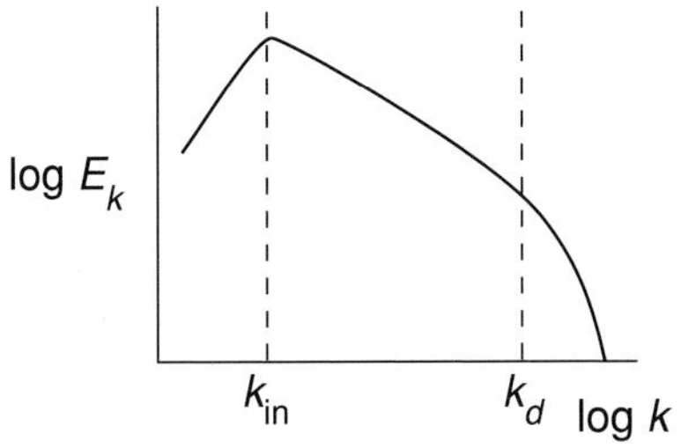

Figure 5.1: A log–log plot of a typical energy spectrum.

and the corresponding microscale Reynolds numbers are

$$\mathrm{Re}_\lambda = \left(\frac{E^2}{\epsilon\nu}\right)^{1/2}, \quad \mathrm{Rm}_\lambda = \left(\frac{E^2}{\epsilon\eta}\right)^{1/2}, \tag{5.27}$$

which are just the square roots of the macroscale Reynolds numbers (5.25).

In Section 5.3.3 we will show that the Reynolds numbers can be expressed as the ratio of the integral scale and the dissipation scale, which allows us to define the Reynolds number also for hyperdiffusion $\alpha > 1$, (2.23).

5.3.2 Phenomenology of the inertial-range spectrum

The power-law spectrum in the inertial range reflects the similarity properties of high-Reynolds number turbulence. The term 'inertial range' indicates that in this spectral range the turbulence develops solely under the influence of the internal nonlinear dynamics without external input (i.e., turbulence drive) or output (i.e., dissipation). In the normal case of a direct cascade the inertial range is loosely defined by

$$k_{\mathrm{in}} \ll k \ll k_d, \tag{5.28}$$

where k_{in} indicates the injection range, where turbulence is excited, and k_d the dissipation range. In practice, the inertial range is taken as the wavenumber range within which the spectrum exhibits a power law behavior. A typical spectrum is plotted schematically in Fig. 5.1. In the presence of an inverse cascade, one finds a second inertial range

$$L_{\mathrm{syst}}^{-1} \ll k \ll k_{\mathrm{in}}, \tag{5.29}$$

the lower limit being determined geometrically by the size of the system. Since there is usually no dissipation at low k, the spectral density of the inversely cascading quantity will partly accumulate at the lowest wavenumbers and partly be reflected, which would finally affect the entire inertial range (5.29). One therefore prefers to study the inverse cascade with a freely propagating spectral front, before the size of the system is reached. For the time being we restrict ourselves to normal cascade conditions. In addition, the theory is usually limited to isotropic turbulence, for which one considers only angle-integrated spectra,

$$E_k = \int d\Omega_{\boldsymbol{k}}\, E_{\boldsymbol{k}}, \quad E = \int_0^\infty dk\, E_k. \tag{5.30}$$

(In experiments one measures the 1D spectrum $E_{k_x} = \int dk_y\, dk_z\, E_{\boldsymbol{k}}$, which is related to the angle-integrated spectrum in the inertial range exhibiting the same power law.)

The dynamics of the turbulence is controlled by the injection of energy at $k \sim k_{\text{in}}$ with a rate ϵ_{in}, which then cascades across the inertial range with the transfer rate ϵ_t and is finally extracted from the system in the dissipation range $k \sim k_d$ with the dissipation rate ϵ_d. For stationary turbulence one has

$$\epsilon_{\text{in}} = \epsilon_t = \epsilon_d := \epsilon. \tag{5.31}$$

This relation holds, to a good approximation, also when the injection rate varies with time, since the rapid dynamics at the small scales in the inertial and dissipation ranges adjusts the spectrum quasi-instantaneously to the slower changes at the large scales.

If we assume that the spectral-transfer process is local, i.e., the mode interactions are dominated by wavenumbers of the same order of magnitude, the energy spectrum of a turbulent fluid can be derived in a simple heuristic way. Divide the inertial range into a discrete number of scales $l_n = k_n^{-1}$,

$$l_0 > l_1 > \cdots > l_N \quad \text{or} \quad k_0 < k_1 < \cdots < k_N, \tag{5.32}$$

where the division is conveniently taken on a logarithmic scale, for instance $l_n = 2^{-n} l_0$, and l_0 is of the order of the integral scale L. A typical turbulent eddy of scale l can be represented by the average difference in velocity δv_l between two points a distance l apart or by the Fourier component v_k with $k = l^{-1}$, where in the following qualitative considerations the precise definition, in particular the vector character of $\boldsymbol{v}$ and $\boldsymbol{x}$, is not important. The time taken for transfer of energy between two neighboring scales l_n and l_{n+1} is given by the distortion, or turnover, time of the eddy δv_{l_n}, or δv_n for simplicity,

$$\tau_n \sim l_n/\delta v_n. \tag{5.33}$$

Since the energy flux is constant across the inertial range,

$$E_n/\tau_n \sim \delta v_n^3/l_n \sim \epsilon, \tag{5.34}$$

one finds the scaling relation

$$\delta v_n \sim \epsilon^{1/3} l_n^{1/3}, \tag{5.35}$$

which is called the Kolmogorov phenomenology or K41 theory (Kolmogorov, 1941a). To obtain the energy spectrum we identify the eddy energy with the band-integrated Fourier spectrum

$$\delta v_n^2 \simeq E_n \simeq \int_{k_n}^{k_{n+1}} E_k \, dk \simeq E_{k_n} k_n,$$

which yields

$$E_k = C_K \epsilon^{2/3} k^{-5/3}, \tag{5.36}$$

the well-known Kolmogorov spectrum (Kolmogorov, 1941a; Obukhov, 1941). The spectrum follows also from purely dimensional considerations on assuming that E_k depends only on the local value k and the energy-transfer rate ϵ,

$$E_k \sim \epsilon^{\alpha} k^{\beta}. \tag{5.37}$$

The exponents α and β are determined by matching the dimensions using $[E_k] = L^3T^{-2}$ and $[\epsilon] = L^2T^{-3}$. The numerical factor C_K, the Kolmogorov constant, is not determined by scaling arguments, but instead requires a dynamic theory such as the Lagrangian-history-direct-interaction approximation (LHDIA) (Kraichnan, 1965a). The Kolmogorov spectrum has been observed in many types of turbulent flows, rather independently of the geometry of the system and the way in which turbulence is generated (see, e.g., Monin and Yaglom, 1975, pp. 467–94), and also the value of the Kolmogorov constant appears to be universal, in particular independent of the Reynolds number. Sreenivasan (1995) collected data from many experiments, which showed that C_K is invariant with a small statistical scatter,

$$C_K = 1.6\text{–}1.7, \tag{5.38}$$

over a broad range of Reynolds numbers, $30 \le \mathrm{Re}_\lambda \le 3 \times 10^4$, and authors of recent high-resolution numerical simulations have found a value in the same range, $C_K = 1.65 \pm 0.05$ for $\mathrm{Re}_\lambda \simeq 500$ (Gotoh and Fukayama, 2001). One should also mention the derivation of $C_K = 1.77$ by Kraichnan (1966) from his LHDIA theory.

In MHD turbulence the Alfvén effect discussed in Section 4.2.2 has been suggested to modify the basic inertial-range scaling (Iroshnikov, 1964; Kraichnan, 1965b). According to this effect, small-scale fluctuations are not independent of the macro-state but are strongly affected by the large-scale magnetic field, which makes them behave approximately as Alfvén waves. Note that the large-scale field is not necessarily a mean field, which we have excluded because of the assumption of isotropy, but rather is the field in the dominant energy-containing eddies. At this point we restrict the discussion to weak velocity–magnetic-field correlation, $\delta z_l^+ \simeq \delta z_l^-$, postponing the case of strong correlation to Section 5.3.4. As discussed in Section 4.2.2, only Alfvén waves moving in opposite directions along the large-scale field interact; hence the interaction time of two wave packets, δz_l^+ and δz_l^-, is the Alfvén time $\tau_A \sim l/v_A$, which is much shorter than the nonmagnetic eddy-distortion time $\tau_l \sim l/\delta z_l$, such that the change in amplitude during one scattering event is small and many such events are needed in order to produce a relative change of order unity. The energy-transfer time, which in hydrodynamic turbulence is just τ_l, is therefore replaced by the longer time

$$T_l \sim (\tau_l)^2/\tau_A. \tag{5.39}$$

This relation can be explained by a simple probabilistic argument as shown in Section 4.2.2. Making the substitution $\tau_l \to T_l$ in (5.34) gives

$$\delta z_l^4 \tau_A / l^2 \sim \epsilon,$$

such that we have the scaling

$$\delta z_l \sim (\epsilon v_A)^{1/4} l^{1/4}, \tag{5.40}$$

which corresponds to the spectrum

$$E_k = C_{IK} (\epsilon v_A)^{1/2} k^{-3/2}, \tag{5.41}$$

the Iroshnikov–Kraichnan (IK) spectrum of MHD turbulence. The coefficient C_{IK} is expected to be different from C_K. The spectrum is less steep than the Kolmogorov spectrum, since the factor τ_l/τ_A, by which the energy-transfer time is longer, increases with decreasing l, and therefore increasingly higher amplitudes are required compared with the hydrodynamic case in order to produce the same transfer of energy. In contrast to the Kolmogorov spectrum, which depends only on the inertial-range quantity ϵ, the MHD energy spectrum depends also on the macroscale quantity v_A and therefore cannot be derived by dimensional analysis without additional assumptions. It should be noted that the difference between the two spectra is quite small, in spite of the rather different transfer mechanisms. The difference becomes more pronounced for the higher-order moments, as will be discussed in Chapter 7.

The energy spectrum can also be obtained in a slightly different way suggested by Kraichnan (1965b). Let τ_k be the decorrelation time of modes with wavenumbers of order k. Then the energy-transfer rate should be proportional to this time, where dimensional analysis gives

$$\epsilon \sim \tau_k E_k^2 k^4. \tag{5.42}$$

In hydrodynamic turbulence τ_k is simply the eddy-distortion time (5.33) or, in terms of the energy spectrum, $\tau_k \sim (k^3 E_k)^{-1/2}$, from which follows the Kolmogorov spectrum (5.36). In the MHD case, decorrelation occurs on the Alfvén time scale, $\tau_k \sim \tau_A \sim (k v_A)^{-1}$, which yields the IK spectrum (5.41).

It is interesting to note that, formally, the same spectral law as the IK spectrum (5.41), with v_A replaced by the large-scale fluid velocity v_0, results from Kraichnan's original direct-interaction approximation (DIA) for hydrodynamic turbulence (Kraichnan, 1959). This is, however, a spurious effect, which arises because the approach does not properly distinguish between advection and distortion of eddies. In hydrodynamics the intrinsic small-scale dynamics should not depend on the large-scale behavior, since a large-scale velocity can be eliminated by a Galilean transformation, whereas in MHD a large-scale magnetic field has a crucial effect on the small scales by strongly coupling velocity and magnetic fluctuations.

When the Alfvén effect controls the inertial-range dynamics, it determines not only the total energy spectrum but also the relative strength of the components E_k^K and E_k^M,

$$E_k^K \simeq E_k^M. \tag{5.43}$$

The equality is, however, only approximate since it results from the interaction of a mode k with the large-scale magnetic field while neglecting the coupling of modes with wavenumbers of similar magnitude, which gives rise to a deviation from the equipartition (5.43) as discussed in Section 5.3.4. Although the arguments leading to the IK energy spectrum (5.41) sound rather convincing, the question of whether the Alfvén effect really dominates the inertial-range dynamics, i.e., whether the inertial-range spectrum is closer to a three-halves law than it is to a five-thirds law, cannot be decided by these arguments but must be checked either experimentally or by appropriate numerical simulations, which will be presented in Section 5.3.5. Since the cross-helicity spectrum has the same dimensions as the energy spectrum, dimensional analysis gives the same spectrum, either $k^{-3/2}$ or $k^{-5/3}$. Observations of turbulence in the solar wind (Section 10.2) reveal a Kolmogorov spectrum both for the energy and for the cross-helicity. Closure-theory computations, however, indicate a steeper spectrum, $H_k^C \sim k^{-2}$ (Grappin *et al.*, 1982) (Section 6.2.2).

Finally, we outline briefly the spectral properties in the inertial range $k < k_{\text{in}}$ of the inverse cascade of the magnetic helicity H_k^M, whose existence is predicted by the absolute equilibrium spectrum (5.18). Since for long wavelengths the Alfvén effect is weak, we apply a local Kolmogorov-type dimensional analysis assuming that we have a constant helicity-transfer rate ϵ_H. With the dimensions $[H_k^M] = L^4T^{-2}$ and $[\epsilon_H] = L^3T^{-3}$, we obtain

$$H_k^M \sim \epsilon_H^{2/3} k^{-2} \tag{5.44}$$

which implies the magnetic-energy spectrum $E_k^M \sim k^{-1}$. The spectrum $H_k^M \sim k^{-2}$ is indeed found in closure-theory computations (Pouquet *et al.*, 1976), see Fig. 6.4.

5.3.3 Anisotropy of MHD turbulence

The phenomenological approaches described in the preceding section tacitly assume the turbulence to be isotropic, only the omnidirectional spectrum being discussed. However, we already know that a mean magnetic field has a strong effect on the turbulence properties, in contrast to a mean flow in hydrodynamic turbulence, which can be eliminated by a Galilean transformation. Since a magnetic field resists bending, but can rather freely be shifted laterally, corresponding to an interchange of field lines, we expect that small-scale modes are primarily excited perpendicular to the magnetic field, $k_\perp \gg k_\parallel$. The anisotropy of MHD turbulence has been discussed by several groups, e.g., Shebalin *et al.* (1983), Sridhar and Goldreich (1994), Goldreich and Sridhar (1995), Montgomery and Matthaeus (1995), Cho and Vishniac (2000), and Milano *et al.* (2001). In the limiting case of a strong mean field $B_0 \gg b$, MHD turbulence becomes essentially 2D, which we shall discuss in more detail in Chapter 8. Here we are concerned with the local anisotropy of the inertial-range fluctuations in globally isotropic turbulence, which arises because of the local presence of the randomly oriented large-scale magnetic field in the energy-containing eddies.

To see the tendency toward the buildup of a spectral anisotropy, consider the weak-turbulence approach, the change of the amplitude of a mode $\boldsymbol{k}_3$ by the scattering of two modes $\boldsymbol{k}_1$ and $\boldsymbol{k}_2$, where the modes are Alfvén waves $z_{\boldsymbol{k}}^{\pm}$ with frequency $\varpi = \pm k_\parallel v_A$. Since in weak-turbulence interactions the frequencies of the modes are not changed, one has, in addition to the wavenumber relation

$$\boldsymbol{k}_3 = \boldsymbol{k}_1 + \boldsymbol{k}_2, \tag{5.45}$$

the frequency condition $\varpi_3 = \varpi_1 + \varpi_2$, i.e.,

$$\pm k_{\parallel 3} = k_{\parallel 1} - k_{\parallel 2}, \tag{5.46}$$

accounting for the effect that only Alfvén waves propagating in opposite directions interact. Comparison of these equations yields that either $k_{\parallel 1} = 0$ or $k_{\parallel 2} = 0$. While the driven mode may have a perpendicular wavenumber exceeding those of the exciting modes, $k_{\perp 3} > k_{\perp 1}, k_{\perp 2}$, corresponding to a transfer of energy to smaller scales, the parallel wavenumber cannot increase, hence there is no direct spectral cascade to large $k_\parallel$.

For fully developed turbulence the weak-turbulence approach is not strictly valid, since the spatial modes $\boldsymbol{k}$ usually have a broad frequency spectrum, which is difficult to treat in a quantitative way. Let us therefore give a simple phenomenological discussion of the spectral anisotropy of fully developed MHD turbulence. Consider an eddy of dimensions $l_\parallel$ and $l_\perp$ parallel and perpendicular to the magnetic field, respectively, i.e., the local field averaged over the size of the eddy. Because of the turbulent transfer to smaller perpendicular scales, $l_\perp$ tends to shrink, making the eddy more elongated, which may finally lead to sheet-like structures limited only by dissipation. In general, this process will, however, be limited by parallel propagation, such that there is the balance of rates

$$\delta z_{l_\perp}/l_\perp \sim v_A/l_\parallel, \tag{5.47}$$

called a critically balanced cascade by Goldreich and Sridhar (1995). Hence the spectral cascade takes place mainly in the $k_\perp$-plane with the energy flux ϵ constant across the inertial range,

$$\epsilon \sim \delta z_{l_\perp}^3/l_\perp. \tag{5.48}$$

Combining this relation with (5.47) gives

$$l_\parallel \sim \frac{v_A}{\epsilon^{1/3}} l_\perp^{2/3} \sim L^{1/3} l_\perp^{2/3}, \tag{5.49}$$

with the integral scale $L = v_A^3/\epsilon$. One thus finds that the spectral anisotropy increases with k,

$$k_\perp/k_\parallel \sim (Lk_\perp)^{1/3}. \tag{5.50}$$

This scaling relation is confirmed by numerical simulations (Cho and Vishniac, 2000). Relation (5.48) is equivalent to a Kolmogorov energy spectrum perpendicular to the local field direction:

$$E(k_\perp) \sim \epsilon^{2/3} k_\perp^{-5/3} \sim (v_A^3/L)^{2/3} k_\perp^{-5/3}. \tag{5.51}$$

The parallel spectrum is enslaved to the perpendicular spectrum. Inserting $k_\perp$ from (5.50) gives

$$E(k_\parallel) \sim \epsilon^{3/2} v_A^{-5/2} k_\parallel^{-5/2} \tag{5.52}$$

Hence, if there is no mean magnetic field, turbulence can be isotropic with a Kolmogorov spectrum, since the parallel contribution is small in the inertial range and falls off more rapidly. In the presence of a mean field B_0, the Kolmogorov spectrum should be measured only in the perpendicular plane, while the amplitude of the parallel field fluctuations is small.

5.3.4 Dissipation scales

The spectral cascade process, i.e., the unidirectional flow of energy (and other directly cascading quantities) in k-space, implies that there is a sink at some wavenumber k_d, at which dissipation becomes dominant, which forms the upper edge of the inertial range, $k_d \sim k_N$ in (5.32). The scale $l_d = k_d^{-1}$ is determined by the condition that the dissipation rate equals the nonlinear transfer rate. In the hydrodynamic case this condition reads

$$\tau_l^{-1} \sim \nu/l^2, \quad \text{with} \quad \tau_l = l/\delta v_l. \tag{5.53}$$

Inserting the Kolmogorov scaling (5.34) gives

$$l_d = (\nu^3/\epsilon)^{1/4} := l_K, \tag{5.54}$$

called the Kolmogorov microscale (Kolmogorov, 1941a) (often denoted by η in the literature, which, however, we do not use here in order to avoid confusion with the resistivity). The dissipation scale can also be regarded as the scale length at which the local Reynolds number $\mathrm{Re}_l = \delta v_l\, l/\nu$ becomes unity.

In MHD turbulence the Alfvén effect weakens the nonlinear energy transfer, $\tau_l \to T_l = (\tau_l)^2/\tau_A$, such that condition (5.53) becomes

$$T_l^{-1} \sim \nu/l^2, \tag{5.55}$$

from which we obtain the corresponding dissipation scale length by inserting the IK scaling (5.40),

$$l_d = (\nu^2 v_A/\epsilon)^{1/3} = l_{IK}. \tag{5.56}$$

Here we assumed that $\nu \sim \eta$ and $\epsilon_\nu \sim \epsilon_\eta$, $\epsilon = \epsilon_\nu + \epsilon_\eta$, where $\epsilon_\nu = \nu \int \omega^2 \, d^3x$ and $\epsilon_\eta = \eta \int j^2 \, d^3x$ are the viscous and Ohmic dissipation rates, respectively. Comparing l_K and l_{IK}, the latter is slightly larger (for v_A and ϵ of order unity) due to the slower transfer of energy, which is balanced already by the weaker dissipation at a larger scale.

If the macroscopic quantities ϵ_{in} and v_A vary with time, such as in decaying turbulence, the energy spectrum changes, too. Since, however, relation (5.31) still holds approximately for $k \gg k_{in}$, we can introduce an invariant normalized spectrum $\widehat{E}(\widehat{k})$ depending on the normalized wavenumber $\widehat{k} = kl_d$. On multiplying (5.36) by $l_K^{-5/3}$ we obtain

$$E(k,t) = \nu^{5/4}\epsilon^{1/4}(t)\widehat{E}(\widehat{k}), \tag{5.57}$$

where, in the inertial range, $\widehat{k} = kl_K < 1$, $\widehat{E}(\widehat{k})$ is the normalized Kolmogorov spectrum

$$\widehat{E}(\widehat{k}) = C_K \, \widehat{k}^{-5/3}, \tag{5.58}$$

while multiplying (5.41) by $l_{IK}^{-3/2}$ gives

$$E(k,t) = \nu v_A(t)\widehat{E}(\widehat{k}), \tag{5.59}$$

where now $\widehat{E}(\widehat{k})$ is the normalized IK spectrum

$$\widehat{E}(\widehat{k}) = C_{IK} \, \widehat{k}^{-3/2} \tag{5.60}$$

in the inertial range $\widehat{k} = kl_{IK}(t) < 1$. In the dissipation range $\widehat{k} > 1$ the spectrum falls off exponentially. The far-dissipation-range ($\widehat{k} \gg 1$) spectrum is predicted from closure theory to have the form

$$\widehat{E}(\widehat{k}) \sim \widehat{k}^{\alpha} \mathrm{e}^{-c\widehat{k}}, \tag{5.61}$$

where in hydrodynamic turbulence $\alpha \simeq 3$ and $c \simeq 7$ (see, for instance, Chen *et al.*, 1993), both depending slightly on the Reynolds number.

As mentioned in Section 5.3.1, the Reynolds number can be expressed in terms of the ratio of the integral and dissipation scales. From (5.24), (5.25), and (5.54) we find

$$\mathrm{Re} = (L/l_K)^{4/3}. \tag{5.62}$$

This relation can also be applied to define the Reynolds number in the case of hyperdiffusion, (2.23),

$$\mathsf{Re}^{(\alpha)} = (L/l_l^{(\alpha)})^{4/3}, \tag{5.63}$$

with the dissipation scale $l_l^{(\alpha)}$ obtained by modifying the balance relation (5.53),

$$\tau_l^{-1} \sim \nu_\alpha / l^{2\alpha},$$

whence, after insertion of the Kolmogorov scaling,

$$l_d^{(\alpha)} = (\nu_\alpha^3/\epsilon)^{1/(6\alpha-2)}. \tag{5.64}$$

The derivation of the corresponding expressions for IK scaling, for instance $\mathsf{Re}\,(=\mathsf{Rm}) = (L/l_{\mathrm{IK}})^{3/2}$, is left to the reader.

5.3.5 Energy spectra in highly aligned turbulence

In the preceding considerations the velocity–magnetic-field correlation was assumed to be small, $\rho_C \ll 1$ or, locally, $z^+ \simeq z^-$. In fact, the $k^{-3/2}$ energy spectrum can be valid only for sufficiently weak correlation. In the opposite case the Elsässer fields $\boldsymbol{z}^\pm$ are more fundamental than $\boldsymbol{v}$ and $\boldsymbol{b}$ or, in terms of the spectral densities, $E_k^\pm$ and the residual spectrum E_k^R,

$$E_k^\pm = \tfrac{1}{4} \int d\Omega_k\, |\boldsymbol{z}_{\boldsymbol{k}}^\pm|^2, \tag{5.65}$$

$$E_k^R = \tfrac{1}{2} \int d\Omega_k\, \boldsymbol{z}_{\boldsymbol{k}}^+ \cdot \boldsymbol{z}_{-\boldsymbol{k}}^- = E_k^K - E_k^M, \tag{5.66}$$

are more fundamental than the total spectral energy $E_k = E_k^K + E_k^M = E_k^+ + E_k^-$ and the cross-helicity $H_k^C = E_k^+ - E_k^-$. Following Grappin *et al.* (1983), we give a phenomenological derivation of the spectra E_k^+ and E_k^-, generalizing the arguments leading to the IK spectrum (5.41) to the case of high alignment. For the Elsässer-field increments $\delta z_l^\pm$ the energy-transfer times are given by (4.55),

$$T^\pm \sim (\tau_l^\pm)^2/\tau_{\mathrm{A}} \sim l v_{\mathrm{A}}/E_l^\mp, \tag{5.67}$$

since $\tau_l^\pm \sim l/\delta z_l^\mp \sim l/(E_l^\mp)^{1/2}$. To obtain the inertial-range spectra $E_k^\pm$, we need expressions for the energy-transfer rates $\epsilon^\pm$. In the strictly local approximation (in k-space) assumed in Section 4.2.2, the energy-transfer rates are found to be equal, (4.42). In reality, however, mode interactions occur in a certain wavenumber band $k/c < k < kc$ with a band parameter $c > 1$. Hence

(4.41) should be replaced by the average over the wavenumber band,

$$\epsilon^{\pm} \simeq kE_k^{\pm}/T_k^{\pm} \simeq kE_k^{\pm}\tau_{\rm A}\langle(\tau_k^{\pm})^{-2}\rangle, \tag{5.68}$$

with $k = l^{-1}$, $E_l \simeq kE_k$, $\tau_{\rm A} \simeq (kv_{\rm A})^{-1}$, and $\tau_k^{\pm} \simeq (k^3 E_k^{\mp})^{-1/2}$. Assuming the validity of power-law spectra $E_k^{\pm} = C^{\pm}k^{-m_{\pm}}$, one can easily evaluate the average in (5.68), for instance

$$\langle(\tau_k^{+})^{-2}\rangle = C^{-}\int_{k/c}^{kc} k^{3-m_-}\,dk \Big/ \int_{k/c}^{kc} dk = F^{+}E_k^{-}k^3,$$

where $F^{+} \simeq c^{3-m_-}/(4-m_-)$ is independent of k. One therefore obtains

$$\epsilon^{\pm} = F^{\pm}E_k^{+}E_k^{-}k^3/v_{\rm A}, \tag{5.69}$$

i.e., $\epsilon^{+} \neq \epsilon^{-}$, since in general $F^{+} \neq F^{-}$, the $\pm$ symmetry of (4.41) being broken, and the spectral indices $m_{\pm}$ satisfy the constraint

$$m_{+} + m_{-} = 3. \tag{5.70}$$

The coefficients $F^{\pm}$ (and the band parameter c) are determined by the condition that the energy-transfer rates $\epsilon^{\pm}$ equal the dissipation rates, which for $\nu = \eta$ reads

$$\epsilon^{\pm} = 2\nu\int_{k_{\rm in}}^{k_d} dk\,k^2 E_k^{\pm}. \tag{5.71}$$

(One can show that the dissipation wavenumbers for E_k^{+} and E_k^{-} are similar, $k_d^{+} \simeq k_d^{-} = k_d$.) From the balance of transfer and dissipation at the dissipation wavenumber, $T_{k_d}^{\pm} = 1/(\nu k_d^2)$, and relation (5.67) the energy spectra become

$$E_k^{\pm} \simeq \nu v_{\rm A}(k/k_d)^{-m_{\pm}}. \tag{5.72}$$

Hence $E_k^{+} \simeq E_k^{-}$ at the dissipation scale, whereas in the inertial range their ratio is very different, $E_k^{+}/E_k^{-} \gg 1$, if $m_{+} > m_{-}$, and so is the ratio of the total energies E^{+}/E^{-}. The difference between ϵ^{+} and ϵ^{-} in (5.69) arises from the different behaviors of the spectra in the inertial range. Inserting (5.72) into (5.71) provides a relation between the injection rates and the spectral indices,

$$\frac{\epsilon^{+}}{\epsilon^{-}} \simeq \frac{3-m_{-}}{3-m_{+}} = \frac{m_{+}}{m_{-}}, \tag{5.73}$$

where (5.70) has been used. Hence, in contrast to the energy ratio, the ratio of the injection rates is finite for $\mathrm{Re} \to \infty$.

Only for uncorrelated turbulence $E_k^+ = E_k^-$, implying that $\epsilon^+ = \epsilon^-$, does the theory predict the IK spectrum (5.41), since only for $m_+ = m_- = \frac{3}{2}$ does the total spectral density $E_k = E_k^+ + E_k^-$ also follow a $k^{-3/2}$ law. In the case of finite alignment, however, for which $m_+ \neq m_-$, E_k does not, in general, obey a power law. Only for extremely high alignment with, say, $E^+ \gg E^-$, does E_k again obey an approximate power law, namely that of the dominant component, $m \simeq m_+$. These results have been confirmed qualitatively by numerical solution of a closure model for MHD (Grappin *et al.*, 1983), as well as by simulations of 2D MHD turbulence (Pouquet *et al.*, 1988; Politano *et al.*, 1989).

As mentioned above, also the residual energy spectrum $E_k^R = E_k^K - E_k^M$ is a fundamental quantity that is expected to follow a power law. Whereas at high k close to the dissipation scale one has quasi-equipartition, $E_k^R \simeq 0$, (5.43), due to the decrease of the nonlocal effect of the large-scale magnetic field with decreasing k, local interactions of eddies of similar wavenumbers become more important, which allows a finite E_k^R. The equation for E_k^R has the following form:

$$\partial_t E_k^R = T_k^{\mathrm{nl}} + T_k^{\mathrm{loc}}, \tag{5.74}$$

omitting dissipation terms. The first term on the r.h.s. represents the Alfvén effect,

$$T_k^{\mathrm{nl}} \sim -E_k^R/\tau_{\mathrm{A}} \sim -E_k^R k v_{\mathrm{A}}, \tag{5.75}$$

which would lead to decay of E_k^R within one Alfvén time if there were no local interactions. From dimensional arguments the latter assume the form

$$T_k^{\mathrm{loc}} \sim \epsilon/k. \tag{5.76}$$

Hence a stationary spectrum arises from the balance $T_k^{\mathrm{nl}} + T_k^{\mathrm{loc}} = 0$,

$$E_k^R \sim (\epsilon/v_{\mathrm{A}})k^{-2}. \tag{5.77}$$

The relative magnitude of the residual spectrum is small, at least in the case of weak velocity–magnetic-field correlation, for which $E_k \sim k^{-3/2}$, being given by $E_k^R/E_k \sim k^{-1/2}$. The sign of E_k^R is not determined by these dimensional considerations. We have, however, seen that, in the absolute equilibrium spectrum, the magnetic energy is always larger than the kinetic energy, (5.21). This tendency is confirmed for dissipative turbulence by numerical solutions of the closure equations (Grappin *et al.*, 1983) and by direct numerical simulations; see Fig. 5.4.

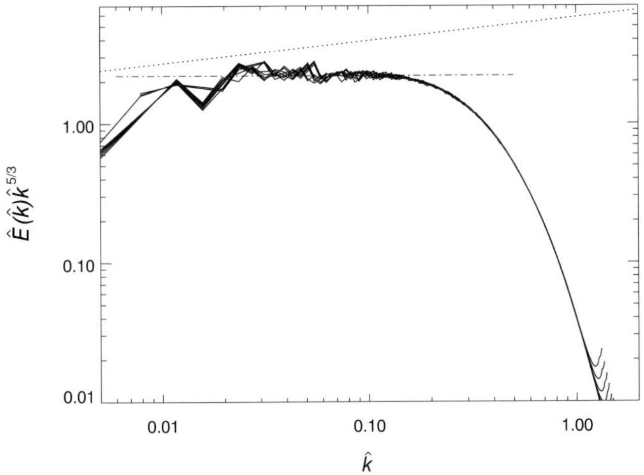

Figure 5.2: A scatter plot of the normalized energy spectrum $\widehat{E}(\widehat{k})$ taken at different times multiplied by $\widehat{k}^{5/3}$, from a direct numerical simulation with $H^M = 0.68 H^M_{\rm max}$. The dash–dotted line indicates a Kolmogorov spectrum with $C_{\rm K} = 2.2$, the dotted line the slope of the IK spectrum $k^{-3/2}$. (From Biskamp and Müller, 2000b.)

5.3.6 Results of numerical simulations

The preceding discussion might evoke the impression that the spectral laws just derived reflect basic principles of physics and must therefore be valid, if only the Reynolds number is sufficiently high. This is true, to a certain extent, of the Kolmogorov spectrum, which follows from simple dimensional analysis and should therefore be rather robust. The Kolmogorov spectrum is indeed observed under very diverse conditions, and deviations from this law seem to occur only if special processes dominate the turbulent dynamics in the inertial range, such as buoyancy forces in thermal convection, which may lead to a steeper energy spectrum, (9.41). Since the IK spectrum is based on such a special process, the Alfvén effect, there is no *a priori* reason why this process should dominate the inertial-range dynamics. A reliable answer can therefore be obtained only from observations or from direct numerical simulations. While observations will be treated in Chapter 10, at this point we discuss the results of numerical studies.

For direct numerical simulations to exhibit a clear inertial range the spatial resolution must be sufficiently high, which is particularly demanding in the 3D case considered here (2D simulation results will be presented in Chapter 8). Most numerical systems of 3D MHD turbulence studied to date were too small for investigators using typically 64^3 Fourier modes, or spatial colocation points, to even discern a power-law spectrum at all, let alone distinguish between the relatively similar spectral indices $\frac{5}{3}$ and $\frac{3}{2}$. We therefore restrict the discussion to some recent work with the relatively high resolution of 512^3 modes (Müller and Biskamp, 2000; Biskamp and Müller, 2000b), involving a number of simulation runs for decaying MHD turbulence.

Studying decaying turbulence has the advantage of avoiding uncontrolled effects arising from the choice of the external forcing, which would have to be applied in order to maintain the turbulence at a steady average level. In a decaying system, on the other hand, spectra are not stationary, such that direct time averaging is not possible. Instead one relies on the self-similarity of the spectrum, which is expressed in normalized form $\widehat{E}(\widehat{k})$ in order to eliminate the variation of the integral quantities, especially of ϵ, as discussed in Section 5.3.3. If the normalization is appropriate, $\widehat{E}(\widehat{k})$ does not vary with time apart from statistical fluctuations.

The initial state is chosen with random phases of the individual modes and specified values of the total energy E, energy ratio $\Gamma = E^K/E^M$, magnetic helicity H^M, and cross-helicity H^C. The dominant wavenumber is taken to be relatively large, $k_0 = 5$, restricting the inertial range to $k > 5$, which is, however, unavoidable in the presence of the inverse cascade dynamics in MHD. Assuming that $k_0 \sim 1$, as is usually done in simulations of hydrodynamic turbulence, would lead to magnetic condensation in the lowest-k state, which would also affect the the inertial-range dynamics. Figure 5.2 shows a scatter plot of the normalized energy spectrum from a run with $\mathrm{Rm}_\lambda \simeq 100$ and finite magnetic helicity. The spectrum has been multiplied by $\widehat{k}^{5/3}$, which very nearly compensates for the slope in the inertial range, the horizontal part of the spectrum extending over almost one wavenumber decade. Compared with the temporal variation of the unnormalized spectra – the curves in Fig. 5.2 are taken from a period of fully developed turbulence, during which the total energy decreases by more than a factor of 2 –, the scatter of the curves in the inertial range $k > k_0$ is small. For the inertial and dissipation ranges the Kolmogorov-type normalization (5.57) is appropriate, the inertial-range spectrum being close to $k^{-5/3}$ and clearly different from the IK power law. The value of the constant $C_K \simeq 2.2$ is somewhat higher than is observed for hydrodynamic turbulence (5.38).

In the small-k range $k < k_0$ the normalized spectrum is not stationary due to the inverse cascade of the magnetic helicity. This should lead to a different spectral law (5.44), which is indeed found in numerical solutions of the closure equations; see Fig. 6.4. The increase of the large-scale magnetic energy is the reason for the decrease of the energy ratio Γ discussed in Section 4.2.3, Γ being primarily determined by the modes $k \lesssim k_0$.

The inertial-range energy spectrum appears to be independent of the magnetic helicity. Figure 5.3 gives a similar scatter plot of the compensated normalized

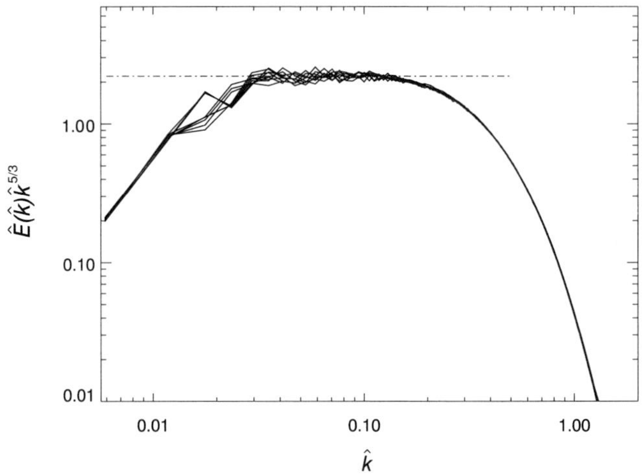

Figure 5.3: A scatter plot of the normalized energy spectrum compensated by $\widehat{k}^{5/3}$ from a simulation run similar to that in Fig. 5.2, but for $H^M = 0$. The dash–dotted line again indicates the Kolmogorov spectrum with $C_K = 2.2$.

energy spectrum for $H^M = 0$, which exhibits the same inertial-range behavior, even the same value of the constant C_K. Since there is no inverse cascade for nonhelical turbulence, the energy spectrum remains self-similar throughout the entire spectral range.

As expected from the discussion in Section 5.3.3, the spectrum does not follow the IK law, and the question of whether the Alfvén effect is completely suppressed arises. Figure 5.4 shows the magnetic- and kinetic-energy spectra, normalized and time-averaged, for the same simulation run as in Fig. 5.2. At small scales $k \sim k_d$ the kinetic energy does indeed approach the magnetic energy, as is required for eddies behaving as Alfvén waves, (5.43). The slight excess of magnetic energy is consistent with the tendency indicated already by the absolute equilibrium spectra (5.21).[3] However, the Alfvén effect does not determine the inertial range dynamics. We shall discuss this dynamics in more detail in Section 7.5 when we treat the higher-order statistical properties of MHD turbulence. There is hence no reason why the prediction of the residual spectrum $E_k^R = E_k^K - E_k^M \sim k^{-2}$, (5.77), should be valid, since it is based on the Alfvén effect. Indeed, the observed spectrum is somewhat steeper, $E_k^R \sim k^{-5/2}$, as shown in Fig. 5.5.

[3] The detailed behavior of the magnetic- and kinetic-energy spectra in the dissipation range $kl_K > 1$ depends on the magnetic Prandtl number $\mathrm{Pr}_m = \nu/\eta$, which is unity in Fig. 5.4. For $\mathrm{Pr}_m > 1$ the ratio E_k^M / E_k^K becomes larger than that shown in Fig. 5.4, whereas for $\mathrm{Pr}_m < 1$ the kinetic-energy spectrum may even exceed the magnetic one for $kl_K > 1$.

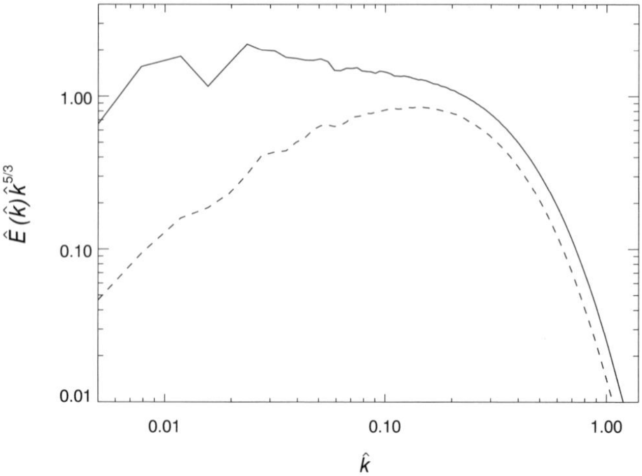

Figure 5.4: Time-averaged normalized spectra of magnetic energy (continuous) and kinetic energy (dashed) compensated by $\widehat{k}^{5/3}$ for the same simulation run as that displayed in Fig. 5.2. (From Biskamp and Müller, 2000b.)

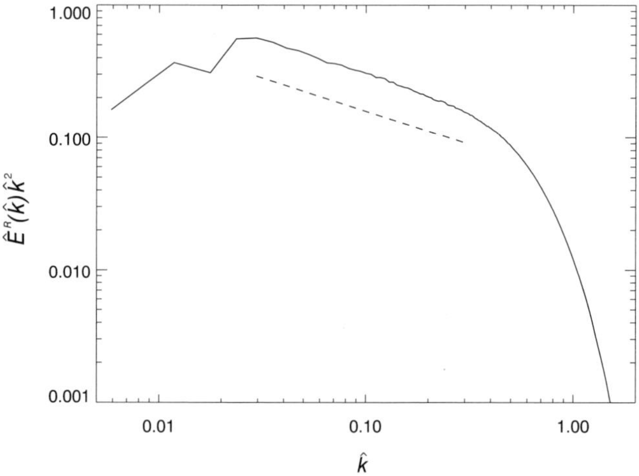

Figure 5.5: The normalized residual spectrum $|E_k^R|$ compensated by k^2 from the simulation run shown in Fig. 5.2. The dashed line indicates a $k^{-5/2}$ law.

Hyperdiffusion and the bottleneck effect

Let us now investigate the effect of hyperdiffusion (2.23) on the shape of the energy spectrum. Since in the presence of normal diffusion the transition from the inertial to the dissipation range is rather gradual (note that the dissipative falloff in the energy spectrum in Fig. 5.2 starts already at $\widehat{k} = kl_d \simeq 0.1$), it is

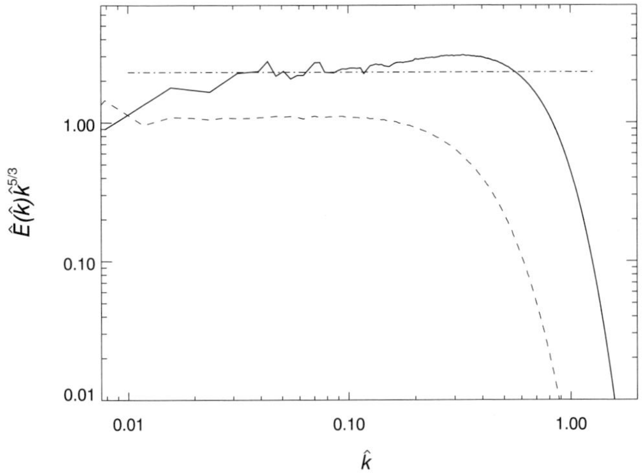

Figure 5.6: A time-averaged normalized energy spectrum multiplied by $\widehat{k}^{5/3}$ for a simulation run with hyperdiffusion $\alpha = 2$ but otherwise the same conditions as in Fig. 5.2. The dash–dotted line gives the Kolmogorov spectrum with the same value of C_K. The dashed line is the 1D spectrum $E(k_z)$. (From Biskamp and Müller, 2000b.)

tempting to introduce higher-order diffusion terms in order to make the transition to the dissipation range more abrupt and thus expand the inertial-scaling range. However, this transition should not be too sharp. The idea of a unidirectional spectral energy flux in the inertial range implies the total absorption of the energy reaching the dissipation scale. Similarly to the case of a light wave impinging on an absorbing medium, where total absorption occurs only if the absorption process is sufficiently gradual and takes place in a layer of width much larger than the wavelength, total absorption of the turbulent energy flux would require very weak diffusivity $\alpha \ll 1$. For normal diffusion $\alpha = 1$ the fraction reflected is still small, but approaches unity for high-order hyperdiffusion $\alpha \gg 1$, which would in fact correspond to an ideal system truncated at $k = k_d$.

As result of a partial reflection, energy accumulates in front of the dissipation range, which is called the *bottleneck effect* (Falkovich, 1994). In hydrodynamic turbulence the energy spectrum has the general form $E_k = \epsilon^{2/3} k^{-5/3} f(k/k_d)$, where the function f describes mainly the exponential falloff in the dissipation range for $k > k_d$. f is, however, not monotonically decreasing, but increases slightly for $k \lesssim k_d$ before falling off, producing a weak hump in the spectrum, which corresponds to an effective flattening of the spectrum in this wavenumber range. The effect has been observed in hydrodynamic turbulence, both experimentally (e.g., She and Jackson, 1993; Saddoughi and

Veeravalli, 1994) and numerically (e.g., She *et al.*, 1993). A mathematical interpretation of this effect is related to a property of the Fourier transformation, namely that a sharply bent structure function in configuration space corresponds to an overshoot in the Fourier spectrum (Lohse and Müller-Groeling, 1995). The bottleneck hump has sometimes been interpreted as a short k^{-1} spectral range in front of the dissipative falloff (e.g., Gagne, 1987). The effect is particularly strong in turbulence simulations using the PPM method (Porter *et al.*, 1998a and 1998b), in which viscosity is not considered explicitly and dissipation occurs on the grid–scale, i.e., sets in abruptly in k-space (see Section 9.2.1).

In the simulations of MHD turbulence with normal diffusion presented above in Fig. 5.2 no bottleneck hump is visible (though a weak hump may appear at still higher Reynolds number). However, on repeating the same run with hyperdiffusion $\alpha = 2$, the effect shows up clearly, as can be seen in Fig. 5.6, where a horizontal part in the range $0.03 \leq \widehat{k} \leq 0.15$ is followed by a hump (the dash–dotted horizontal line indicates the Kolmogorov spectrum with the same constant $C_{\mathrm{K}} = 2.2$ as that observed in the normal-diffusion run. (The dashed line is the normalized 1D spectrum $E(k_z) = E_{k_z} + E_{-k_z}$, which is lower by a factor $\frac{3}{5}$; it does not exhibit a hump, since, as a consequence of taking the integral over k_x and k_y, dissipation already affects smaller wavenumbers k_z, making the transition to the dissipation range gentler.)

In the presence of a strong mean field B_0 the spectrum is globally anisotropic. While the spectrum parallel to the field falls off steeply, since it is difficult to excite those modes involving bending of the strong field, the perpendicular spectrum is somewhat flatter and roughly consistent with a $k^{-3/2}$ law. A general discussion of how the statistical properties depend on the strength of the mean field is given in Section 7.5.2, where we show the transition from the globally isotropic case B_0 to that of large B_0, which converges to the behavior in a purely 2D system.

6
Two-point-closure theory

In the derivation of spectral laws presented in Chapter 5 we used only certain general properties of the turbulence, in particular the integral invariants, which lead to the spectral cascades. (Only the Alfvén effect resulting in the IK spectrum is based on a specific dynamic process of the MHD system.) Though phenomenological arguments, especially dimensional analysis, are often very powerful and robust, since they represent basic physical principles, they only predict a few scaling laws but cannot, for instance, specify proportionality factors, such as the Kolmogorov constant and the sign of the residual energy spectrum. Morover, these arguments provide little insight into the turbulence *dynamics*. Such properties must be treated by a statistical theory derived from the basic fluid equations. Here the most practical approach is two-point closure theory. An alternative method, renormalization-group (RNG) theory, which was originally developed in the context of the theory of critical phenomena, has also been applied to hydrodynamic turbulence (e.g., Yakhot and Orszag, 1986) and MHD turbulence (Fournier *et al.*, 1982; Camargo and Tasso, 1992), but there is still a considerable degree of arbitrariness and even inconsistency. We shall therefore not discuss RNG theory any further but restrict the treatment in this chapter to closure theory.

In Chapter 4 we introduced the one-point-closure approximation consisting of the equations for the average fields and some phenomenological expressions for the correlation functions appearing in these equations, which is appropriate for describing large-scale inhomogeneous-turbulence processes. To study intrinsic small-scale properties, for which correlation functions are of primary interest, one has to go one step further in the hierarchy of moment equations. In Section 6.1 we discuss briefly the various two-point-closure approximations that have been studied in the literature, of which the EDQNM model is the most widely used. This model is then applied in Section 6.2 to MHD turbulence, where we analyze the two most interesting cases of helical turbulence and of correlated turbulence, considering, in particular, the effects of nonlocal mode interactions. Section 6.3 deals with the limitations and shortcomings of closure theory.

6.1 Quasi-normal-type approximations

6.1.1 The problem of closure

Let us write the fluid equations in a general form valid for many types of nonlinear system,

$$\frac{du_i}{dt} + \nu_i u_i = \sum_{jk} M_{ijk} u_j u_k, \tag{6.1}$$

where in the case of incompressible MHD $\{u_i\} = \{\boldsymbol{v}_k, \boldsymbol{b}_k\}$, $\{\nu_i\} = \{\nu k^2, \eta k^2\}$, and the coupling coefficients M_{ijk} are functions of the wave vectors $\boldsymbol{k}$, $\boldsymbol{p}$, $\boldsymbol{q}$ forming a triangle, $\boldsymbol{k}+\boldsymbol{p}+\boldsymbol{q} = 0$. Multiplying (6.1) by u_j and the corresponding equation for u_j by u_i, and then adding and averaging the results gives the equation for the two-point correlation function,

$$\frac{d}{dt}\langle u_i u_j\rangle + (\nu_i + \nu_j)\langle u_i u_j\rangle = \sum_{mn} \big(M_{imn}\langle u_j u_m u_n\rangle + M_{jmn}\langle u_i u_m u_n\rangle\big). \tag{6.2}$$

The equation for the triple-correlation function is obtained in an analogous way,

$$\frac{d}{dt}\langle u_i u_j u_k\rangle + (\nu_i + \nu_j + \nu_k)\langle u_i u_j u_k\rangle = \sum_{mn} \big(M_{imn}\langle u_j u_k u_m u_n\rangle + M_{jmn}\langle u_i u_k u_m u_n\rangle + M_{kmn}\langle u_i u_j u_m u_n\rangle\big). \tag{6.3}$$

We thus arrive at an infinite hierarchy of moment equations replacing the single dynamic equation (6.1). This apparent incongruity is due to the fact that the solution of the hierarchy is equivalent to an infinite set of solutions of the dynamic equation for different initial conditions or, in the case of stationary turbulence, to a solution followed over an infinitely long time. While in general the solution of the 'simple' dynamic equation is very complicated, with rapid spatial and temporal variations, the solution of the statistical equations can be smooth and stationary.

To be of practical use, the hierarchy must be closed, resulting in a set of approximate equations for a finite number of functions. Since, in the most interesting case of high Reynolds number, the nonlinear term in (6.1) is much larger than the linear term, finite-series approximations, neglecting for instance correlations of order $n > N$, are not appropriate. A natural starting point for an approximate treatment is the observation, which is also found in numerical simulations, that the components of the fields have probability distribution functions very close to Gaussian, or normal. This property is rather generally expected for stochastic systems of many degrees of freedom as a consequence of the central-limit theorem,[1] for instance for the molecules in a gas, which assume a Maxwellian velocity distribution.

The statistical properties of a random variable χ become particularly clear on considering its cumulants. The nth cumulant $C^{(n)}$ is defined as the nth-order derivative of the logarithm of the characteristic function $\varphi(s) = \langle e^{is\chi} \rangle$ taken at the origin,

$$C^{(n)} = (d^n/ds^n) \ln \varphi(s)|_{s=0}, \tag{6.4}$$

which can be written in terms of the moments $\langle \chi^m \rangle$, $m \leq n$ (see, for instance, Lumley 1970, Chapter 2). In the case of zero mean the first four cumulants are

$$C^{(1)} = 0, \quad C^{(2)} = \langle \chi^2 \rangle, \quad C^{(3)} = \langle \chi^3 \rangle, \quad C^{(4)} = \langle \chi^4 \rangle - 3\langle \chi^2 \rangle^2. \tag{6.5}$$

For a set of random variables $\chi_1, \ldots, \chi_n$ with the multivariate Gaussian distribution (5.13),

$$p(\chi_1, \ldots, \chi_n) = (2\pi)^{-n/2} |A^{-1}|^{1/2} \exp\left(-\tfrac{1}{2} \sum_{i,j}^{n} A_{ij} \chi_i \chi_j \right),$$

the characteristic function can be given analytically,

$$\varphi(s_1, \ldots, s_n) = \left\langle \exp\left(i \sum_j s_j \chi_j \right) \right\rangle = \exp\left(-\tfrac{1}{2} \sum_{ij} A_{ij}^{-1} s_i s_j \right), \tag{6.6}$$

where A_{ij}^{-1} is the inverse of the matrix A_{ij} and $|A^{-1}|$ its determinant. A_{ij}^{-1} equals the second-order moments, $\langle \chi_i \chi_j \rangle = A_{ij}^{-1}$. From the definition (6.4) one finds immediately that, for a Gaussian distribution, all cumulants of order $n > 2$ vanish; in particular, one has

$$\langle \chi_i \chi_j \chi_k \rangle = 0, \tag{6.7}$$

$$\langle \chi_i \chi_j \chi_k \chi_l \rangle = \langle \chi_i \chi_j \rangle \langle \chi_k \chi_l \rangle + \langle \chi_i \chi_k \rangle \langle \chi_j \chi_l \rangle + \langle \chi_i \chi_l \rangle \langle \chi_k \chi_j \rangle. \tag{6.8}$$

While odd-order moments vanish, even-order moments consist of the sum of products of second-order moments corresponding to all possible pairings, as illustrated in Fig. 6.1. Hence, apart from the second order one, the cumulants represent the non-Gaussian effects of a general probability distribution. For a turbulence field, it is clear that, even if the velocity components considered at one point in space are Gaussian, this cannot be true for the joint probability

[1] In its simplest form the central-limit theorem states that the sum of a large number of identically distributed stochastic variables assumes a Gaussian, or normal, distribution.

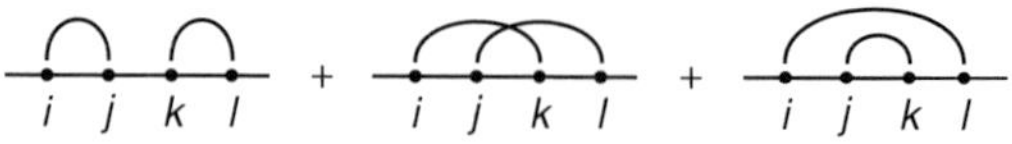

Figure 6.1: Pairings of the fourth-order moment.

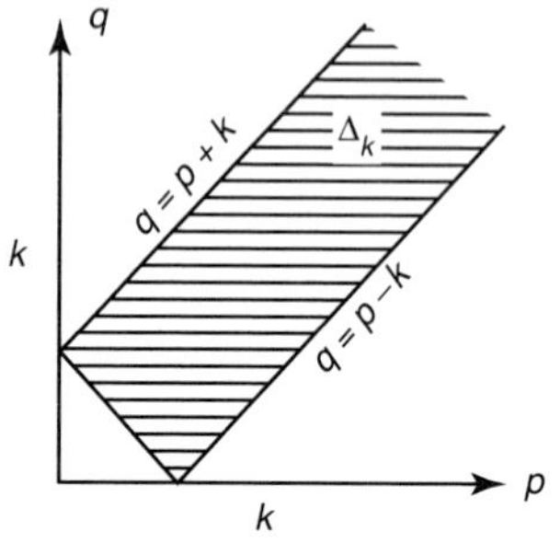

Figure 6.2: The integration domain in the p, q-plane.

distribution involving different spatial arguments, since otherwise the triple correlation in (6.2) would vanish, eliminating all nonlinear interactions.

6.1.2 The quasi-normal approximation

One therefore assumes only that the joint probability distributions are *close* to normal. Here the simplest approximation is to neglect all cumulants of order $n > 3$, which implies (6.8), and to compute the triple correlations from the evolution equation (6.3) using (6.8). This closure of the hierarchy is the *quasi-normal approximation*, or *cumulant-discard method*, which was first proposed by Millionshtchikov (1941). The corresponding spectral equations for isotropic hydrodynamic turbulence were derived by Proudman and Reid (1954). Integrating (6.3) and inserting the result into (6.2) gives the nonlinear evolution equation for the modal intensity $U_k = \langle |u_k|^2 \rangle$, the equal-time correlation function,

$$\partial_t U_k + 2\nu_k U_k = \int_0^t dt' \int_{\Delta_k} dp\, dq\, G_{kpq}(t,t') \frac{a_{kpq}}{q} \big[k^2 U_p(t') U_q(t') - p^2 U_k(t') U_q(t') \big]. \tag{6.9}$$

Here Δ_k indicates integration over the subdomain of the p, q-plane such that $\boldsymbol{k}, \boldsymbol{p}, \boldsymbol{q}$ form a triangle, given in Fig. 6.2, $a_{kpq} = xy + z^3$, where x, y, z are the cosines of the angles in the triangle facing the sides k, p, q, respectively, and $G_{kpq}(t, t')$ is the linear response function, which in the quasi-normal approximation is simply Green's function of the l.h.s. of (6.3),

$$G_{kpq}(t,t') = \mathrm{e}^{-(\nu_k + \nu_p + \nu_q)(t-t')}. \tag{6.10}$$

The r.h.s. of (6.9) contains two terms; the first one, the mode-coupling term, acts as a source or sink of turbulence energy at wavenumber k, while the second one, which is proportional to U_k, represents a turbulent friction. More details of the properties of the quasi-normal equations for hydrodynamic turbulence can be found in Leslie's book (1973), which gives a general review of closure theory.

While it is difficult to judge the value of the quasi-normal approximation on purely theoretical grounds, numerical solutions have shown that the quasi-normal equations do not provide a consistent theory of fluid turbulence. Ogura (1963) showed that the spectral intensity U_k develops negative values, which contradicts a basic requirement on U_k as the expectation value of a nonnegative quantity. Since negative U_k occur in the main, energy-containing, part of the spectrum, the theory is also unacceptable from a practical point of view. This unphysical feature is caused by the buildup of excessive levels of the triple correlation in (6.3) due to the excessively long memory time $\sim\nu_k^{-1}$ in the time integral in (6.9) which, for a mode in the inertial range, is much longer than the corresponding eddy-turnover time. Hence the viscous damping ν_k should be replaced by a turbulent eddy damping $\widetilde{\nu}_k$. A statistical theory accounting for this effect is the direct-interaction approximation (DIA) of Kraichnan (1959), or its Lagrangian variant (Kraichnan, 1965a), which corrects an inconsistency of the former Eulerian version mentioned in Section 5.3.2. Formally, the DIA amounts to supplementing (6.9) by a second nonlinear equation for the response function G instead of using expression (6.10). The Lagrangian DIA is a fundamental two-point turbulence theory with no adjustable parameters, which gives good agreement with experimental measurements, predicting, in particular, the $k^{-5/3}$ energy spectrum with the correct value of the Kolmogorov constant. However, the equations are very complex, too unwieldy for practical applications. Interest has therefore shifted to somewhat more phenomenological models, which are sufficiently simple to allow numerical solutions for high Reynolds numbers, but still satisfy the important consistency conditions.

6.1.3 The eddy-damped quasi-normal Markovian approximation (EDQNM)

Instead of calculating the response function self-consistently from a nonlinear equation as in the DIA, we introduce a phenomenological eddy-damping rate, which determines the relaxation time of the triple correlation in (6.3). The simplest expression follows from dimensional arguments (note that ν_k has the dimension T^{-1}):

$$\widetilde{\nu}_k \sim \epsilon^{1/3} k^{2/3} \sim (E_k k^3)^{1/2}, \tag{6.11}$$

which was suggested by Orszag (1970). $\widetilde{\nu}_k^{-1}$ gives the order of magnitude of the turnover time of an eddy of scale k^{-1}, which is also called the eddy-distortion, or scrambling, time. Here the Kolmogorov spectrum (5.36) was used to replace ϵ by the local quantity E_k. For a steeply decreasing spectrum, such as in the initial stage of the development of turbulence from a smooth state, expression (6.11) is, however, not suitable, since $\widetilde{\nu}_k$ would decrease for increasing k while, as a typical dynamic frequency, it should increase. An expression satisfying this requirement was introduced by Pouquet *et al.* (1975),

$$\widetilde{\nu}_k \sim \left(\int_0^k p^2 E_p \, dp \right)^{1/2} . \tag{6.12}$$

Replacing the viscous-damping rate in (6.10) by the eddy-damping rate $\widetilde{\nu}_k$ gives the renormalized response function

$$\widetilde{G}_{kpq}(t, t') = \exp\left(- \int_{t'}^{t} d\tau \left[\widetilde{\nu}_k(\tau) + \widetilde{\nu}_p(\tau) + \widetilde{\nu}_q(\tau) \right] \right). \tag{6.13}$$

There is, of course, a considerable degree of arbitrariness about the way in which the 'naked' quasi-normal expression (6.10) can be 'dressed', or renormalized to give the expression (6.13) representing only the simplest and most straightforward form. An important property is the symmetry in k, p, and q, which guarantees the detailed-balance relations for each triad interaction discussed in Section 5.2, and hence conservation of the ideal invariants of the dynamic equations.

While the use of the eddy viscosity shortens the response time and thus avoids the buildup of artificially large triple correlations, it is still not assured that the spectral intensity remains positive if it is so initially, i.e., that the solution of (6.9) is realizable in the sense that it corresponds to the statistical average with a nonnegative probability distribution. A simple way to satisfy this condition is to replace the time argument t' in the spectral intensities inside the integral over time in (6.9) by the outer time t, a process called *Markovianization*. Since the effective time domain in the integral is the eddy interaction time $\widetilde{\nu}_k^{-1}$, which is the shortest time scale on which the spectral intensity U_k can possibly change, Markovianization should not change the evolution $U_k(t)$ significantly. Equation (6.9) now takes the form (omitting wavenumber indices and integrals)

$$\frac{dU}{dt} + 2\nu U(t) = \theta(t) U(t) U(t), \tag{6.14}$$

where $\theta(t) = \int_0^t \widetilde{G}(t, t') \, dt'$ is the triple-correlation, or triad, relaxation time. The Realizability of the Markovian equation (6.14) can be proved by showing that it constitutes the *exact* ensemble average of the spectral intensity of a model dynamic system. The model amplitude equation has the form of a Langevin equation,

$$\frac{d}{dt}\widehat{u}_k = -\alpha_k \widehat{u}_k + \widehat{q}_k, \tag{6.15}$$

where $\widehat{u}_k$ is a stochastic variable, which is, in general, *not* identical with the turbulence modes u_k but has the same variance $\langle \widehat{u}_k^2 \rangle = U_k$, $\widehat{q}_k(t)$ is a white-noise stochastic variable independent of $\widehat{u}_k$, and $\alpha_k(t)$ is a nonstochastic damping rate. $\widehat{q}_k$ and α_k can be chosen such that the equation for the average spectral intensity derived from the Langevin equation coincides with the Markovian equation (6.14). Turbulence-model equations of this type are called eddy-damped quasi-normal Markovian (EDQNM) (see, for instance, Orszag, 1977). We should mention that Markovianization is sufficient but not necessary to guarantee realizability. Kraichnan's DIA is realizable, though it is not Markovian. On the other hand, the Lagrangian version of the DIA, which eliminates the spurious mode interactions of largely disparate scales, has not been shown to be realizable, and probably is not in a strict sense.

6.2 The EDQNM theory of MHD turbulence

The ideas of statistical closure theory have also been applied to MHD turbulence, in particular the EDQNM approximation, and several numerical studies of these equations have been performed. Even more than in the case of hydrodynamic turbulence there is considerable freedom in choosing the form of the eddy damping. Pouquet *et al.* (1976) suggested the following expression, which was also used in later studies:

$$\begin{aligned}\widetilde{\nu}_k &= a_S\left(\int_0^k q^2(E_q^K + E_q^M)\,dq\right)^{1/2} + a_A k\left(2\int_0^k E_q^M\,dq\right)^{1/2} + (\nu+\eta)k^2 \\ &= \widetilde{\nu}_k^S + \widetilde{\nu}_k^A + \nu_k^D.\end{aligned} \tag{6.16}$$

The first term, $\widetilde{\nu}_k^S$, represents the nonlinear eddy-distortion rate analogous to the eddy damping (6.12). The second term, $\widetilde{\nu}_k^A \sim k(E^M)^{1/2} = kv_A = \tau_A^{-1}$, corresponds to the Alfvén effect, which leads to relaxation of the triple correlations (6.3) within an Alfvén time due to propagation of Alfvén waves along the large-scale field as discussed in Section 4.2.2. Note that, for a spectrum k^{-n} that is not too steep, with $n < 3$, $\widetilde{\nu}_k^S$ is independent of the lower limit and mainly determined by the spectrum at the upper limit; hence it is a local effect, whereas $\widetilde{\nu}_k^A$ is a nonlocal effect determined by the field in the energy-containing eddies at $q \ll k$. The third term, ν_k^D, comprises the collisional dissipation effects. Definite values of the coefficients were chosen by

Pouquet *et al.*; a_S is related to the Kolmogorov constant, which gives $a_S \simeq 0.3$, while an explicit calculation of the relaxation rate of the triple correlation for a Gaussian large-scale field gives $a_A = 1/\sqrt{3}$, but, in view of the arbitrariness of the *Ansatz* (6.16), the coefficients should be considered essentially free numerical parameters of order unity.

The EDQNM closure model consists of the equations for five spectral functions, the kinetic and magnetic energies E_k^K and E_k^M, the kinetic and magnetic helicities H_k^K and H_k^M, and the cross-helicity H_k^C. The model has been derived and discussed in detail for the most interesting special situations, namely helical turbulence with vanishing velocity–magnetic-field correlations and correlated turbulence with vanishing kinetic and magnetic helicities.

6.2.1 Helical turbulence

Magnetic turbulence in astrophysical systems is, in general, governed by two effects, buoyancy and rotation, which naturally lead to helical flows and twisted field lines or, in formal parlance, kinetic and magnetic helicity. These properties play a crucial role in building up large-scale magnetic structures. Though closure theory, which in its most common form is limited to homogeneous and isotropic turbulence, does not describe the global inhomogeneous features of such systems, it provides a valuable tool for understanding the basic growth mechanism of large-scale fields. To simplify the formalism and facilitate physical understanding we restrict our consideration to the case of vanishing velocity–magnetic-field correlation, following Pouquet *et al.* (1976). If $H_k^C = 0$ initially, it will remain zero. The spectral equations are of the general form

$$\partial_t U_k^{(i)} + 2\nu_k^{(i)} U_k^{(i)} - f_k^{(i)} = \int_\Delta dp\, dq\, \theta_{kpq} \sum_{j,l=1}^{4} \left(b_{kpq}^{(jl)} U_k^{(j)} U_p^{(l)} + c_{kpq}^{(jl)} U_p^{(j)} U_q^{(l)} \right), \tag{6.17}$$

where the $U_k^{(i)}$ are the spectral intensities E_k^K, E_k^M, H_k^K, and H_k^M, the coefficients $b_{kpq}^{(i)}$ and $c_{kpq}^{(i)}$ are geometrical factors depending on the wave-vector triad and the cosines of the angles, $\nu_k^{(i)} = \nu k^2$ or ηk^2, and $f_k^{(i)}$ are appropriate forcing terms, which are applied if a stationary state should be maintained. For the explicit algebraic forms of the equations, which are rather complex, the reader is referred to the original paper. The relaxation time of the triple correlations is given by (6.13),

$$\theta_{kpq}(t) = \int_0^t \widetilde{G}(t,t')\, dt' \simeq \frac{1 - e^{-(\widetilde{\nu}_k + \widetilde{\nu}_p + \widetilde{\nu}_q)t}}{\widetilde{\nu}_k + \widetilde{\nu}_p + \widetilde{\nu}_q}, \tag{6.18}$$

since the spectral intensities, and hence the $\widetilde{\nu}_k$ terms, can be assumed constant during the time θ. The EDQNM equations (6.17) satisfy the same conservation relations as the original fluid equations, *viz.*, conservation of total energy and magnetic helicity in the absence of dissipation and forcing,

$$\frac{d}{dt}\int_0^\infty dk\,(E_k^K + E_k^M) = 0, \quad \frac{d}{dt}\int_0^\infty H_k^M\,dk = 0. \tag{6.19}$$

They also preserve the properties $|H_k^K| \le E_k^K k$ and $|H_k^M| \le E_k^M/k$ required by the definition of the helicities, if the forcing terms satisfy these relations.

We first want to show the consequences of the closure model (6.17) for the inertial-range energy spectrum. The r.h.s. of the equation, called the transfer function T_k, represents the nonlinear interaction, which conserves energy. Hence

$$\frac{d}{dt}\int_0^k E_p\,dp\Big|_{\mathrm{nl}} = \int_0^k T_p\,dp = -\epsilon \tag{6.20}$$

is the energy flux from the range $p < k$ into the range $p > k$, which is constant in the inertial range. Assuming that the major contribution to the integral comes from the region $p \sim q \sim k$, the coefficients in the equation for $U_k^{(i)} = E_k^{K,M}$ are $b \sim c \sim k$; hence we find the scaling relation, see (5.42),

$$\epsilon \sim \theta_k k^4 E_k^2. \tag{6.21}$$

For a weak magnetic field, for which $\theta \sim (\widetilde{v}_k^S)^{-1} \sim (k^3 E_k)^{-1/2}$ from (6.16) and (6.18), one obtains the Kolmogorov spectrum $E_k \sim \epsilon^{2/3}k^{-5/3}$, whereas for a strong magnetic field, for which $\theta \sim (\widetilde{v}_k^A)^{-1} \sim (kv_A)^{-1}$, one finds the IK spectrum $E_k \sim (\epsilon v_A)^{1/2}k^{-3/2}$. Hence there should be a transition from a $k^{-5/3}$ behavior for $k < k_0$ to a flatter spectrum $k^{-3/2}$ for $k > k_0$, which occurs at $k_0 \sim \epsilon/v_A^3$. In the case of roughly equal magnetic and kinetic energies, $E^K \sim E^M \sim B^2$, the transition wavenumber is of the order of the inverse integral scale $k_0 \sim L^{-1}$, (4.45); hence the entire inertial range should exhibit the IK spectrum. Figure 6.3 shows a numerical solution of the EDQNM equations for kinetically driven nonhelical turbulence, which confirms these scaling arguments.

To obtain more insight into the turbulence dynamics, it is useful to distinguish between local and nonlocal mode interactions. Local means that the wavenumbers of a triad are of the same order, i.e., we have $q \le ap$ and $k \ge p/a$ for $k \le p \le q$, where $a \sim 1$, say $a = \sqrt{2}$, such that the maximum ratio of the largest to the smallest wavenumber is 2. Nonlocal interactions occur between strongly disparate wavenumbers. These are on the one hand the action of small-scale, i.e., large-wavenumber, fluctuations on k,

$$p \sim q \gg k, \tag{6.22}$$

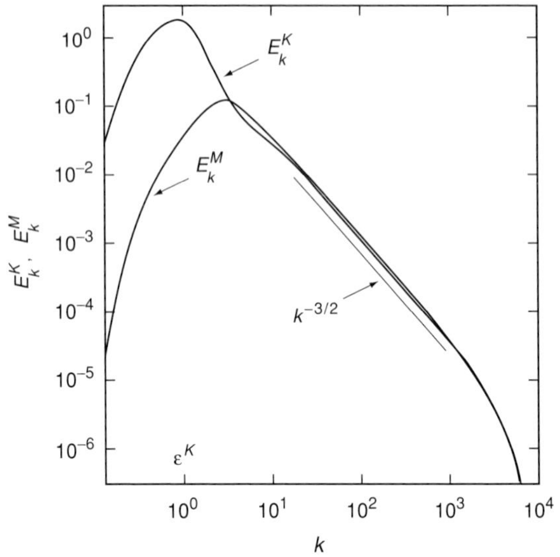

Figure 6.3: Stationary kinetic- and magnetic-energy spectra from a numerical solution of the EDQNM equations (6.17) driven kinetically at large scales $k \sim 1$, $f_k^K = ck^4 \exp(-2k^2)$. The spectra exhibit the $k^{-3/2}$ behavior in the inertial range $10 < k < 10^3$. While the kinetic energy is larger in the forcing range, there is a slight excess of magnetic energy in the inertial range. (From Pouquet *et al.*, 1976.)

which leads to the concept of eddy diffusivities. On the other hand we have the effect of large scales, i.e., small wavenumbers, acting on k,

$$q \ll k \sim p \quad \text{or} \quad p \ll k \sim q. \tag{6.23}$$

Nonlocal interactions are particularly important in MHD turbulence. In the nonhelical case these are mainly effects of the latter type (6.23) connected with the Alfvén effect, whereby the large-scale magnetic field couples kinetic and magnetic fluctuations, such that $E_k^K \simeq E_k^M$. In helical MHD turbulence, the most important nonlocal effect is of the former type (6.22), the generation of large-scale fields by small-scale helicity, called the turbulent dynamo effect, which has been introduced in Section 4.1. By expanding the EDQNM equations (6.17) Pouquet *et al.* derived expressions for the nonlocal processes due both to small-scale modes and to large-scale modes.

Let us first consider the nonhelical case, in which the Alfvén effect is the strongest nonlocal effect, which follows from the equations

$$\partial_t E_k^K \Big|_{\text{nl}} = -k\Gamma_k(E_k^K - E_k^M), \quad \partial_t E_k^M \Big|_{\text{nl}} = k\Gamma_k(E_k^K - E_k^M), \tag{6.24}$$

where

$$\Gamma_k = \tfrac{4}{3} k \int_0^{ak} \theta_{kkq} E_q^M \, dq \sim (E^M)^{1/2} \simeq v_{\rm A}, \tag{6.25}$$

on inserting θ and using $\widetilde{v}_k^{\rm A} \gg \widetilde{v}_k^S$. These results, which quantify the phenomenological *Ansatz* (5.75), show that the kinetic- and magnetic-energy spectra relax to equipartition in a time of the order of the Alfvén time.

In the presence of small-scale helicities H_k^K and H_k^M the most important nonlocal effect is the destabilization of large-scale magnetic modes, an effect corresponding to the inverse cascade of magnetic helicity, which is described by the terms

$$\partial_t E_k^M \Big|_{\rm nl} = \alpha_k^R k^2 H_k^M, \quad \partial_t H_k^M \Big|_{\rm nl} = \alpha_k^R E_k^M, \tag{6.26}$$

with

$$\alpha_k^R = -\tfrac{4}{3} \int_{k/a}^{\infty} \theta_{kqq} (H_q^K - q^2 H_q^M) \, dq. \tag{6.27}$$

Hence for $\alpha_k^R > 0$ the small-scale residual helicity destabilizes the large-scale magnetic fields. (There is no corresponding destabilizing nonlocal effect on the kinetic quantities E_k^K and H_k^K.) The first term in α_k^R corresponds to the α_t-effect in mean-field electrodynamics (4.29), i.e., kinematic dynamo theory, while the second term results from the small-scale Lorentz force on the velocity. This term can be derived in a similar way to the α_t-effect in Section 4.1.3. Assume that, at some time $t = 0$, the velocity field vanishes but there are small-scale magnetic fluctuations $\widetilde{\boldsymbol{b}}$ about a large-scale field $\langle \boldsymbol{B} \rangle$. On integrating the equation of motion

$$\partial_t \widetilde{\boldsymbol{v}} \simeq \langle \boldsymbol{B} \rangle \cdot \nabla \widetilde{\boldsymbol{b}}, \tag{6.28}$$

we obtain

$$\widetilde{\boldsymbol{v}}(t) \simeq \langle \boldsymbol{B} \rangle \cdot \nabla \int_0^t \widetilde{\boldsymbol{b}}(t') \, dt'. \tag{6.29}$$

Substituting for $\widetilde{v}$ in the electromotive force $\boldsymbol{\epsilon} = \langle \widetilde{\boldsymbol{v}} \times \widetilde{\boldsymbol{b}} \rangle$, (4.5), gives the magnetic contribution to the latter:

$$\boldsymbol{\epsilon}^M = \tfrac{1}{3} \langle \boldsymbol{B} \rangle \int_0^t \langle \widetilde{\boldsymbol{b}} \cdot (\nabla \times \widetilde{\boldsymbol{b}}') \rangle \, dt' = \tfrac{1}{3} \tau^M \langle \widetilde{\boldsymbol{b}} \cdot \widetilde{\boldsymbol{j}} \rangle \langle \boldsymbol{B} \rangle, \tag{6.30}$$

where we again assumed that we have isotropy of the small-scale fluctuations and τ^M is the magnetic correlation time, which approximately equals

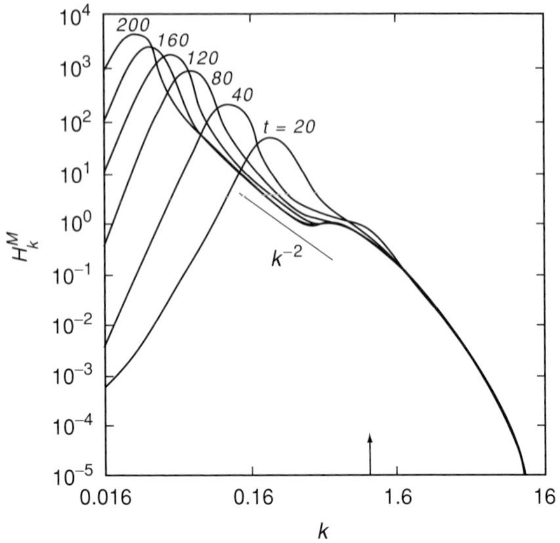

Figure 6.4: The inverse cascade of the magnetic helicity. Solutions of the EDQNM equations with injection both of H_k^K and of H_k^M in the range $k \sim 1$ are shown. (From Pouquet *et al.*, 1976.)

the velocity-correlation time in (4.29), $\tau^M \simeq \tau$, because of the Alfvén effect. We thus find the total electromotive force $\boldsymbol{\epsilon} = \alpha^R \langle \boldsymbol{B} \rangle$ with

$$\alpha^R = \alpha^V + \alpha^M \propto \langle (\widetilde{\boldsymbol{v}} \cdot \widetilde{\boldsymbol{\omega}} - \widetilde{\boldsymbol{b}} \cdot \widetilde{\boldsymbol{j}}) \rangle = \int dq \, (H_q^K - q^2 H_q^M). \qquad (6.31)$$

Numerical solutions of the EDQNM equations confirm the development of the inverse cascade of the magnetic helicity, an example being given in Fig. 6.4 for a case with injection both of magnetic and of kinetic helicity in the range $k \sim 1$. The spectral slope is consistent with the prediction of dimensional analysis, $H_k^M \sim k^{-2}$, (5.44).

Actually, the inverse-cascade process results from the competition of the dynamo mechanism (6.26), driven by the residual helicity α^R, and the Alfvén effect (6.24). As the residual helicity at wavenumber k leads to an increase of magnetic field at a larger scale, say $\frac{1}{2}k$, the latter exerts an increasingly strong effect on the fields at the primary wavenumbers k, transforming them into Alfvén waves, which tends to switch off the driving at those wavenumbers, $\alpha_k^R \to 0$. This can be seen most directly from the basic form of the electromotive force $\boldsymbol{\epsilon} = \langle \widetilde{\boldsymbol{v}} \times \widetilde{\boldsymbol{b}} \rangle$, (4.5), which vanishes for Alfvén waves $\widetilde{\boldsymbol{v}} = \pm \widetilde{\boldsymbol{b}}$. However, only the magnetic helicity is amplified by the dynamo process, such that the residual helicity is still finite at $\frac{1}{2}k$, driving the inverse cascade to even larger scales, say $\frac{1}{4}k$, and so forth. Hence the inverse cascade, the generation of

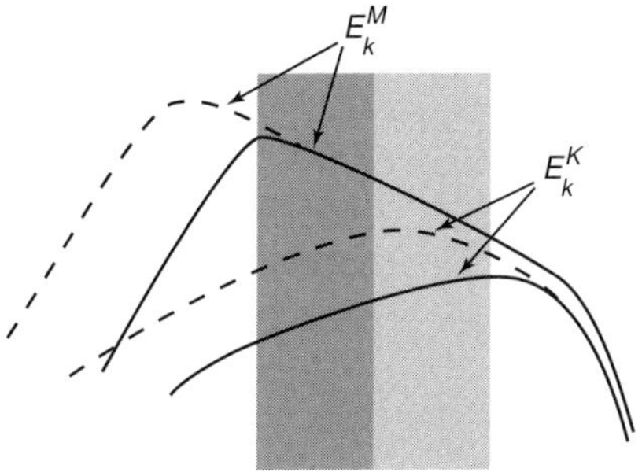

Figure 6.5: A schematic drawing illustrating the mechanism of the inverse magnetic cascade. The continuous lines are the energy spectra at time t_0; the dashed lines are those at a later time t_1. The spectral region of most influence in driving the cascade at t_0 is lightly shaded; that at t_1 is more darkly shaded.

large-scale magnetic fields, is mainly due to helical fields at only somewhat smaller scales, while the dynamo action at much smaller scales is quenched by velocity–magnetic-field equipartition caused by the Alfvén effect. The mechanism, which is illustrated schematically in Fig. 6.5, has essentially been confirmed by direct numerical simulation (Brandenburg, 2001).

6.2.2 Correlated turbulence

Observations in the solar wind reveal a turbulent behavior, whereby velocity and magnetic-field fluctuations are often strongly correlated, $\widetilde{\boldsymbol{v}} \simeq \pm\widetilde{\boldsymbol{b}}$. While we shall consider solar-wind turbulence in more detail in Chapter 10, at this point we want to discuss the properties of homogeneous correlated turbulence in the framework of closure theory, following mainly the work by Grappin *et al.* (1982, 1983). Since here we are not interested in the inverse magnetic cascade, we assume that we have vanishing magnetic and kinetic helicities, such that the system is now described by the kinetic and magnetic energies E_k^K and E_k^M, and the cross-helicity H_k^C. As discussed in Section 5.3.5, for correlated turbulence a treatment in terms of the Elsässer fields z^+ and z^- is more appropriate; hence we are dealing with the spectral intensities E_k^+, E_k^-, $E_k^+ + E_k^- = E_k$, $E_k^+ - E_k^- = H_k^C$, and the residual energy spectrum $E_k^R = E_k^K - E_k^M$, which gives the correlation of the Elsässer fields, (5.66). Grappin *et al.* (1982) used the EDQNM approximation to derive a set of equations for these spectral functions, which have the same structure as that given in (6.17); for the detailed form, see the original paper. The model satisfies the ideal conservation laws of the primitive MHD equations, and the same expression (6.16) for the eddy damping is used.

We first study the development of a correlated turbulent system and show that, as suggested by the phenomenological argument in Section 4.2.2, the correlation indeed increases with time. To simplify matters we assume the presence of a sufficiently large field $\boldsymbol{B}_0$, whose direction is random and isotropically distributed, which simplifies the EDQNM equations to the following equations for E_k^+ and E_k^-:

$$\partial_t E_k^\pm + (\nu + \eta)k^2 E_k^\pm = \int_{\Delta_k} dp\, dq\, \theta_{kpq} M_{kpq} \left(k^2 E_p^\pm E_q^\mp - p^2 E_k^\pm E_q^\mp\right), \tag{6.32}$$

while the strong Alfvén effect enforces equipartition,

$$E_k^R = E_k^K - E_k^M = 0, \tag{6.33}$$

and reduces the eddy-decorrelation time to the Alfvén time,

$$\theta_{kpq} = \theta_{kpq}^{\mathrm{A}} = \sqrt{3}/\big[(k+p+q)v_{\mathrm{A}}\big].$$

The coupling coefficient is $M_{kpq} = [k/(pq)](1-y^2)(1+z^2)$, where, as before, y and z are the cosines of the angles facing the triangle sides p and q. Note that there is only cross-coupling of E_k^+ and E_k^-, reflecting the corresponding property of the primitive equations. The system (6.32) is symmetric in E_k^+ and E_k^-, such that a difference in the evolution of the two fields results only from the initial conditions (or the forcing).

For the correlation to grow, the transfer to small scales must be slower for H_k^C than for E_k. We know from the large-k behavior of the ideal spectra (5.17) and (5.19) that both E_k and H_k^C exhibit a direct cascade, i.e., nonlinear interactions generate spectral fluxes to small scales and hence efficient dissipation. However, these processes are not identical for these two quantities, as has already been indicated in the forms of the dissipation terms (4.32) and (4.33), which are negative definite only for the energy. The corresponding property in the nonlinear interactions becomes apparent in the closure equations (6.32) on rewriting these in terms of the kinetic energy $E_k^K (=E_k^M)$ and the cross-helicity H_k^C, using the relations in Table 6.1:[2]

$$\begin{aligned}\partial_t E_k^K + (\nu+\eta)k^2 E_k^K = \int_{\Delta_k} dp\, dq\, \theta_{kpq}^{\mathrm{A}} M_{kpq}\Big[&k^2(E_p^K E_q^K - \tfrac{1}{4}H_p^C H_q^C) \\ &- p^2(E_k^K E_q^K - \tfrac{1}{4}H_k^C H_q^C)\Big],\end{aligned} \tag{6.34}$$

$$\begin{aligned}\partial_t H_k^C + (\nu+\eta)k^2 H_k^C = \int_{\Delta_k} dp\, dq\, \theta_{kpq}^{\mathrm{A}} M_{kpq}\Big[&k^2(H_p^C E_q^K - E_p^K H_q^C) \\ &- p^2(H_k^C E_q^K - E_k^K H_q^C)\Big],\end{aligned} \tag{6.35}$$

[2] The definitions in Table 6.1 differ slightly from those given by Grappin *et al.* (1982).

Table 6.1: *Definitions of the moments and invariants in terms of* v, b *and* z^+, z^- *and their interrelationship; the corresponding* k*-spectra satisfy the same relations as the moments*

Fields	$\mathbf{z}^{\pm} = \mathbf{v} \pm \mathbf{b}$	$\mathbf{v} = (\mathbf{z}^+ + \mathbf{z}^-)/2$, $\mathbf{b} = (\mathbf{z}^+ - \mathbf{z}^-)/2$
Second-order moments	$E^{\pm} = \frac{1}{4}\langle(\mathbf{z}^{\pm})^2\rangle$ $E^R = \frac{1}{2}\langle \mathbf{z}^+ \cdot \mathbf{z}^- \rangle = E^K - E^M$ $E^{\pm} = \frac{1}{2}(E^K + E^M \pm H^C)$ $E = E^+ + E^-$	$E^K = \frac{1}{2}\langle \mathbf{v}^2 \rangle$, $E^M = \frac{1}{2}\langle \mathbf{b}^2 \rangle$ $H^C = \langle \mathbf{v} \cdot \mathbf{b} \rangle = E^+ - E^-$ $E^{K,M} = \frac{1}{2}(E^+ + E^- \pm E^R)$ $E = E^K + E^M$
Ideal invariants	E^+, E^-	E, H^C

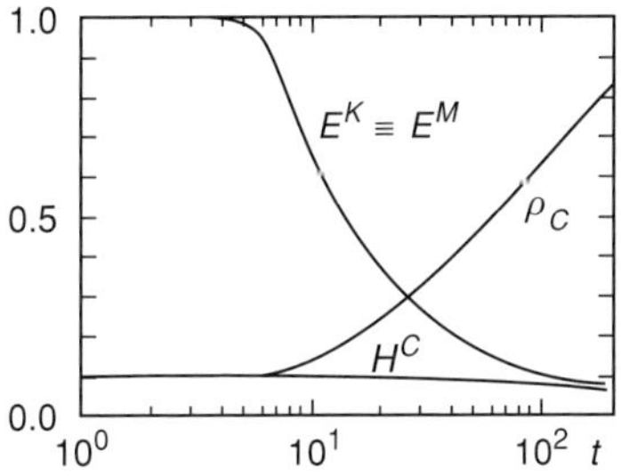

Figure 6.6: Fast decay of energy E and slow decay of cross-helicity H^C leading to growth of the correlation $\rho_C = H^C/E$, from a numerical solution of the closure equations (6.32) for Reynolds number Re $= 10^4$. (From Grappin *et al.*, 1982.)

The equation for H_k^C shows that the nonlinear term, and hence the spectral transfer, vanishes when $H_k^C \propto E_k$, whereas the nonlinear term in the equation for E_k^K does not, in general. One consequence of the structure of the nonlinear term in (6.35) is that the cross-helicity spectrum develops alternating ranges of positive and negative values, as found in the numerical solution of the closure equations, wherein the final stationary H_k^C-spectrum exhibits opposite signs in the dissipation range and the inertial range, resulting in a steeper spectrum than for the energy, typically $H_k^C \sim k^{-2}$ compared with $E_k \sim k^{-3/2}$. Consequently, the rate of dissipation of the cross-helicity is smaller than that of the energy, which leads to the increase of the correlation $\rho_C = H^C/E$, (4.61). As an example, Fig. 6.6 shows the decay of energy and the growth of the correlation from a numerical solution of (6.32).

In Section 5.3.5 we derived the energy spectra $E^{(\pm)} \sim k^{-m_\pm}$ for correlated turbulence using a phenomenological model of the spectral energy transfer, (5.68), where the main result was that, instead of the IK result for uncorrelated

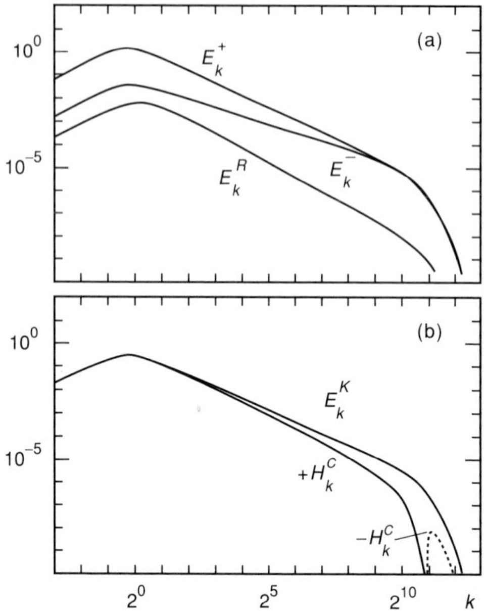

Figure 6.7: Energy spectra obtained from solution of the EDQNM closure equations, for correlation $\rho_C = 0.88$. The Reynolds number is $\mathrm{Re} = \mathrm{Rm} = 1.5 \times 10^6$. (a) $E_k^\pm$ and E_k^R, which exhibit almost perfect power laws $m_+ = 1.8$, $m_- = 1.2$, and $m_R = 2$. (b) E_k and H_k^C. Since $E_k^R \ll E_k$, one has $E_k^K \simeq E_k^M$, which is still nearly a power law with spectral index $\simeq 1.7$. The cross-helicity spectrum is steeper and does not follow a clear power law. (From Grappin *et al.*, 1983.)

turbulence, $m_+ = m_- = \frac{3}{2}$, the spectral indices are in general different, only their sum being constant, $m_+ + m_- = 3$, (5.70). Numerical solutions of the closure equations by Grappin *et al.* (1983) confirm the phenomenological results. Instead of the reduced model (6.32) discussed before, the authors solved the full EDQNM equations in order to compute also the residual spectrum E_k^R. Stationary spectra were obtained with constant dissipation rates ϵ^+ and ϵ^-, by applying random forces at large scales. Figure 6.7 gives the spectra $E_k^\pm$, E_k^R as well as E_k, H_k^C for a case of strong, but not extreme, correlation $\rho_C = 0.88$, for which one finds the spectral indices $m_+ = 1.8$, $m_- = 1.2$ and, for the residual spectrum, $m_R = 2$. Increasing the correlation makes the difference $m_+ - m_-$ larger; the results are in qualitative agreement with the phenomenological relation (5.73). The residual-spectral index $m_R = 2$ is practically independent of the correlation, confirming the basic mechanism of the competition between the Alfvén effect and local mode interaction, which does not involve the correlation.

6.3 Shortcomings of closure approximations

The two-point-closure approximation is a convenient approach by which to describe the temporal evolution of the various Fourier spectra, or the correlation functions in configuration space. In fact, it is the only statistical theory available to date which, starting from the primitive fluid equations, allows a quantitative description of turbulence. A self-consistent closure model such as Kraichnan's Lagrangian DIA provides numerical values of certain dimensionless constants, above all the Kolmogorov constant in the energy spectrum, which are not determined by dimensional analysis. Since, in practice, however, closure models, in particular EDQNM models, contain adjustable parameters, the Kolmogorov constant remains undetermined. The merits of closure studies consist of corroborating the validity of mechanisms and processes that have been suggested from general arguments, such as the tendency toward alignment and the inverse cascade of the magnetic helicity, as well as spectral scaling laws for various quantities, as we have shown in the previous section.

The major deficiency of two-point-closure theory is the inability to deal with spatial structures such as vortex filaments and current and vorticity sheets. It is now well acknowledged that real fluid turbulence consists of an intricate interplay of random and coherent effects, of which intermittency, the increasing sparseness of small-scale eddies, is a characteristic consequence. Closure theory assumes that the joint probability distributions are close to Gaussian, which is an expression of maximum randomness. As will be discussed in detail in Chapter 7, in real turbulence probability distributions of field increments become highly non-Gaussian at small distances, indicating the tendency toward structure formation.

This inability of closure theory to treat small scales correctly may even affect the behavior of the integral quantities. Closure models tend to overemphasize the strength of the nonlinear interactions, ignoring local rearrangments of the fields, which often give rise to a depletion of nonlinearity as discussed in Section 3.1. In this way the EDQNM equations for 2D MHD predict the presence of a finite-time singularity of the mean-square vorticity and current density (Pouquet, 1978), whereas direct numerical simulations and analytical considerations clearly indicate that these quantities grow only exponentially (Section 3.1.2).

A general problem of closure approximations is the probabilistic constraints, the most important one being the condition that the mean square of a random variable cannot have a negative value. This property is usually not preserved when one closes the hierarchy of moment equations by some *ad hoc* assumption, a well-known example being the quasi-normal approximation discussed in

Section 6.1.2, which gives rise to massively negative parts in the energy spectrum. In view of the complexity of the closure equations a direct proof of statistical consistency is usually not possible; one can, however, show in some cases that the equations are equivalent to the exact moment equations of some stochastic model, which guarantees realizability, for instance the EDQNM approximation, for which the equivalent model follows a Langevin equation (6.15). The relationship to such a basic stochastic model also indicates that closure approximations overemphasize the random character of the turbulence.

In the MHD case a specific difficulty with closure theory arises because of the inherent anisotropy of the turbulence. As we have discussed in Section 5.3.3, the small-scale dynamics in MHD turbulence is dominated by motions perpendicular to the large-scale magnetic field $k_\perp \gg k_\parallel$,

$$k_\parallel v_A \lesssim k_\perp \delta z_l,$$

which reduces the influence of the Alfvén effect. Though closure theory, in principle, allows one to deal with anisotropic, even inhomogeneous, systems, the equations become rather unwieldy, to such an extent that practical studies are limited to homogeneous isotropic systems. In MHD turbulence this implies, in particular, that the Alfvén frequency is kv_A, which is much higher than the (perpendicular) eddy-scrambling time and thus dominates the dynamics, for instance in the eddy damping (6.16) leading to the IK energy spectrum $k^{-3/2}$, whereas direct numerical simulations produce a Kolmogorov spectrum. Also the derivation of the residual spectrum $E_k^R \sim k^{-2}$, (5.74)–(5.77), which makes use of the Alfvén effect, appears to be doubtful. On the other hand, the arguments using the equipartition $E_k^V \sim E_k^M$ are still correct, which seems to be a general property of the MHD equations that is not tied to the dominance of rapidly propagating Alfvén waves.

7
Intermittency

Turbulence is usually associated with the idea of self-similarity, which means that the spatial distribution of the turbulent eddies looks the same on any scale level in the inertial range. This is a basic assumption in the Kolmogorov phenomenology K41 and, on the same lines, the IK phenomenology introduced in Section 5.3.2. It is, however, well known that this picture is not exactly true, since it ignores the existence of small-scale structures, which cannot be distributed in a uniform space-filling way. In fact, in a real turbulence field experiments as well as numerical simulations show that smaller eddies, or higher frequencies, become increasingly sparse, or intermittent, which apparently violates self-similarity. This chapter deals with the various aspects of intermittency.

Section 7.1 gives a brief introduction. We illustrate the concept of self-similarity by some simple examples and clarify the notion of intermittency, distinguishing between dissipation-range and inertial-range intermittency. Section 7.2 deals with structure functions, in particular the set of inertial-range scaling exponents, which are convenient parameters for a quantitative description of the statistical distribution of the turbulence scales. We discuss the important constraints on these exponents imposed by basic probabilistic requirements. Since experiments and, even more so, numerical simulations deal with turbulence of finite, often rather low, Reynolds number, the scaling range may be quite short, or even hardly discernable, especially for higher-order structure functions, which makes determination of the scaling exponents difficult. The scaling properties can, however, be substantially improved by making use of the extended self-similarity (ESS), which often provides surprisingly accurate values of the relative scaling exponents. A crucial assumption to connect dissipation-range and inertial-range quantities is Kolmogorov's refined similarity hypothesis. We also investigate the probability distribution function (p.d.f.) of the field increments. In Section 7.3 some exact relations for the third-order

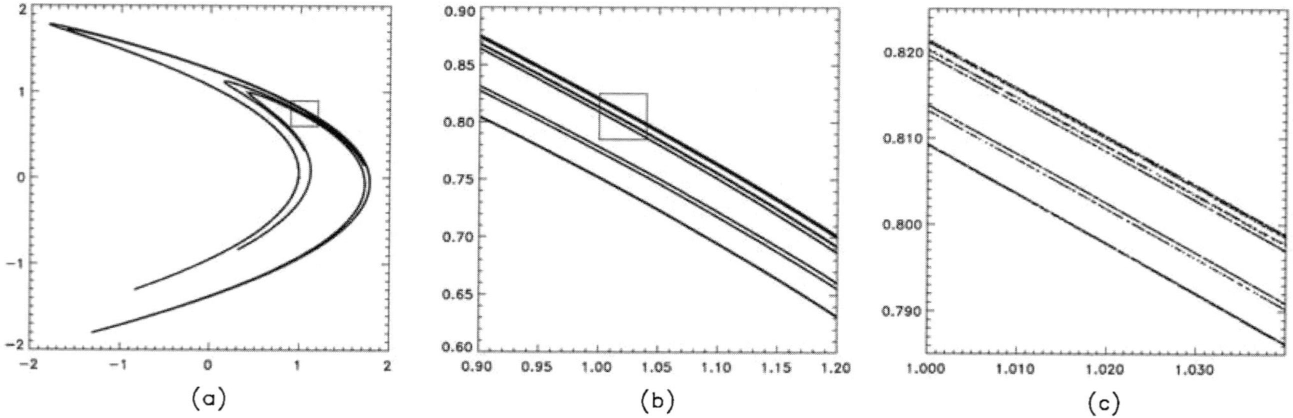

Figure 7.1: Self-similarity of the Hénon attractor (a); (b) is the amplification of the square in (a), (c) that of the square in (b). The figure contains 10^6 iterations of the Hénon map $x_{n+1} = \alpha - x_n^2 + \beta y_n, \; y_{n+1} = x_n$ (Hénon, 1976) with $\alpha = 1.4$ and $\beta = 0.3$.

structure functions in hydrodynamic and MHD turbulence are derived, which serve as a gauge of the relative scaling exponents. Section 7.4 deals with some intermittency models, which have been developed in order to interpret the observations. We describe in some detail the log-normal model and the log-Poisson model. Finally, in Section 7.5 we discuss more specifically the intermittency properties of MHD turbulence, which are obtained mainly from numerical simulations.

7.1 Self-similarity versus intermittency

A system is called self-similar or scale-invariant if it is reproduced by magnification of some part of it. Self-similarity is a widespread phenomenon encountered in physics and biology, but one should keep in mind that, while mathematical systems can be exactly self-similar, in nature this property exists only for a certain scale range. Well-known systems exhibiting discrete self-similarity are fractals, mathematical objects with a noninteger spatial dimension. Examples are fractal curves such as the boundary of Koch's snow flake, which is a special case of a fractal 'coastline', the Cantor-set-like structure of a strange attractor such as the attractor of Hénon's map, Fig. 7.1, and the fascinating Julia sets resulting from mappings $z_{n+1} = f(z_n)$ in the complex plane; for an introduction to chaos theory see Schuster (1988). In the case of a fractal curve self-similarity follows in a deterministic way from the local construction prescription, whereas in a dissipative map the position of the individual iteration points on the attractor is random.

A dynamic system more closely connected to turbulence exhibiting a continuous self-similarity is Brownian motion. A Brownian particle, for instance a colloidal particle immersed in a liquid or a smoke particle suspended in air, performs incessant irregular motions, which are caused by the continual bombardment by the molecules of the surrounding medium. Let $x(t)$ be the position of the particle. The displacement $\Delta x(s,t) = x(t) - x(s)$ during an interval $t-s$, which is assumed to be long compared with the time between molecular impacts, can be regarded as the sum of a large number of small displacements. Hence, by virtue of the central-limit theorem (see footnote on page 115), $\Delta x(s,t)$ exhibits a normal probability distribution,

$$p[\Delta x(s,t)] = \frac{1}{\sqrt{2\pi|t-s|}} \exp\left(-\frac{\Delta x(s,t)^2}{2|t-s|}\right).$$

Since this is valid for any interval $t - s$, it follows that the stochastic motions of the particle look similar on all time scales. This self-similarity persists as long as $t - s$ is large compared with the molecular collision time.

When the fluctuations of a quantity on a certain scale are not distributed in a (statistically) uniform way but become increasingly sparse in time or space with decreasing scale size, the behavior is called intermittent.[1] It is convenient to represent the fluctuations on scale l by the high-pass-filter $v_K^>$, (4.18). Though the high-pass filter contains all scales $l < K^{-1}$, for a spectrum steeper than k^{-1}, which is true in most cases of interest, the contribution of the high wavenumbers $k \gg K$ is negligible, such that the high-pass-filtered quantity represents the scale range $l \lesssim K^{-1}$. A practical measure of intermittency is the *flatness*[2] of the filtered quantity,

$$F_K = \frac{\langle (v_K^>)^4 \rangle}{\langle (v_K^>)^2 \rangle^2}. \tag{7.1}$$

A simple model of intermittent behavior is to assume that fluctuations are present only during a certain fraction δ_K of space or time, such that the averages in (7.1) are proportional to δ_K and thus

$$F_K \sim \delta_K^{-1}.$$

Hence the flatness increases with increasing sparseness of the fluctuations. Note that, for a Gaussian probability distribution, the flatness has the constant value of 3; see (6.5). Since the basic assumption in the closure approximations discussed in Chapter 6 is the proximity to a Gaussian behavior, these theories essentially neglect the effects of intermittency. Larger values of the flatness arise when the tails of the distribution are more strongly populated, emphasizing the importance of 'large events', which is characteristic of an intermittent behavior.

In fluid turbulence intermittency is easily demonstrated for the dissipative scales. Figure 7.2 shows (a) the distribution of the current density in a 2D MHD turbulent state, which is compared with (b) a purely random state with the same energy spectrum, $E_k \sim k^{-3/2}$, and the same dissipation length l_d, $j_k = -k^2\psi_k = k^{-3/4}\mathrm{e}^{-kl_d + i\alpha_k}$, where α_k are random phases. Whereas in the turbulent state the dissipative eddies form well-defined structures, *viz.*, current sheets, which are clearly separated from each other, the dominating 'dissipative' small-scale eddies in the random state are space filling. It is interesting to see the difference between the two cases in the cross-sectional

[1] This form of intermittency should be distinguished from the phenomenon of intermittency which occurs in certain low-order dynamic systems on the route to chaos (see Pomeau and Manneville, 1980).

[2] Occasionally also called *kurtosis*.

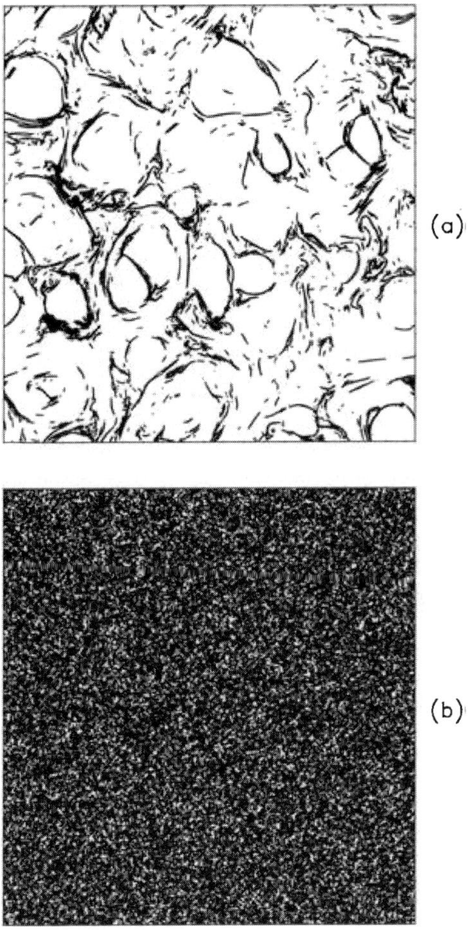

Figure 7.2: Contour plots of the current density. (a) A 2D MHD turbulence state. (b) The corresponding random state with the same energy spectrum and dissipation length.

profiles shown in Fig. 7.3, which were taken along the diagonals of both contour plots in Fig. 7.2. The striking feature in the trace across the turbulent state is the occurrence of large events mentioned above, whereas the behavior in the random state is statistically uniform. Also the p.d.f.s differ in a characteristic way, as can be seen in Fig. 7.4; that of the random state exhibits a Gaussian distribution as expected, whereas the p.d.f. of the turbulent state is far from Gaussian, with strongly flaring flanks corresponding to the occurrence of events of large amplitude. These features, the sparseness of the distribution of small scales, the occurrence of large events, and the strongly non-Gaussian

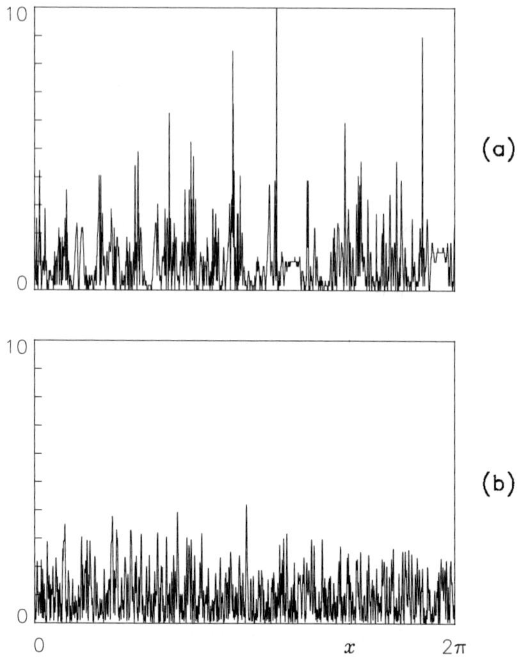

Figure 7.3: Cross-sectional profiles of $|j|$ taken along the diagonals of the corresponding contour plots in Fig. 7.2. The units on the vertical axis are the respective mean values.

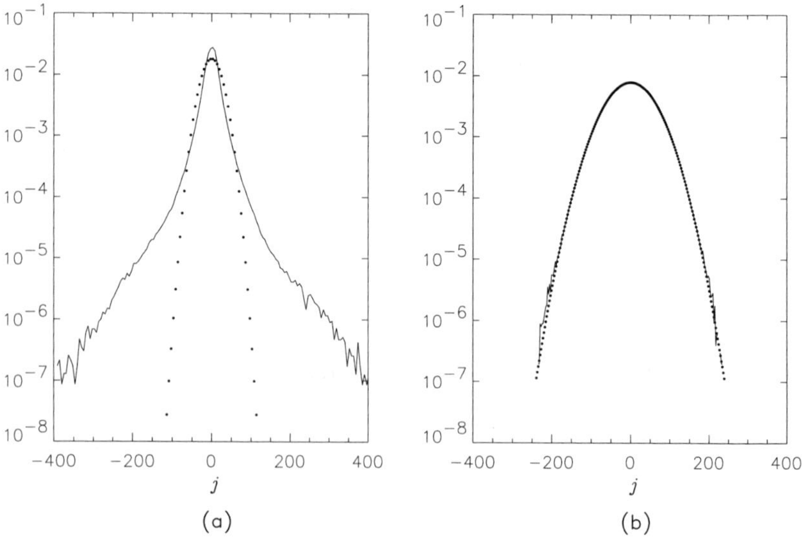

Figure 7.4: P.d.f.s of the current density, (a) of the turbulence state in Fig. 7.2(a) and (b) of the random state in Fig. 7.2(b). The dotted curves are the corresponding Gaussian distributions with the same variance.

shape of the p.d.f. of the velocity increments, are observed in fluid-turbulence experiments; see, for instance, Anselmet *et al.* (1984). Hence it is obvious that real turbulence, simulated or observed, is intermittent on the dissipative scales.

A more important, but also more intricate, question concerns intermittency on the inertial-range scales. Is self-similarity lost only on reaching the dissipation range, or does this loss occur gradually throughout the inertial range, where a scaling behavior is expected? At this point it is useful to emphasize the difference between scaling and self-similarity. Since the inertial range is characterized by the absence of an intrinsic length scale and a local cascade dynamics – modes interact most efficiently when their wavenumbers are of the same order –, quantities such as the energy spectrum, or the structure functions in general, are expected to exhibit a power-law behavior. Self-similarity, or scale-invariance, requires, in addition, that the exponents of these power laws follow a linear relation. The scaling properties of turbulence will be the central issue in the subsequent sections.

7.2 Structure functions

In the discussion of turbulence phenomenology in Chapter 5 we talk of 'turbulent scales', meaning turbulent eddies of a certain size, an expression that relates to a pictorial description of turbulence. For the following quantitative treatment a more precise definition is needed. We consider the increment of the velocity, or some other field,

$$\delta \boldsymbol{v}(\boldsymbol{x}, \boldsymbol{l}) = \boldsymbol{v}(\boldsymbol{x} + \boldsymbol{l}) - \boldsymbol{v}(\boldsymbol{x}) \tag{7.2}$$

between two points separated by $\boldsymbol{l}$, which gives a natural coordinate frame for the vector $\delta \boldsymbol{v}$. We thus have the longitudinal component,

$$\delta v_{\parallel} = \delta \boldsymbol{v}(\boldsymbol{x}, \boldsymbol{l}) \cdot \boldsymbol{l}/l \tag{7.3}$$

and two transverse, or lateral, components $\delta \boldsymbol{v}_{\perp}$. The statistical distribution of velocity eddies of size l is described by the moments of the increments called *structure functions*, both longitudinal and lateral,

$$S^{(n)}(l) = \langle [\delta v_{\parallel}(l)]^n \rangle, \quad U^{(n)}(l) = \langle [\delta v_{\perp}(l)]^n \rangle, \tag{7.4}$$

where we assume that the turbulence is homogeneous and isotropic, such that the moments depend only on the distance l, and both tranverse moments are equal. We will mostly be concerned with the longitudinal structure functions. The set of moments $S^{(n)}(l)$ is equivalent to the p.d.f. of the velocity increments, enhanced tails corresponding to the occurrence of events of relatively large amplitude, see Fig. 7.4(a), which are necessarily sparsely distributed.

The structure functions are related to the correlation functions

$$\langle v_i(\boldsymbol{x}_1)v_j(\boldsymbol{x}_2)\cdots v_k(\boldsymbol{x}_n)\rangle,$$

in particular the longitudinal correlation functions

$$\langle v_\parallel(\boldsymbol{x})^{n-1}v_\parallel(\boldsymbol{x}+\boldsymbol{l})\rangle = C^{(n)}(l),$$

the simplest relations being

$$S^{(2)}(l) = 2[C^{(2)}(0) - C^{(2)}(l)], \quad S^{(3)}(l) = 6C^{(3)}(l), \tag{7.5}$$

which can easily be verified. On the same lines one finds relations between lateral structure functions and correlation functions of the lateral velocity components and, more generally, among lateral, longitudinal, and mixed functions, using the incompressiblity condition $\nabla\cdot\boldsymbol{v} = 0$; for details see Monin and Yaglom (1975).

The second-order structure functions and correlation functions are related to the (angle-integrated) energy spectrum $E(k)$, (5.30). One has, in particular,

$$C^{(2)}(l) = 2\int_0^\infty \left(-\frac{\cos(kl)}{(kl)^2} + \frac{\sin(kl)}{(kl)^3}\right)E(k)\,dk \tag{7.6}$$

and, using (7.5),

$$S^{(2)}(l) = 4\int_0^\infty \left(\frac{1}{3} + \frac{\cos(kl)}{(kl)^2} - \frac{\sin(kl)}{(kl)^3}\right)E(k)\,dk. \tag{7.7}$$

If the energy spectrum follows a power law $E(k) \sim k^{-\alpha}$, it is easy to see that $S^{(2)} \sim l^{\alpha-1}$, since (7.7) becomes

$$S^{(2)} \sim l^{\alpha-1}\int_0^\infty \left(\frac{1}{3} + \frac{\cos\kappa}{\kappa^2} - \frac{\sin\kappa}{\kappa^3}\right)\kappa^{-\alpha}\,d\kappa, \tag{7.8}$$

where the integral converges for $1 < \alpha < 3$.

7.2.1 Scaling exponents

While higher-order correlation functions have little practical importance, higher-order structure functions play a crucial role in turbulence theory. The special interest in the structure functions arises from their scaling properties, in contrast to the correlation functions, which do not in general exhibit a scaling behavior. Thus one finds in the inertial range

$$S^{(n)}(l) = a_n l^{\zeta_n}. \tag{7.9}$$

One should note the difference between $S^{(n)}(l)$ as a function of l for a particular n and as a function of n for a particular scale l. The latter determines the full statistical distribution of structures at one scale, whereas the former compares some particular feature at different scales. The scaling property (7.9) guarantees that the full information (within the inertial range) is contained in the set of scaling exponents ζ_n together with the values of $S^{(n)}$ at some $l = l_0$ to fix the coefficients a_n in (7.9). Self-similarity requires that the mean turbulent state at some scale can be mapped onto that at a different scale by a simple scale factor, which implies that the scaling exponents are linear in n,

$$\zeta_n = cn, \quad c > 0. \tag{7.10}$$

Note that the more general linear relation $\zeta_n = an + b$, $b \neq 0$, does *not* correspond to a self-similar behavior, but describes an intermittent system, for instance in the β-model (Frisch *et al.*, 1978). The Kolmogorov phenomenology K41, $\delta v(l) \sim l^{1/3}$, (5.35), implies the linear relation

$$\zeta_n = n/3, \tag{7.11}$$

whereas for the IK phenomenology, (5.40), one has

$$\zeta_n = n/4. \tag{7.12}$$

Even in the most general intermittent case the exponents are not completely free but satisfy certain relations, which follow from probabilistic constraints. These imply that ζ_n is a monotonically growing concave function of n. Let us briefly sketch the proof. For any three positive even integers n_1, n_2, n_3 with $n_1 \leq n_2 \leq n_3$, one has the following inequalities:

(a) the convexity (or better, concavity) constraint

$$(n_3 - n_1)\zeta_{n_2} \geq (n_3 - n_2)\zeta_{n_1} + (n_2 - n_1)\zeta_{n_3}; \tag{7.13}$$

(b) the monotonicity constraint

$$\zeta_{n_1} \leq \zeta_{n_2}. \tag{7.14}$$

Relation (a) follows from Hölder's inequality (see, e.g., Loève, 1963),

$$\langle(\delta v)^{n_2}\rangle^{n_3-n_1} \leq \langle(\delta v)^{n_1}\rangle^{n_3-n_2}\langle(\delta v)^{n_3}\rangle^{n_2-n_1}. \tag{7.15}$$

On inserting the scaling relation (7.9) and noting that, for even orders n_i, the proportionality factors a_n are positive, comparison of the exponents of l yields the inequality (7.13) (for sufficiently small l). The monotonicity constraint can be proved by showing that, in the opposite case, $\zeta_{n_1} > \zeta_{n_2}$, the amplitude of the velocity fluctuation would not be bounded in the limit $l \to 0$, which contradicts the requirement that, for $\mathrm{Re} \to \infty$, the turbulence energy remains finite.

These arguments do not apply to structure functions of odd order, which not only may have negative coefficients, but also could even change sign within, or at the boundaries of, the scaling range. The proof will, however, remain valid for odd orders when the structure functions are defined with the absolute value of the argument,

$$S'^{(n)}(l) = \langle |\delta v(l)|^n \rangle, \tag{7.16}$$

which in general exhibit a more distinct scaling behavior for finite Reynolds number and are therefore often considered instead of the original functions $S^{(n)}(l)$. In a system with regular scaling properties both S and S' are expected to exhibit the same exponents,

$$S'^{(n)}(l) = a'_n l^{\zeta_n}, \quad a'_n > 0. \tag{7.17}$$

7.2.2 Extended self-similarity (ESS)

Structure functions have been measured for hydrodynamic turbulence under various different conditions, ranging from small table-size experiments through large-wind-tunnel turbulence to atmospheric turbulence (e.g., Anselmet *et al.*, 1984), and similar observations of magnetic turbulence have been conducted by satellite in the solar wind, to which we shall come in Chapter 10. Structure functions are also investigated in numerical-simulation studies of relatively high resolution, e.g., Fig. 7.5. A common feature of the observations is that, even at what appear to be high Reynolds numbers, the scaling properties of the structure functions are far from perfect, in particular for higher orders, such that fitting a straight line to the log–log plots of the data becomes rather arbitrary. The inertial range is usually defined by the scaling range where Kolmogorov's four-fifths law, $S^{(3)}(l) \propto l$, which is discussed in more detail in Section 7.3.1, is approximately satisfied. This range is quite short, if it is visible at all, except for very high Reynolds numbers, say, $\mathrm{Re} > 10^5$.

Using this exact relation for $S^{(3)}$, one can consider $S^{(3)}(l)$ as a dynamic length and write the structure functions within the inertial range in terms of $S^{(3)}$, or $S'^{(3)}$,

$$S'^{(n)} = c_n \left| S^{(3)} \right|^{\zeta_n} = c'_n \left[S'^{(3)} \right]^{\zeta_n}, \tag{7.18}$$

with different coefficients but the same exponents as in (7.9). In the following we restrict our consideration to even orders n, for which $S^{(n)} = S'^{(n)}$. The unexpected feature of the representation (7.18), which was discovered by Benzi *et al.* (1993), is that $S^{(n)}(S'^{(3)})$ exhibits a substantially broader scaling range

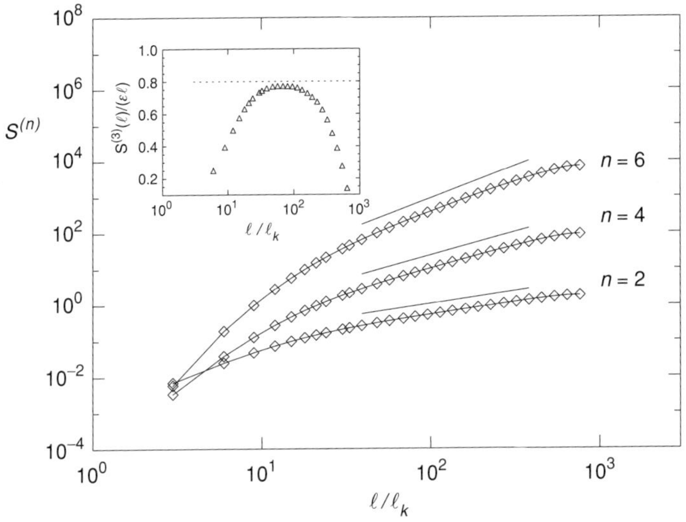

Figure 7.5: Structure functions $S^{(n)}$ of the longitudinal velocity increments in Navier–Stokes turbulence for $n = 2$, 4, and 6. Data are shifted for clarity (by factors of 10^2 for $n = 4$ and 10^4 for $n = 6$). The straight lines indicate the power laws $\zeta_2 = 0.69$, $\zeta_4 = 1.27$, and $\zeta_6 = 1.77$ obtained by the ESS technique. The insert shows the extent to which Kolmogorov's four-fifths law (7.29) is approached. (From Celani and Biskamp, 1999.)

than do the original functions $S^{(n)}(l)$, covering nearly the entire span of $S^{(3)}$-values and reaching into the dissipation range. This is the more surprising, since, by construction, (7.18) is valid only in the inertial range. It thus appears that the deficiencies in the scaling properties of $S^{(n)}(l)$ and $S^{(3)}(l)$ somehow compensate each other in the implicit form $S^{(n)}(S'^{(3)})$. The practical advantage of this phenomenon called *extended self-similarity* (ESS)[3] is dramatic, leading to a precision in determining the scaling exponents that had not even been dreamed of before. Some examples are given in Benzi's paper and many more can be found in the subsequent turbulence literature. ESS is also extensively used in numerical simulations of turbulence; see, for instance, the caption of Fig. 7.5.

The property of ESS is rather generally valid. The scaling exponents thus obtained practically do not depend on the Reynolds number and thus allow information on high-Reynolds-number turbulence to be gained from studies at much lower Reynolds numbers (However, since the ESS scaling properties degrade with increasing order n, only a finite number of exponents can be determined with sufficient accuracy at a given Reynolds number.) ESS can

[3] This term, which is now generally used, is somewhat misleading since, as we have seen, turbulence is not self-similar. A more appropriate name would be 'extended scaling property'.

also be made to work when one is considering $S^{(n)}(S'^{(m)})$ with $m \neq 3$ by substituting l from the inverse of $S'^{(m)}(l)$, which yields the relative scaling exponents $\xi_{m,n} = \zeta_n/\zeta_m$,

$$S^{(n)} \sim l^{\zeta_n} = (l^{\zeta_m})^{\xi_{m,n}} \sim (S'^{(m)})^{\xi_{m,n}}.$$

ESS has also been found in systems other than Navier–Stokes turbulence, for instance in thermal convection, (2.37) and (2.43), with the density fluctuation related to the temperature fluctuation by (2.41) or (2.42). In this system there is another exact relation, Yaglom's four-thirds law, $\langle \delta v_l\, \delta T_l^2 \rangle \propto l$ (Section 7.3.2), such that the structure functions $\langle (\delta v\, \delta T^2)^{n/3} \rangle \sim l^{z_n}$ can again be written in terms of the third-order function,

$$\langle (\delta v\, \delta T^2)^{n/3} \rangle \sim \left| \langle \delta v\, \delta T^2 \rangle \right|^{z_n} \sim \Big(\langle |\delta v\, \delta T^2| \rangle \Big)^{z_n}. \tag{7.19}$$

Here, too, the ESS property is realized with high quality, similarly to the behavior in Navier–Stokes turbulence. This is, however, *not* true for other structure functions such as those of the temperature, for instance $\langle \delta T^n \rangle$ written in terms of $\langle \delta v\, \delta T^2 \rangle$ does not exhibit an extended scaling range. A similar behavior is found in 2D MHD turbulence (Section 8.2.4).

7.2.3 The refined similarity hypothesis

Experiments and numerical simulations show that the dissipation of energy occurs in a highly intermittent manner, an example being given in Fig. 7.2(a), i.e., the local dissipation rate

$$\epsilon(\boldsymbol{x}) = \tfrac{1}{2}\nu \sum_{i,j} (\partial_i v_j + \partial_j v_i)^2$$

varies rapidly in space and time. This throws some doubt on the strict validity of the K41 phenomenology (5.35), which connects the local velocity fluctuation at scale l with the average energy-dissipation rate $\epsilon = \langle \epsilon(\boldsymbol{x}) \rangle$, an inconsistency already noted by Landau shortly after the publication of the K41 theory. On the basis of Landau's criticism, Obukhov (1962) proposed that $\langle \epsilon \rangle$ in the K41 scaling relation should be replaced by the spatial average over a volume of scale l,

$$\epsilon_l = \frac{1}{V_l} \int_{V_l} \epsilon(\boldsymbol{x})\, dV, \tag{7.20}$$

where the precise shape of the volume is not important. Hence, instead of (5.35), one should write

$$\delta v_l \sim \epsilon_l^{1/3} l^{1/3}. \tag{7.21}$$

The scaling law now has a clear mathematical meaning, that of a stochastic relation between two random variables, ϵ_l and δv_l. To make Obukhov's suggestion more quantitative, Kolmogorov (1962) introduced the *refined similarity hypothesis* relating the moments of $\delta v(l)$ to those of ϵ_l,

$$S_{\parallel}^{(n)}(l) = \langle [\delta v_{\parallel}(l)]^n \rangle = d_n \langle \epsilon_l^{n/3} \rangle l^{n/3}. \tag{7.22}$$

Since we are mostly interested in the scaling properties, we shall usually refer to (7.21) as the refined similarity hypothesis. Denoting the scaling exponents of ϵ_l by μ_n,

$$\langle \epsilon_l^n \rangle \sim l^{\mu_n}, \tag{7.23}$$

insertion of (7.9) into (7.22) gives the relation

$$\zeta_n = \tfrac{1}{3}n + \mu_{n/3}. \tag{7.24}$$

The refined similarity hypothesis connects dissipation-range and inertial-range physics and provides the basis for the verification of the intermittency models discussed in Section 7.4, which deal with the statistics of ϵ_l, while experiments measure only δv_l. Comprehension and validation of the hypothesis is therefore crucial for the theory. Its probabilistic essence is that ϵ_l has the same statistics as $\delta v_l^3/l$ or, in other words, the p.d.f. $p(V_l)$, $V_l = \delta v_l/(\epsilon_l l)^{1/3}$, is universal, independent both of l and of the Reynolds number. In particular, the coefficients d_n in (7.22) should be independent of Re. A direct check of (7.21) or (7.22) can, in principle, be obtained from numerical simulation by investigating the correlation between V_l and ϵ_l. Although in the first study, by Hosokawa and Yamamoto (1992), this correlation was found to be weak, later simulations at significantly higher Reynolds number showed that the hypothesis is reasonably well satisfied, at least for low-order moments of $p(V_l)$; see Borue and Orszag (1996). Wang *et al.* (1996), who conducted a detailed quantitative analysis, found that, in particular, $p(V_l)$ is indeed universal and close to Gaussian in the inertial range.

The refined similarity hypothesis has also been examined in a number of turbulence experiments by various authors (e.g., Praskovsky, 1992; Thoroddsen and Van Atta, 1992; Gagne *et al.*, 1994), who claimed to have verified the hypothesis. However, this claim only applies to low-order moments, since the experimentally measured quantity is not $\epsilon(\boldsymbol{x})$ itself, but rather the 1D surrogate

$\epsilon'(x) = 15\nu(\partial_x v_x)^2$ and the corresponding 1D average $\epsilon'_l = \int_l \epsilon'(x)\,dx/l$, instead of the true local dissipation rate. Even for isotropic turbulence, for which the mean values of the surrogate and the true dissipation rate are equal, $\langle\epsilon(x)\rangle = \langle\epsilon'(x)\rangle$, the probability distributions of ϵ_l and ϵ'_l are very different.

7.3 Exact turbulence relations

The phenomenological approaches such as K41 presented in Chapter 5, and the more general framework of structure-function scaling discussed in the preceding Section 7.2, are based solely on certain symmetry properties of the turbulence without direct reference to the dynamics such as the Navier–Stokes equations. On the other hand, closure theory, which was treated in Chapter 6, starts from the exact statistical turbulence equations but introduces severe approximations, which reduce the theory again to a more or less phenomenological model of the spectral quantities. There are, however, a few exact relations derived from the dynamic equations, which require only homogeneity and isotropy of the turbulence and are extremely helpful when we want to check the validity of assumptions and to gauge free parameters. In this section we shall consider three exact relations, namely Kolmogorov's four-fifths law for Navier–Stokes turbulence, Yaglom's four-thirds law for scalar convection, and a relation obtained recently for certain third-order structure functions in MHD turbulence.

7.3.1 *Kolmogorov's four-fifths law*

In the equation of refined similarity (7.22) a special behavior occurs for $n = 3$, where on the r.h.s. $\langle\epsilon_l\rangle = \epsilon$ is just the mean dissipation rate, which is independent of the statistics of ϵ_l, such that we have the simple scaling relation $S^{(3)}(l) = a\epsilon l$. This equation, which is indeed exact with $a = -\frac{4}{5}$, was first obtained by Kolmogorov (1941b), whose derivation is, however, too concise to be easily verifiable. Algebraically more detailed analyses can be found in Landau and Lifshitz (1959, p. 123) and in Monin and Yaglom (1975, pp. 121 and 396), and a generalization to anisotropic turbulence is included in the book by Frisch (1995, p. 76), who also gives a discussion of its physical meaning and implications. The central point of the proof in these references is the von Kármán–Howarth equation (von Kármán and Howarth, 1938), from which the four-fifths law follows immediately. Let us briefly sketch the derivation. Consider the Navier–Stokes equation (2.45) written in tensorial form,

$$\partial_t v_i + \partial_k(v_k v_i) = -\partial_i p + \nu\,\partial_{kk} v_i, \quad \partial_i v_i = 0,$$

where summation over repeated indices is implied. (An alternative proof using only elementary vector analysis has been given by Rasmussen, 1999.) On multiplying this equation by $v'_i = v_i(\boldsymbol{x}')$ and the equation for v'_i by v_i, summing and averaging, one obtains

$$\partial_t \langle v_i v'_i \rangle + \partial_k \langle v_k v_i v'_i \rangle + \partial'_k \langle v'_k v'_i v_i \rangle = -\langle v'_i \, \partial_i p \rangle - \langle v_i \, \partial'_i p' \rangle + \nu(\partial_{kk} + \partial'_{kk})\langle v_i v'_i \rangle. \quad (7.25)$$

In homogeneous turbulence the two-point moments depend only on the difference vector $\boldsymbol{l} = \boldsymbol{x}' - \boldsymbol{x}$, such that the derivatives ∂_i and ∂'_j can be replaced by $-\partial_{l_i}$ and ∂_{l_j}, respectively. Assuming that we have isotropy in addition, the velocity correlation functions in (7.25) become functions of $l = |\boldsymbol{l}|$ and can be expressed in terms of the parallel correlations $C^{(2)}(l)$ and $C^{(3)}(l)$, (7.5), which we do not want to write down here; we refer the interested reader to Chapter 7 of the classical treatise by Monin and Yaglom (1975). Using these relations in (7.25), some straightforward algebra gives the von Kármán–Howarth equation

$$\partial_t C^{(2)}(l,t) = \frac{1}{l^4} \, \partial_l l^4 \, [C^{(3)}(l,t) + 2\nu \, \partial_l C^{(2)}(l,t)]. \quad (7.26)$$

The pressure terms in (7.25) vanish because of incompressibility. For l in the inertial range, $l \ll L$, the correlation functions relax rapidly, such that they do not depend explicitly on t but only through the dissipation rate ϵ, $C(l,t) = C[l, \epsilon(t)]$. Hence the time derivative in (7.26) can be expressed in terms of the dissipation rate,

$$\partial_t C^{(2)}(l) \simeq \partial_t C^{(2)}(0) = \tfrac{1}{3} \, \partial_t \langle v_i v_i \rangle = -\tfrac{2}{3}\epsilon.$$

On substituting the correlation functions by the structure functions using the relations (7.5), the von Kármán–Howarth equation assumes the form

$$-\frac{2}{3}\epsilon = \frac{1}{l^4}\frac{d}{dl} l^4 \left(\frac{1}{6} S^{(3)}(l) - \nu \, \frac{dS^{(2)}(l)}{dl} \right), \quad (7.27)$$

or, after integration,

$$-\frac{4}{5}\epsilon l = S^{(3)}(l) - 6\nu \, \frac{dS^{(2)}(l)}{dl}. \quad (7.28)$$

In the inertial range $l_K \ll l \ll L$ the viscous term is negligible; hence

$$S^{(3)}(l) = \langle (\delta v_\parallel(l))^3 \rangle = -\tfrac{4}{5}\epsilon l, \quad (7.29)$$

which is Kolmogorov's four-fifths law. One consequence of this relation is that the *skewness* S is negative in the inertial range,

$$S = \frac{S^{(3)}(l)}{[S^{(2)}(l)]^{3/2}} < 0. \tag{7.30}$$

Since the mean velocity increment vanishes, $\langle \delta v_\| (l) \rangle = 0$, (7.29) implies that the p.d.f. $p(\delta v_\|)$ is skewed, with negative $\delta v_\|$ occurring less frequently than positive values but reaching larger amplitudes.

Since (7.29) is valid in the inertial range for homogeneous isotropic Navier–Stokes turbulence, it is often used to define the extent of the inertial range in experimental observations (e.g., Antonia *et al.*, 1997) and numerical simulations of turbulence, an example being given in Fig. 7.5. It is interesting to note that this range is typically shorter than the scaling range in the (angle-integrated or omnidirectional) energy spectrum. This difference arises partly because the structure functions have a similar character to the unidirectional spectrum, which is more strongly affected by dissipation and hence appears to fall off at smaller k, partly because of the bottleneck effect (Section 5.3.6) in the spectrum, which appears to stretch out the inertial range in the latter. The four-fifths law also forms the basis of the ESS *Ansatz* (7.18), which allows one to determine the scaling exponents ζ_n with comparatively high precision, as discussed in Section 7.2.2.

7.3.2 *Yaglom's four-thirds law*

A similar exact relation to the four-fifths law in Navier–Stokes turbulence was derived by Yaglom (1949) for thermal convection. Here the temperature follows a simple advection–diffusion equation, (2.44), which we give here in tensorial form,

$$\partial_t \theta + \partial_i (v_i \theta) = \kappa\, \partial_{ii} \theta, \quad \text{with} \quad \partial_i v_i = 0, \tag{7.31}$$

writing θ instead of T, since this may be any passive scalar such as the concentration of smoke particles in the atmosphere or of some pollutant in a river. Yaglom's relation can be derived directly without recourse to the equation for the correlation function $\langle \theta \theta' \rangle$, the analog of the von Kármán–Howarth equation (7.26). Consider (7.31) both for $\theta = \theta(\boldsymbol{x})$ and for $\theta' = \theta(\boldsymbol{x}')$, which we write in the forms

$$\partial_t \theta + \partial_i v_i (\theta - \theta') = \kappa\, \partial_{ii} (\theta - \theta'), \tag{7.32}$$

$$\partial_t \theta' + \partial_i' v_i' (\theta' - \theta) = \kappa\, \partial_{ii}' (\theta' - \theta), \tag{7.33}$$

using the independence of $\boldsymbol{x}$ and $\boldsymbol{x}'$. Subtraction of the first from the second, multiplication by $2(\theta' - \theta)$ and averaging gives

$$\partial_t \langle (\delta\theta)^2 \rangle + \partial_i \langle v_i (\delta\theta)^2 \rangle + \partial'_i \langle v'_i (\delta\theta)^2 \rangle = 2\kappa \langle \delta\theta \, (\partial_{ii} + \partial'_{ii}) \delta\theta \rangle,$$

where $\delta\theta = \theta' - \theta$. Using homogeneity, the average quantities depend only on $\boldsymbol{l} = \boldsymbol{x}' - \boldsymbol{x}$, such that the derivatives ∂_i and ∂'_i can be replaced by $-\partial_{l_i}$ and ∂_{l_i}, respectively. The equation now takes the simple form

$$\begin{aligned} \partial_t \langle (\delta\theta)^2 \rangle + \partial_{l_i} \langle \delta v_i \, (\delta\theta)^2 \rangle &= 2\kappa \langle \delta\theta \, (\partial_{ii} + \partial'_{ii}) \delta\theta \rangle \\ &= 2\kappa \, \partial_{l_i l_i} \langle \theta^2 \rangle - 4\kappa \langle \partial_{l_i} \theta \, \partial_{l_i} \theta \rangle. \end{aligned} \tag{7.34}$$

Assuming also that isotropy applies, one has $\langle \delta v_i \, (\delta\theta)^2 \rangle = \langle \delta v_\parallel \, (\delta\theta)^2 \rangle l_i / l$. At scales small compared with the integral scale, $l \ll L$, the first term on the l.h.s. of (7.34) is small, since

$$\langle (\delta\theta)^2 \rangle = 2\Big(\langle (\theta)^2 \rangle - \langle \theta\theta' \rangle \Big) \propto l^2,$$

while the second term on the r.h.s. is the scalar dissipation rate,

$$4\kappa \langle \partial_{l_i} \theta \, \partial_{l_i} \theta \rangle = 4\epsilon_\theta,$$

which remains finite. Hence (7.34) becomes

$$\frac{1}{l^{D-1}} \frac{d}{dl} l^{D-1} \left(\langle \delta v_\parallel \, (\delta\theta)^2 \rangle - 2\kappa \, \frac{d \langle (\delta\theta)^2 \rangle}{dl} \right) = -4\epsilon_\theta, \tag{7.35}$$

using the relation $\partial_{l_i} [l_i f(l)/l] = l^{-(D-1)} (d/dl)(l^{D-1} f)$ and D is the spatial dimension. Integration gives

$$\langle \delta v_\parallel \, (\delta\theta)^2 \rangle - 2\kappa \, \frac{d \langle (\delta\theta)^2 \rangle}{dl} = -\frac{4}{D} \epsilon_\theta l. \tag{7.36}$$

In the inertial range $l_\kappa \ll l \ll L$, where the diffusion term is negligible, we thus find

$$\langle \delta v_\parallel \, (\delta\theta)^2 \rangle = -\frac{4}{D} \epsilon_\theta l, \tag{7.37}$$

which is called Yaglom's four-thirds law for scalar turbulence (Yaglom, 1949), implying that $D = 3$. Here l_κ is the scalar dissipation scale, which can formally be connected to the Kolmogorov microscale by some function of the Prandtl number $\mathrm{Pr} = \nu/\kappa$, $l_\kappa = f(\mathrm{Pr}) l_K$. Both Kolmogorov's four-fifths law and Yaglom's four-thirds law have been verified in experiments (Fulachier and Dumas, 1976) and in numerical simulations (see Fig. 7.6).

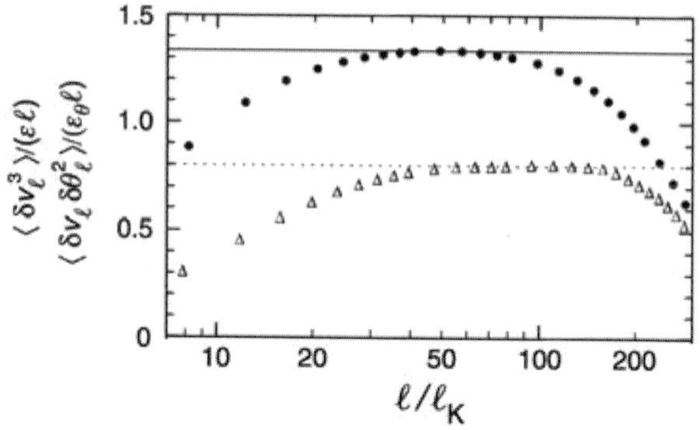

Figure 7.6: Verifications of Kolmogorov's four-fifths law and Yaglom's four-thirds law obtained from 3D simulation of passive scalar convection by hydrodynamic turbulence. Plotted are $-\langle[\delta v_\parallel(l)]^3\rangle/(\epsilon l)$ (triangles) and $-\langle\delta v_\parallel(l)\,[\delta\theta(l)]^2\rangle/(\epsilon_\theta l)$ (full circles). (From Chen and Cao, 1997.)

7.3.3 *Four-thirds law in MHD turbulence*

When they are written in terms of the Elsässer fields (2.106), the structure of the MHD equations is formally similar to that of a passive scalar (7.31). The $z^\pm$-fields are, of course not passive, but coupled to each other; however, as a matter of fact, we did not use the passivity of θ in the derivation of Yaglom's law. It is, instead, the vector character of the Elsässer fields which prevents a simple scaling relation of the structure functions of the parallel field increments analogous to Kolmogorov's four-fifths law, as will be discussed below. We first derive an equation similar to Yaglom's law for the scalar product of the increment vectors $\delta z_i^\pm\,\delta z_i^\pm$, where again summation over repeated indices is meant. Proceeding as before, we write the MHD equations for z_i^+ in a form analogous to (7.32) and (7.33), subtract one of the equations from the other and multiply by δz_i^+. Averaging, assuming that we have homogeneity, yields

$$\partial_t\langle\delta z_i^+\,\delta z_i^+\rangle + \partial_{l_j}\langle\delta z_j^-\,\delta z_i^+\,\delta z_i^+\rangle = 2\eta\,\partial_{l_j l_j}\langle\delta z_i^+\,\delta z_i^+\rangle - 4\eta\langle\partial_j z_i^+\,\partial_j z_i^+\rangle. \tag{7.38}$$

The pressure terms vanish because of incompressibility, $\partial_i z_i^\pm = 0$, and $\nu = \eta$ is assumed for simplicity. In the case of isotropic turbulence we have again $\langle\delta z_j^-\delta z_i^+\delta z_i^+\rangle = \langle\delta z_\parallel^-\,\delta z_i^+\delta z_i^+\rangle l_j/l$. For scales $l \ll L$ the time derivative on the l.h.s. is negligible, while the third term on the r.h.s. can be expressed in terms of the dissipation rate ϵ^+. Equation (7.38) now becomes

$$\frac{1}{l^{D-1}}\frac{d}{dl}l^{D-1}\left(\langle\delta z_\parallel^-\,\delta z_i^+\,\delta z_i^+\rangle - 2\eta\,\frac{d}{dl}\langle\delta z_i^+\,\delta z_i^+\rangle\right) = -4\epsilon^+. \tag{7.39}$$

In the inertial range, where the η-term on the l.h.s. is negligible, integration gives

$$\langle \delta z_\parallel^- \, \delta z_i^+ \, \delta z_i^+ \rangle = -\frac{4}{D}\epsilon^+ l, \tag{7.40}$$

and the analogous equation with + and − interchanged. Relation (7.40) was obtained by Politano and Pouquet (1998b) (the corresponding relation for hydrodynamic turbulence had previously been derived by Antonia *et al.*, 1997). It is formally very similar to Yaglom's law for scalar-field convection and also to Kolmogorov's law (7.29) for hydrodynamic turbulence. The main difference compared with the latter is that the MHD relation (7.40) contains both longitudinal and lateral structure functions,

$$\langle \delta z_\parallel^- \, \delta z_i^+ \, \delta z_i^+ \rangle = \langle \delta z_\parallel^- \, (\delta z_\parallel^+)^2 \rangle + 2\langle \delta z_\parallel^- \, (\delta z_\perp^+)^2 \rangle.$$

In the hydrodynamic limit $\mathbf{z}^\pm = \mathbf{v}$ the MHD relation reduces to the four-fifths law. On substituting for $\langle \delta v_\parallel \, (\delta v_\perp)^2 \rangle$ using the relation

$$\langle \delta v_\parallel \, (\delta v_\perp)^2 \rangle = \frac{1}{6l}\frac{d}{dl} l \langle (\delta v_\parallel)^3 \rangle,$$

one recovers (7.29).

One can also derive an equation for the third-order structure function involving only longitudinal field components, which, because of the form of the nonlinear term, will necessarily contain both fields, z^+ and z^-. However, the simple direct approach used to obtain Yaglom's law and the MHD relation (7.40) does not work in this case. Instead Politano and Pouquet (1998a) proceeded as in the proof of Kolmogorov's law by deriving first the corresponding MHD version of the von Kármán–Howarth equation, from which follows the relation

$$\langle \delta z_\parallel^-(\mathbf{l}) \, [\delta z_\parallel^+(\mathbf{l})]^2 \rangle - 2\langle [z_\parallel^+(\mathbf{x})]^2 z_\parallel^-(\mathbf{x}') \rangle = -\frac{8}{D(D+2)}\epsilon^+ l. \tag{7.41}$$

In contrast to the four-fifths law in hydrodynamics, this relation couples the structure function to the third-order correlation function and thus does not directly provide a scaling law for the former. Politano and Pouquet (1998b) showed that this relation is, of course, compatible with the scaling relation (7.40).

Relation (7.40) suggests that MHD turbulence is governed by similar dynamics to hydrodynamic turbulence, which is characterized by the approximate validity of the Kolmogorov phenomenology $\delta v_l \sim l^{1/3}$ with moderate corrections due to intermittency, which essentially rules out the IK phenomenology $\delta z_l \sim l^{1/4}$ based on the interaction of Alfvén waves along the local field. A

Kolmogorov-like behavior is, indeed, found in isotropic, or nearly isotropic, 3D MHD turbulence (Section 7.5). In 2D MHD turbulence, however, corresponding to the presence of a strong mean field, conditions seem to be more complicated. While spectra and structure functions of the Elsässer fields exhibit anomalous, strongly intermittent scaling properties, the scaling exponents of the mixed functions $\langle|\delta z_{\|}^{\mp}\, \delta z_i^{\pm}\, \delta z_i^{\pm}|^{n/3}\rangle$ are similar to those in hydrodynamic turbulence given by the She–Lévêque formula (7.64) (Section 8.2.4).

7.4 Phenomenological models of intermittency

Sobering experience from many years in turbulence theory does not foster much hope for a first-principles theory of intermittency, such as a derivation of the scaling exponents from the dynamic equations. Instead the only viable approach seems to be phenomenological modeling, for which the exact results outlined in the preceding section serve as guidelines and gauge. We discuss in some detail two different approaches, namely Kolmogorov's log-normal theory, the first intermittency model proposed, which seems to have a clear physical basis but is inconsistent with basic probabilistic requirements, and the more recent log-Poisson theory, whose physical meaning is less transparent but which describes the experimental results very well. Turbulence seems to lend itself naturally to fractal theory, which has been a fashionable subject in turbulence research. These studies led to several noted intermittency models, such as the β-model, which is the simplest fractal model (Frisch *et al.*, 1978), and its IK variant (Biskamp, 1993a), and more general multifractal models, such as the random-β model (Benzi *et al.*, 1984) and the p-model (Meneveau and Sreenivasan, 1987) or the corresponding IK variant (Carbone, 1993). These models contain a number of parameters, which must be adjusted to fit the experimental results. While the β-model with a linear law $\zeta_n = \frac{1}{3}n(1 - \delta) + \delta$ is ruled out, the p-model with the choice $p = 0.7$ gives very good agreement. However, though multifractal theory certainly conveys an impression of elegance and mathematical rigor, it permits little insight into the *physics* of turbulent dynamics. We will therefore not consider these models in any detail. An introduction to the fractal theory of turbulence can be found in Frisch's treatise on turbulence (Frisch, 1995).

7.4.1 The log-normal model

Most intermittency theories refer to the spatial distribution of the dissipation rate ϵ_l, (7.20), which is then connected with the corresponding field increments δv_l, or $\delta z_l^{\pm}$ in MHD, by the refined similarity hypothesis (7.21). The first intermittency theory was developed by Obukhov (1962) and Kolmogorov (1962), who claimed the p.d.f. $p(\epsilon_l)$ has a specific form, a log-normal distribution. To derive this result, one assumes that there is a simple cascade process of self-similar eddy fragmentation (Yaglom, 1966). Consider a cube of width l_0 and mean dissipation rate ϵ_0, where l_0 can be chosen as the integral scale L and $\epsilon_0 = \langle\epsilon\rangle$, which we divide into 2^D cubes of width $l_1 = l_0/2$. The dissipation rate ϵ_1 in these cubes is a random variable with the mean value $\langle\epsilon_1\rangle = \epsilon_0$. The l_1-cubes are then further subdivided, each into 2^D cubes of width $l_2 = l_1/2$, and this process is continued N times until the dissipation scale l_d is reached, $l_N = l_d$. Consider a scale l_j in the inertial range,

$$l_0 \gg l_j = l_0/2^j \gg l_d. \tag{7.42}$$

The random variable ϵ_j can be written as a product

$$\epsilon_j = \epsilon_0 \chi_1 \cdots \chi_j, \tag{7.43}$$

where $\chi_i = \epsilon_i/\epsilon_{i-1}$ are dimensionless random variables. Hence $\ln(\epsilon_j/\epsilon_0) = \chi_1 + \cdots + \chi_j$ is a sum of random variables. One now assumes that the χ_i are independent, which reflects the random character of the fragmentation process, and identically distributed, because of the self-similarity of this process, with mean value a and variance σ. It follows from the central-limit theorem that the p.d.f. of $\ln \epsilon_j$ approaches a Gaussian distribution for $j \gg 1$ with mean value $a_j = ja$ and variance $\sigma_j^2 = j\sigma^2$. Hence the p.d.f. of ϵ_j approaches the log-normal distribution

$$p(\epsilon_j) = \frac{1}{\epsilon_j} \frac{1}{\sqrt{2\pi\sigma_j^2}} \exp\left(-\frac{(\ln \epsilon_j - a_j)^2}{2\sigma_j^2}\right). \tag{7.44}$$

The requirement that the mean value at each level j is independent of the level of subdivision, $\langle\epsilon_j\rangle = \epsilon_0$, gives the relation $a_j = -\sigma_j^2/2$.

Since j is proportional to $\ln(L/l_j)$, one can write $\sigma_j^2 = \mu \ln(L/l_j)$, introducing the parameter μ. Hence the statistics of ϵ_j depends, if only logarithmically, on the global scale L. From (7.44) the moments can be obtained,

$$\begin{aligned} \langle\epsilon_l^n\rangle &= \epsilon_0^n \exp\left[\tfrac{1}{2}n(n-1)\sigma_l^2\right] \\ &= \epsilon_0^n (L/l)^{\mu_n}, \end{aligned} \tag{7.45}$$

with

$$\mu_n = \tfrac{1}{2}\mu n(n-1). \tag{7.46}$$

Here l_j has been replaced by the continuous variable l. Since in the nonintermittent K41 theory $\langle\epsilon_l^n\rangle = \epsilon_0^n$, i.e., $\mu = 0$, μ is called the intermittency parameter, the free parameter in the log-normal theory, which must be determined by comparison with experiments. Since experimentally only the statistics of the field increments δv_l is measured, one applies the refined similarity hypothesis with (7.24) connecting the corresponding scaling exponents. Inserting (7.45) gives

$$\zeta_n = \frac{n}{3} - \mu \frac{n(n-3)}{18}. \tag{7.47}$$

For $n = 2$ we have, in particular, $S^{(2)}(l) \sim \epsilon^{2/3} l^{2/3} (L/l)^{-\mu/9}$; hence

$$E_k \sim \epsilon^{2/3} k^{-5/3} (kL)^{-\mu/9}. \tag{7.48}$$

While the μ-correction to the Kolmogorov spectrum seems to be too small to be reliably measurable, the corrections to the higher-order moments become substantial for large n. The intermittency parameter was originally introduced as the scaling exponent of the correlation function

$$\langle \epsilon(\boldsymbol{x}) \epsilon(\boldsymbol{x} + \boldsymbol{l}) \rangle \sim l^{-\mu}. \tag{7.49}$$

Since $\langle \epsilon(\boldsymbol{x}) \epsilon(\boldsymbol{x} + \boldsymbol{l}) \rangle \simeq \langle \epsilon_l^2 \rangle \sim l^{\mu_n}$, we have $\mu_2 = -\mu$, which is connected with the scaling of the sixth-order structure function by the refined similarity hypothesis,

$$\mu = 2 - \zeta_6. \tag{7.50}$$

Both experiments (e.g., Anselmet *et al.*, 1984) and numerical simulations of Navier–Stokes turbulence (e.g., Vincent and Meneguzzi, 1991) give $\mu \simeq 0.2$. With this value the scaling exponents ζ_n, (7.47), agree well with the experimental results up to $n \simeq 10$. The subsequent flattening and falloff beyond the maximum at $n = 16$ is, however, not physical because it contradicts the basic probabilistic requirement of monotonicity (7.14). Hence the log-normal model does not provide a consistent statistical description of turbulence. On the other hand, it has been observed, both in experiments (e.g., Monin and Yaglom, 1975, Chapter 25) and in simulations (Biskamp, 1995), that, for l in the inertial range, $p(\epsilon_l)$ obeys a log-normal law reasonably well. The origin of this somewhat paradoxical behavior can be traced to the weak convergence of $p(\epsilon_l)$ to the log-normal distribution with increasing Reynolds number, such that the moments do not tend to those of the log-normal distribution. Mathematically this behavior is related to the property that the p.d.f. is not uniquely determined by the moments if the latter increase too rapidly (see, e.g., Feller, 1966, Chapter 7; also Paladin and Vulpiani, 1987, Appendix A). Hence it is more practical and reliable to construct a model for the moments rather than the probability distribution.

7.4.2 The log-Poisson model

A major success in describing intermittency in hydrodynamic turbulence was achieved by She and Lévêque (1994), who proposed a rather simple, though somewhat unconventional, model, which, at least formally, contains no adjustable parameters. The excellent agreement with experimental results is, indeed, surprising, especially since the statistical errors in the experimental values are very small due to the use of ESS (Section 7.2.3). The theory deals with the moments of ϵ_l. She and Lévêque introduced the hierarchy of relative moments

$$\epsilon_l^{(n)} = \langle \epsilon_l^{n+1} \rangle / \langle \epsilon_l^n \rangle, \tag{7.51}$$

which are dominated by increasingly large values in the wings of the p.d.f. $p(\epsilon_l)$. The extremes are $n = 0$, at which $\epsilon_l^{(0)} = \epsilon$ is the mean dissipation rate, and $n \to \infty$, at which $\epsilon_l^{(\infty)}$ represents the effect of the strongest dissipative events in the far tail of the p.d.f.. The main assumption in the theory is a relation connecting consecutive orders of $\epsilon_l^{(n)}$,

$$\epsilon_l^{(n+1)} = A_n (\epsilon_l^{(n)})^\beta (\epsilon_l^{(\infty)})^{1-\beta}, \quad 0 < \beta < 1, \tag{7.52}$$

where the coefficients A_n are independent of l. The parameter β is characteristic of the intermittency of the turbulence; the limit $\beta \to 1$ corresponds to nonintermittent turbulence, whereas the opposite limit $\beta \to 0$ represents an extremely intermittent state, in which the dissipation is concentrated in one singular structure. Assumption (7.52) may appear to be rather *ad hoc*, but it has an interesting probabilistic interpretation, as was pointed out by Dubrulle (1994). Instead of the relative moments $\epsilon_l^{(n)}$ she considered the nondimensional dissipation rate

$$\pi_l = \epsilon_l / \epsilon_l^{(\infty)}, \tag{7.53}$$

the moments of which are seen from (7.52) to obey the recursion relation

$$\langle \pi_l^{n+2} \rangle = A_n \frac{\langle \pi_l^{n+1} \rangle^{\beta+1}}{\langle \pi_l^n \rangle^\beta}. \tag{7.54}$$

The solution can easily be found,

$$\langle \pi_l^n \rangle = B_n \left(\frac{\langle \pi_l \rangle}{B_1} \right)^{(1-\beta^n)/(1-\beta)}, \tag{7.55}$$

where the coefficients B_n are functions of the A_m, $m \leq n$. The p.d.f. $p(\pi_l)$ corresponding to the moments (7.55) is a generalized Poisson distribution for the variable $Y = \log \pi_l / \log \beta$,

$$p(\pi_l)\, d\pi_l = F_G(Y)\, dY, \quad F_G(Y) = \int e^{-\lambda} \frac{\lambda^Z}{Z!} G(Y - Z)\, dZ, \qquad (7.56)$$

i.e., $p(\pi_l)$ is a log-Poisson distribution. The parameter λ is the expectation value of the Poisson distribution and G is an arbitrary probability density function, $G > 0$, $\int G(Y)\, dY = 1$.[4] We can easily show that the p.d.f. (7.56) does indeed lead to the moments (7.55). On substituting for π_l, one has

$$\begin{aligned} \langle \pi_l^n \rangle &= \int dY e^{-\lambda + nY \ln \beta} \int dZ \frac{\lambda^Z}{Z!} G(Y - Z) \\ &= \int e^{-\lambda + nZ \ln \beta} \int dY' \, e^{nY' \ln \beta} G(Y') \\ &= B_n \int dZ \, e^{-\lambda} \frac{(\lambda \beta^n)^Z}{Z!} = B_n e^{-\lambda(1-\beta^n)}, \end{aligned} \qquad (7.57)$$

where

$$B_n = \int dY' \, \beta^{nY'} G(Y'). \qquad (7.58)$$

From (7.57) for $n = 1$ we obtain the value of λ,

$$\lambda = -\frac{1}{1 - \beta} \ln\left(\frac{\langle \pi_l \rangle}{B_1} \right). \qquad (7.59)$$

Insertion into (7.57) gives (7.55) for the moments. The function $G(Y)$, which determines the coefficients B_n, can be chosen to account for the geometry of the turbulence set-up; in particular one finds $B_n = 1$ for $G(Y) = \delta(Y)$, whereas the scaling exponents depend only on the parameter β. The actual values of β as well as the scaling exponent x of $\epsilon_l^{(\infty)}$,

$$\epsilon_l^{(\infty)} \sim l^{-x}, \qquad (7.60)$$

[4] The integral in (7.56) should, properly speaking, be written as a sum, since the variable Z in the Poisson distribution takes only integer values,

$$p_X(n) = \frac{X^n}{n!} e^{-X}, \quad \sum_{n=0}^{\infty} p_X(n) = 1, \quad \langle n \rangle = X.$$

However, in the most interesting regime λ is large, such that for the main part of the distribution the argument Z may be considered continuous.

are intrinsic properties of the turbulence and thus have to be obtained from physical arguments. Equations (7.53) and (7.55) give the scaling exponents μ_n, $\langle \epsilon_l^n \rangle \sim l^{\mu_n}$,

$$\mu_n = -nx + x\frac{1-\beta^n}{1-\beta}. \tag{7.61}$$

The scaling exponents can also be written in terms of a third parameter C_0, the co-dimension of the most intermittent turbulent structures, which is connected with the parameters x and β. From the definition (7.51) and the scaling of $\epsilon_l^{(\infty)}$, (7.60), it follows that

$$\lim_{n\to\infty} \mu_{n+1} - \mu_n = x; \quad \text{hence} \quad \lim_{n\to\infty} \mu_n = -nx + c.$$

Since the probability of finding an object of linear size l and dimension d in a D-dimensional unit box is $\sim l^{D-d}$, $\langle \epsilon_l^n \rangle \sim l^{-nx} l^{D-d}$ for $n \gg 1$, we find $c = D - d = C_0$, the co-dimension of the dissipative eddies. Comparison with (7.61), where $\lim_{n\to\infty} \mu_n = -nx + x/(1-\beta)$, yields the relation

$$C_0 = \frac{x}{1-\beta}, \tag{7.62}$$

which was derived by Politano and Pouquet (1995). Note that the results (7.61) and (7.62) are independent of the specific type of turbulence.

In hydrodynamic turbulence the dissipative eddies are vortex filaments, i.e., 1D structures; hence the co-dimension is $C_0 = 2$.[5] The choice of x is less obvious. She and Lévêque assumed that the most intermittent structures exhibit a regular Kolmogorov scaling, $\epsilon_l^{(\infty)} \sim E/\tau_l \sim l^{-2/3}$; hence from (7.60) $x = \frac{2}{3}$ and therefore $\beta = \frac{2}{3}$ using (7.62),

$$\mu_n = -\frac{2}{3}n + 2\left[1 - \left(\frac{2}{3}\right)^n\right]. \tag{7.63}$$

To connect μ_n with the scaling exponents of the structure functions of the velocity increments, one applies the refined similarity hypothesis, in particular (7.24). These assumptions yield the celebrated She–Lévêque formula

$$\begin{aligned}\zeta_n^{\mathrm{SL}} &= \frac{n}{3}(1-x) + C_0(1-\beta^{n/3}) \\ &= \frac{n}{9} + 2\left[1 - \left(\frac{2}{3}\right)^{n/3}\right],\end{aligned} \tag{7.64}$$

[5] It should be emphasized that C_0 need not be an integer, since the dissipative structures may exhibit a fractal dimension.

which, as mentioned before, fits the experimental and computational results exceptionally well.

The success of the model, however, tends to make us forget a certain arbitrariness of the assumptions in its derivation, for instance the choice of the scaling parameter x. Moreover, Novikov (1994) raised a mathematical argument against the validity of the She–Lévêque theory (7.64) as a fundamental statistical theory of turbulence. Novikov claimed that, if asymptotically $\lim_{n\to\infty}(-\mu_n/n) < 1$ or, equivalently, $\lim_{n\to\infty}(\zeta_n/n)$ is finite, i.e., ζ_n increases linearly for large n, the probability distribution $p(\epsilon_l)$ should exhibit a gap, which seems to be rather implausable and hence would contradict the validity of the log-Poisson distribution (7.56), a point that is still under discussion. It would, however, be difficult to determine the asymptotic behavior of the scaling exponents experimentally, as was shown by Nelkin (1995), so the consequences of Novikov's objection cannot easily be verified.

7.5 Intermittency in MHD turbulence

7.5.1 Log-Poisson models for MHD turbulence

We now consider the intermittency properties of MHD turbulence, applying the framework outlined in the preceding section. In the discussion of the energy spectrum in Chapter 5, two types of turbulence phenomenology have been introduced, Kolomogorov's theory K41 with $\delta v_l \sim l^{1/3}$ for a local spectral cascade process, and the IK scaling $\delta v_l \sim l^{1/4}$ based on the Alfvén effect, the coupling of small-scale velocity and magnetic-field fluctuations by the large-scale field. To obtain the intermittency corrections to these basic scalings in the log-Poisson framework, we have to know the structure of the dissipative eddies in order to determine their co-dimension C_0. As outlined in Section 3.1.2, ideal MHD tends to form current–vorticity sheets. With finite resistivity these become unstable, breaking up into small elements. Though there have been suggestions that these elements might have a filamentary character, being current–vorticity filaments similar to the vortex filaments in Navier–Stokes turbulence shown in Fig. 7.7, numerical simulations of fully developed MHD turbulence clearly demonstrate that the smallest eddies are micro-current sheets; see Fig. 7.8. (This is not true for kinetically driven turbulence in magnetoconvection, in which the magnetic field is primarily dragged around passively and stretched into thin flux tubes, Section 9.3.4.) Hence $d = 2$, i.e., $C_0 = 1$. Assuming that basic Kolmogorov scaling applies, as suggested by the energy spectrum Fig. 5.2 or 5.3, we have $x = \frac{2}{3}$ and $\beta = \frac{1}{3}$,

$$\mu_n^{\mathrm{MHD}} = -\frac{2}{3}n + 1 - \left(\frac{1}{3}\right)^n . \tag{7.65}$$

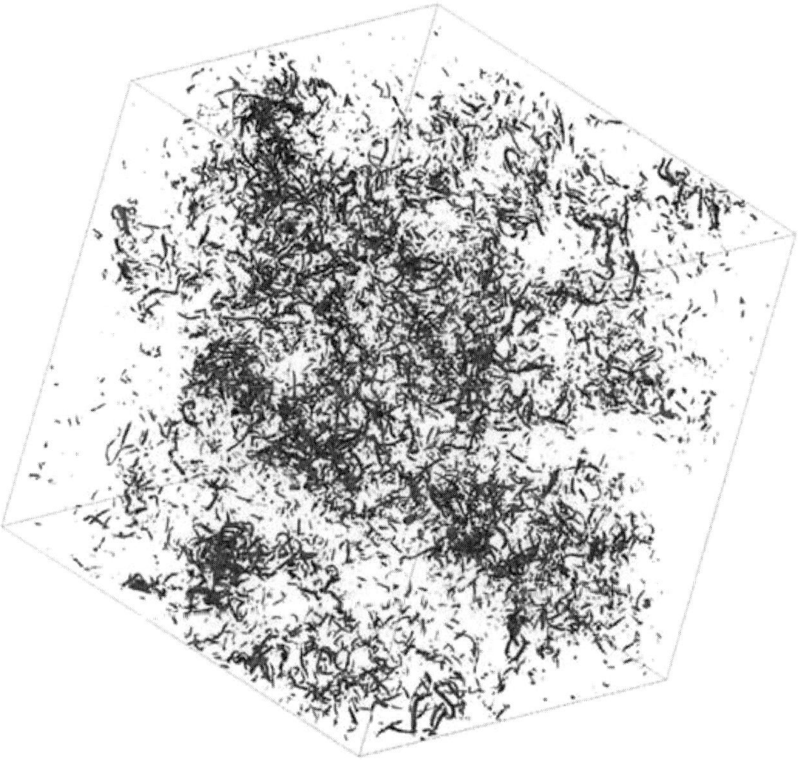

Figure 7.7: Isosurfaces of the vorticity in Navier–Stokes turbulence, demonstrating the filamentary character of the dissipative eddies. (From Biskamp and Müller, 2000b.)

Applying the refined similarity hypothesis (7.21) yields the structure-function scaling exponents (Horbury and Balogh, 1997; Müller and Biskamp, 2000)

$$\zeta_n^{\mathrm{MHD}} = \frac{n}{9} + 1 - \left(\frac{1}{3}\right)^{n/3}. \tag{7.66}$$

In particular we have $\zeta_3^{\mathrm{MHD}} = 1$, which is consistent with the exact relation (7.40).

If, on the other hand, the IK phenomenology were appropriate, the energy-transfer time would be given by (5.39), $T_l \sim l^{1/2}$; hence $x = \frac{1}{2}$ and, since we still have $C_0 = 1$ and $\beta = \frac{1}{2}$,

$$\mu_n^{\mathrm{IK}} = -\frac{n}{2} + 1 - \left(\frac{1}{2}\right)^n. \tag{7.67}$$

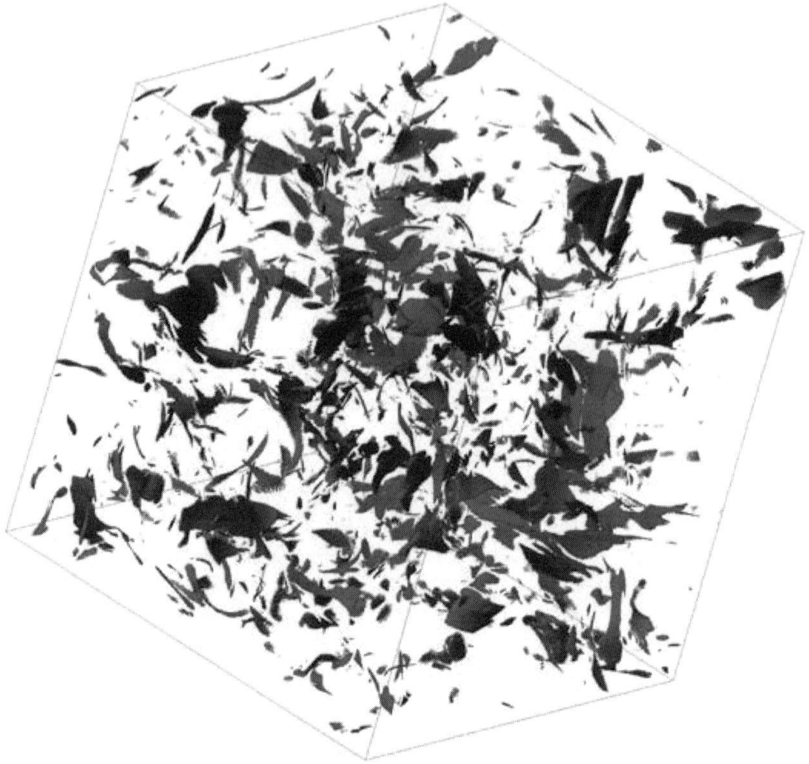

Figure 7.8: Isosurfaces of the current density in MHD turbulence. The smallest tubulence eddies are sheet-like, in contrast to the filamentary structure in Navier–Stokes turbulence shown in Fig. 7.7. (From Biskamp and Müller, 2000b.)

Instead of Kolmogorov's refined similarity hypothesis (7.21) we should now use an analogous generalization of the IK relation (5.40),

$$\delta z_l \sim (\epsilon_l v_{\mathrm{A}})^{1/4} l^{1/4}, \tag{7.68}$$

which gives the the scaling exponents of the structure functions of the z-fields,

$$\zeta_n^{\mathrm{IK}} = \frac{1}{4}n + \mu_{n/4}^{\mathrm{IK}} = \frac{n}{8} + 1 - \left(\frac{1}{2}\right)^{n/4}. \tag{7.69}$$

This is the IK version of the log-Poisson model, which was suggested by Grauer *et al.* (1994) and Politano and Pouquet (1995). The three different log-Poisson models ζ_n^{SL}, ζ_n^{MHD}, and ζ_n^{IK} are plotted in Fig. 7.9. While ζ_n^{MHD} corresponds to a more intermittent turbulent state than does ζ_n^{SL}, turbulence following $\zeta_4^{\mathrm{IK}} = 1$ would be significantly less strongly intermittent. Which of these models is more

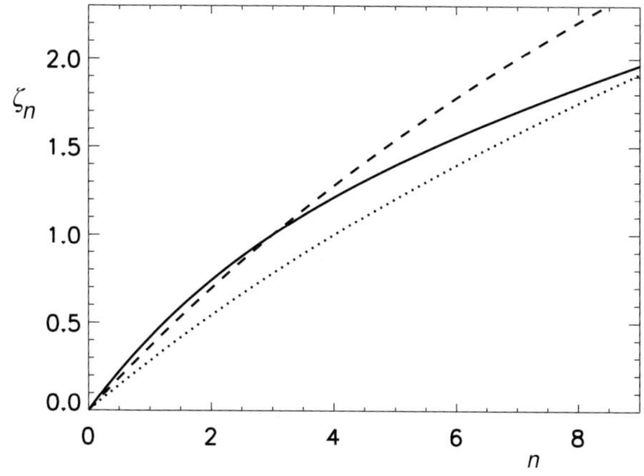

Figure 7.9: Comparison of different log-Poisson intermittency models, namely the MHD log-Poisson model ζ_n^{MHD}, (7.65) (continuous), the hydrodynamic log-Poisson model, the She–Lévêque model, ζ_n^{MHD}, (7.64) (dashed), and the IK model, (7.69) (dotted).

appropriate to describe intermittency in MHD turbulence should be decided by comparison with numerical simulations of sufficiently high Reynolds number. The discussion of the intrinsic local anisotropy of MHD turbulence outlined in Section 5.3.3 which favors a Kolmogorov scaling of the dominant inertial-range modes with $k_\perp \gg k_\parallel$, together with the exact relation (7.40) for the third-order structure functions, seems to rule out the IK model; no correponding exact relation for fourth-order structure functions appears to exist. The numerical results are presented in the subsequent section.

7.5.2 The effect of the mean magnetic field

In Section 5.3.3 we discussed the intrinsic spectral anisotropy of MHD turbulence, which exists also in a globally isotropic system, as small-scale modes develop mainly perpendicularly to the local magnetic field. We now investigate the spatial structure of MHD turbulence more generally by analyzing the scaling properties of the structure functions of the Elsässer fields $S_\parallel^{(n)\pm}(l)$ and $S_\perp^{(n)\pm}(l)$, where the subscripts indicate that the displacement $\boldsymbol{l}$ is either parallel or perpendicular to the magnetic field $\langle \boldsymbol{b} \rangle_{V_l}$ averaged over a local cubic box of volume $V_l = l^3$. In addition, we admit a constant mean magnetic field B_0, since magnetic fluctuations usually occur about a mean fields, which is often much higher than the r.m.s. fluctuating field amplitude $\overline{b} = \langle b^2 \rangle^{1/2}$. The scaling exponents obtained from several simulation runs with mean fields $B_0/\overline{b} = 0, 2.5,$ and 10 are summarized in Fig. 7.10. The scaling exponents of the perpendicular structure functions are always smaller than those of the parallel structure functions, with

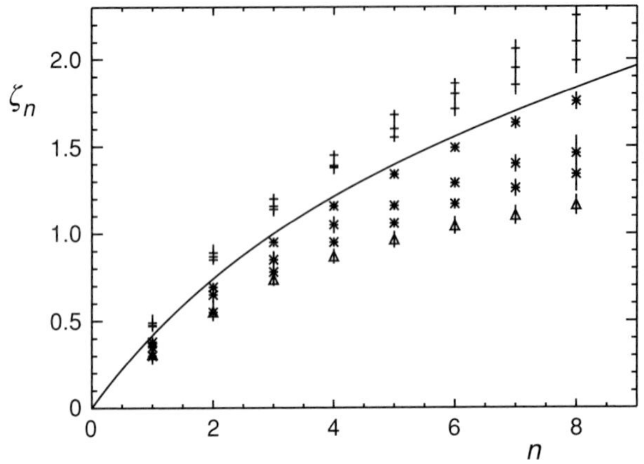

Figure 7.10: Scaling exponents ζ_n^+ and ζ_n^- of the perpendicular (x) and parallel (+) structure functions of the Elsässer fields for three strengths of the mean field, $B_0/\overline{b} = 0, 2.5,$ and 10. The line indicates the MHD log-Poisson model $\zeta_n^{\rm MHD}$, which reproduces the perpendicular exponents well in the globally isotropic case $B_0 = 0$. While the parallel exponents increase with increasing B_0, the perpendicular exponents decrease, approaching the values observed in 2D MHD turbulence (triangles). (From Müller *et al.*, 2002.)

the difference increasing with B_0. In the globally isotropic case $B_0 = 0$ the perpendicular exponents are well described by the MHD log-normal, whereas for large B_0 they approach the values observed in 2D MHD turbulence (see Section 8.2.4), indicated by the triangles in Fig. 7.10, which is consistent with the idea that 3D turbulence degenerates to 2D turbulence in the limit $B_0 \to \infty$. The parallel dynamics becomes weaker with increasing mean field, which is reflected in the increase of the scaling exponents.

8

Two-dimensional turbulence

In the previous chapters we assumed, explicitly or tacitly, that turbulence is isotropic, in particular that there is no mean magnetic field, so that spectra depend only on the modulus of the wave vector k and structure functions only on the distance l between two points. We also assumed that the kinetic and magnetic energies have similar magnitudes, which implies that the magnetic field is distributed in a space-filling way. However, even if the turbulence is globally isotropic, the local dynamics is not, differing strongly between the directions parallel and perpendicular to the *local* magnetic field. As discussed in Section 5.3.3, the small-scale fluctuations are dominated by perpendicular modes, and Alfvén waves propagating parallelly are only weakly excited, which gives rise to the observed Kolmogorov energy spectrum $k^{-5/3}$ instead of the IK spectrum $k^{-3/2}$.

In nature magnetic turbulence often occurs about a mean magnetic field, just as hydrodynamic turbulence occurs about a mean flow. However, whereas the latter can be eliminated by transforming to a moving coordinate system, the presence of a mean magnetic field has a strong effect on the turbulent dynamics. If this field is much larger than the fluctuation amplitude, turbulence becomes essentially 2D in the plane perpendicular to the field, since the stiffness of field lines suppresses magnetic fluctuations, Alfvén waves, with short wavelengths along the field. Hence turbulent motions tend to simply displace field lines without bending them. To deal with this situation quantitatively we derive a set of equations for a plasma embedded in a strong magnetic field $\boldsymbol{B}_0 = B_0 \boldsymbol{e}_z$. This is essentially an expansion of the MHD equations in the small parameter ϵ measuring the relative magnitude of the perpendicular, or poloidal, field. Gradient scales along the field are much longer than perpendicular scales, since the latter do not involve bending. To be definite we assume that

$$l_\perp / l_\parallel \sim B_\perp / B_0 \sim \epsilon. \tag{8.1}$$

The dynamics will lead to perpendicular equipartition (order-of-magnitude-wise),

$$v_\perp \sim B_\perp, \quad p \sim v_\perp^2,$$

therefore the variation of the axial field is small and pressure balance $\delta(B_z)^2 \simeq 2B_0\,\delta B_z \sim p$ yields

$$\delta B_z/B_0 \sim \epsilon^2.$$

It follows that the poloidal field is divergence-free to order ϵ^2,

$$\nabla_\perp \cdot \boldsymbol{B}_\perp = -\partial_z B_z \sim \epsilon^3,$$

and can be written in terms of the axial component of the vector potential A_z, or the flux function $\psi = -A_z$,

$$\boldsymbol{B}_\perp = \nabla \times A_z \boldsymbol{e}_z = \boldsymbol{e}_z \times \nabla\psi.$$

Since the axial forces are small, $\partial_z p \sim j_\perp B_\perp \sim \epsilon^3$, the axial velocity is small, $v_z/v_\perp \sim \epsilon$. Integration of Faraday's law (2.15) gives

$$\partial_t \boldsymbol{A} = \boldsymbol{v} \times \boldsymbol{B} - \nabla\chi + \eta\,\nabla^2 \boldsymbol{A} \tag{8.2}$$

with $\boldsymbol{B} = \nabla \times \boldsymbol{A}$ and the scalar potential χ. The perpendicular component of the vector potential generates the axial field, $B_z = (\nabla_\perp \times \boldsymbol{A}_\perp)\cdot\boldsymbol{e}_z$. Since the variation of B_z is small, so is $\partial_t \boldsymbol{A}_\perp$, such that the l.h.s. of (8.2) as well as the resistive term can be neglected, which gives

$$(\boldsymbol{v} \times \boldsymbol{B})_\perp = \boldsymbol{v} \times B_0 \boldsymbol{e}_z = \nabla_\perp \chi,$$

so that the poloidal velocity is

$$\boldsymbol{v}_\perp = \boldsymbol{e}_z \times \nabla\phi, \tag{8.3}$$

where we introduce the streamfunction $\phi = \chi/B_0$. Equation (8.3) implies that the poloidal flow is incompressible, $\nabla \cdot \boldsymbol{v}_\perp = 0$. The axial component of (8.2) now becomes

$$\partial_t \psi + \boldsymbol{v}_\perp \cdot \nabla\psi = B_0\,\partial_z\phi + \eta\,\nabla^2\psi. \tag{8.4}$$

The equation for the streamfunction is obtained by considering the axial component of the vorticity equation (2.28). Since $\boldsymbol{j}\cdot\nabla B_z \ll \boldsymbol{B}\cdot\nabla j_z$, one finds

$$\partial_t \omega + \boldsymbol{v} \cdot \nabla \omega - \boldsymbol{b} \cdot \nabla j = B_0 \, \partial_z j + \nu \, \nabla^2 \omega, \tag{8.5}$$

with

$$\boldsymbol{v} = \boldsymbol{e}_z \times \nabla \phi, \quad \boldsymbol{b} = \boldsymbol{B}_\perp = \boldsymbol{e}_z \times \nabla \psi,$$
$$\omega = \omega_z = \nabla^2 \phi, \quad j = j_z = \nabla^2 \psi.$$

Equations (8.4) and (8.5) are called the (lowest-order) reduced MHD equations (Kadomtsev and Pogutse, 1974; Strauss, 1976). The coupling of the poloidal dynamics to the axial direction occurs through the linear terms $\propto B_0 \, \partial_z$. Since gradient scales along the mean field are much longer than those across it, this coupling is effective only in elongated systems such as long flux tubes or a slender toroidal plasma column in a laboratory device, whereas for systems of comparable axial and poloidal dimensions, in particular in the cubic box usually chosen in numerical turbulence simulations, the dynamics is expected to be 2D.

In hydrodynamics 2D turbulence has received much attention in spite of the fact that, because of the absence of the vortex-stretching term – the vorticity is simply advected (apart from viscous damping) –, the 2D case differs fundamentally from its fully 3D counterpart. The motivation for this interest is twofold. On the one hand 2D turbulence models, within certain limits, the dynamics in the atmosphere and the oceans and therefore has important practical applications; on the other hand it represents the simplest system of chaotic incompressible fluid motions and thus serves as a convenient paradigm in turbulence theory. Hence it is useful to start this chapter with a brief introduction to 2D hydrodynamic turbulence (Section 8.1).

Two-dimensional MHD turbulence is considered in Section 8.2. Addition of the Lorentz force provides a source term in the vorticity equation, leading to amplification of vorticity even in 2D, which makes 2D and 3D turbulence much more alike in MHD than they are in hydrodynamics. We compare the main properties of the ideal systems, the absolute equilibrium distributions of the invariants and the resulting cascade directions, which indicate that also in 2D the energy performs a direct energy cascade. The decay of 2D MHD turbulence is dominated by the selective decay of the mean-square magnetic potential. The energy spectrum is found to follow an IK law scaling rather than the Kolmogorov law which prevails in 3D MHD turbulence. Numerical simulations provide rather accurate values of the structure-function scaling exponents, which show that MHD turbulence is more intermittent in 2D than it is in 3D.

8.1 Two-dimensional hydrodynamic turbulence

Restricting fluid motions to 2D, $\partial_z = 0$, the Navier–Stokes equation (2.46) reduces to an advection–diffusion equation for the vorticity component $\omega_z = \omega$,

$$\partial_t \omega + \boldsymbol{v} \cdot \nabla \omega = \nu \nabla^2 \omega, \tag{8.6}$$

where

$$\boldsymbol{v} = \boldsymbol{e}_z \times \nabla \phi, \quad \omega = \nabla^2 \phi. \tag{8.7}$$

This means that, apart from diffusive decay, the vorticity of a fluid element is constant; there is no source term corresponding to the vortex-stretching effect $\boldsymbol{\omega} \cdot \nabla \boldsymbol{v}$ in 3D turbulence. In fact 2D motions are much more constrained, since any functional $f(\omega)$ is ideally conserved. It therefore appears that this approximation cannot lead to genuine turbulence in the classical sense and should hence have little practical importance.

This is, however, not true since approximately 2D fluid motions are realized under various conditions both in nature and in the laboratory. The best known example is the large-scale dynamics in rotating planetary atmospheres. If rotation is sufficiently fast or, quantitatively speaking, if the Rossby number $\mathrm{Ro} = v/(2\Omega L)$ is sufficiently small, $\mathrm{Ro} \ll 1$, v being a typical horizontal velocity and L a typical horizontal scale size, inertia and viscous terms in the Navier–Stokes equation

$$\partial_t \boldsymbol{v} + \boldsymbol{v} \cdot \nabla \boldsymbol{v} = -\nabla p/\rho_0 + 2\boldsymbol{v} \times \boldsymbol{\Omega} + \nu \nabla^2 \boldsymbol{v} \tag{8.8}$$

are negligible, such that the pressure gradient and Coriolis force balance each other to lowest order,

$$\nabla p = 2\rho_0 \boldsymbol{v} \times \boldsymbol{\Omega}. \tag{8.9}$$

This is called the geostrophic approximation. It implies that the horizontal velocity $\boldsymbol{v}_h$ can be expressed in terms of a streamfunction as in (8.7) with $\phi = p/(2\Omega\rho_0)$ and is hence divergence-free, $\nabla \cdot \boldsymbol{v}_h = 0$. Taking the curl of (8.9),

$$\nabla \times (\boldsymbol{v} \times \boldsymbol{\Omega}) = \boldsymbol{\Omega} \cdot \nabla \boldsymbol{v} = 0,$$

we find that the velocity does not vary along $\boldsymbol{\Omega}$; the fluid motion consists of cylindrical columns parallel to $\boldsymbol{\Omega}$ (Taylor–Proudman columns, see Chandrasekhar, 1961). The temporal evolution of the horizontal flow is determined by the 2D Navier–Stokes equation (8.6). However, this is, in general, true only for scales larger than the vertical width d of the layer in the atmosphere, to which the dynamics is restricted, whereas at smaller scales, $l < d$, turbulence becomes 3D except in the case of stable stratification, for which vertical motions are damped. Thus 2D turbulence describes the main features of the development of atmospheric cyclones and anticyclones as well as oceanic circulation. In the laboratory, 2D flows may be realized at scales $l > d$ in a fluid confined between horizontal plates a distance d apart. To reduce the effect of the boundary layers at the plates, which tend to generate 3D turbulence and thus give rise to strong dissipation of energy, rapidly rotating systems are used (e.g., Hopfinger *et al.*, 1982). The same effect of two-dimensionalization is achieved by a vertical magnetic field in a conducting fluid (e.g., Sommeria, 1986). To suppress the vertical dynamics one also uses stably stratified systems of two fluid layers, with the lighter fluid on top of the heavier one (e.g., Paret and Tabeling, 1998). An interesting alternative approach allowing one to eliminate the friction at the confining boundaries is to generate flows in thin liquid soap films (Couder and Basdevant, 1986; Gharib and Derango, 1989; Rutgers, 1998).

While the experimental realization of quasi-2D flows is certainly interesting, the most convenient tools for studying 2D turbulence are direct numerical simulations, with which considerably higher Reynolds numbers can be reached than is possible in laboratory devices. In this section we present a brief overview of 2D turbulence, discussing separately the properties of the turbulence in the energy and the enstrophy cascade. For a more conceptual treatment the reader is referred to the review article by Kraichnan and Montgomery (1980).

8.1.1 Properties of the ideal system

We mentioned above that equation (8.6) has infinitely many constants of motion, namely the vorticity of each fluid element. In turbulence theory, however, as discussed in Section 5.2, the quadratic integral invariants are most important, which are the energy E and the mean-square vorticity, or enstrophy, usually denoted by Ω (not to be confused with the angular velocity, which we denote by the same symbol),

$$E = \tfrac{1}{2}\int v^2\, d^2x = \tfrac{1}{2}\sum_{\boldsymbol{k}} k^2 |\phi_{\boldsymbol{k}}|^2, \tag{8.10}$$

$$\Omega = \int \omega^2\, d^2x = \sum_{\boldsymbol{k}} k^4 |\phi_{\boldsymbol{k}}|^2. \tag{8.11}$$

Following the analysis in Section 5.2, we derive the energy spectrum $E_k = \frac{1}{2}|v_{\boldsymbol{k}}|^2$ corresponding to the absolute equilibrium distribution of a truncated ideal system with modes in the band $k_{\min} \leq k \leq k_{\max}$. From the Gibbs probability distribution

$$\rho_{\mathrm{G}} = Z^{-1} \exp(-\alpha E - \beta \Omega)$$

one obtains, as discussed in Section 5.2.1,

$$E_{\boldsymbol{k}} = \frac{1}{2(\alpha + \beta k^2)}, \tag{8.12}$$

or the angle-integrated spectrum

$$E_k = 2\pi k E_{\boldsymbol{k}} = \frac{\pi k}{\alpha + \beta k^2}. \tag{8.13}$$

The parameters α and β are determined by the values of the invariants E and Ω. E_k must be positive, i.e., $\alpha + \beta k^2 > 0$ in the truncated spectral range. Of the three possible cases, $\{\alpha, \beta > 0\}$, $\{\alpha > 0, \beta < 0\}$, and $\{\alpha < 0, \beta > 0\}$, the last, with negative α, is the most interesting one, $|\alpha| < \beta k_{\min}^2$. Here the absolute equilibrium energy spectrum is peaked at the smallest k, indicating that there is an inverse turbulence cascade in a dynamically evolving dissipative system. The enstrophy spectrum $\Omega_k = k^2 E_k$ is peaked at high k; hence it should exhibit a direct cascade, as would also be expected from the conservation of energy in a dissipative system, see below (8.14), since E_k and Ω_k are tightly connected. Because of the traditional meaning of the coefficient α in the Gibbs distribution as the inverse of the temperature, such states are also called negative-temperature states. Numerical simulations show that the inverse energy cascade is a characteristic property of 2D turbulence, dominating the dynamics, as will be discussed in more detail below.

Comparison with the equilibrium results in 3D hydrodynamic turbulence (5.22) (see, in particular, the discussion in Section 5.2.2) indicates that the cascade properties in 2D differ fundamentally from those in the 3D case. A summary of the cascade properties is given in Table 8.1.

8.1.2 The decay of enstrophy

Switching on viscosity, the ideal invariants E and Ω decay, but their decay rates are greatly different. From (8.6) we obtain the relations

$$\frac{dE}{dt} = -\nu\Omega, \quad \frac{d\Omega}{dt} = -2\nu \int (\nabla\omega)^2 \, d^2x. \tag{8.14}$$

While E decays very slowly, $dE/dt = O(\nu)$, since Ω can only decrease, the decay of Ω is much more rapid; $d\Omega/dt$ may remain finite for $\nu \to 0$, as large vorticity gradients are formed. Hence we are dealing with a case of

Table 8.1: *Ideal invariants and cascade directions in Navier–Stokes and MHD turbulence in 3D and 2D*

Theory	3D		2D	
Navier–Stokes	E_k^K	direct	E_k^K	inverse
	H_k^K	direct	Ω_k	direct
MHD	E_k	direct	E_k	direct
	H_k^C	direct	H_k^C	direct
	H_k^M	inverse	A_k	inverse

selective decay, which determines the enstrophy-decay law in a similar way to that discussed for the energy-decay laws in MHD turbulence in Section 4.2.3. We introduce the length scale L by taking the relation

$$L^2\Omega = E = \text{constant}. \tag{8.15}$$

Assuming that the macroscopic dynamics of the turbulence decay is independent of the value of the viscosity, L can depend only on E and the enstrophy-decay rate $\epsilon_\Omega = -d\Omega/dt$. From dimensional arguments one has

$$\epsilon_\Omega \sim E^3/L^3.$$

Hence, inserting L from (8.15),

$$\frac{d\Omega}{dt} \sim -\Omega^{3/2}, \tag{8.16}$$

which has the similarity solution $\Omega \sim t^{-2}$ and therefore $L \sim t$. This result was first obtained by Batchelor (1969), though by invoking a slightly different argument making use of the spectral properties in the enstrophy-cascade range. Numerical simulations of decaying turbulence confirm the decay scaling (8.16) only during the first phase of the decay process, while later on the decay is slowed down by formation of coherent structures, whose interaction dominates the turbulence dynamics (McWilliams, 1990; Carnevale *et al.*, 1992). Because of the nonlocality of the spectral transfer discussed below, finite-box-size effects become important even if the dominant wavenumber is still much larger than unity. Results of recent numerical studies (Chasnov, 1997; Das *et al.*, 2001) show that, at sufficiently large Reynolds number and box size the decay does indeed proceed in a self-similar way, $\Omega \sim t^{-1}$, $L \sim t^{1/2}$.

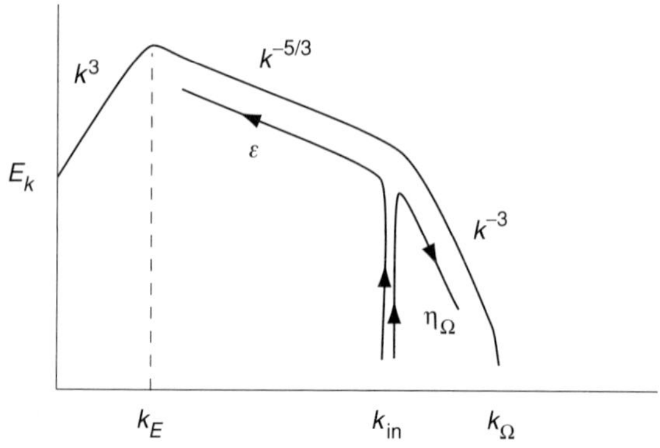

Figure 8.1: A schematic drawing of the energy spectrum in 2D hydrodynamic turbulence and the cascade directions of E_k and Ω_k when energy is injected at wavenumber $k_{\rm in}$.

8.1.3 The phenomenology of the dual cascade

The opposite cascade directions of energy and enstrophy lead to the following picture of the spectral dynamics of 2D hydrodynamic turbulence, which is essentially due to Kraichnan (1967) and Batchelor (1969) and is illustrated in Fig. 8.1. When energy – and enstrophy, since they are intrinsically connected – is injected at some intermediate wavenumber $k_{\rm in}$, enstrophy flows to large k, at which it is finally dissipated, while energy flows in the inverse direction, forming a front that propagates to ever smaller k, forming eddies of increasing size, since there is no natural dissipation process at small k until some, essentially geometrical, limit is reached. As a consequence of the different cascade dynamics one expects different power laws of the energy spectra in the spectral ranges $k < k_{\rm in}$ and $k > k_{\rm in}$. Let us first consider the latter range, which is dominated by the enstrophy cascade. Since in the inertial range the energy spectrum is not affected by viscosity, it depends only the wavenumber k and the enstrophy-transfer rate ϵ_Ω, which is constant across the inertial range and equal to the enstrophy-dissipation rate for stationary conditions. Making the *Ansatz* $E_k \sim \epsilon_\Omega^\alpha k^\beta$, the exponents are obtained by comparing dimensions using $[E_k] = L^3T^{-2}$ and $[\epsilon_\Omega] = T^{-3}$, which gives

$$E_k = C_\Omega \epsilon_\Omega^{2/3} k^{-3}, \quad k_{\rm in} < k < k_\Omega, \tag{8.17}$$

where C_Ω is a proportionality factor analogous to the Kolmogorov constant and k_Ω is the enstrophy-dissipation wavenumber. Strictly speaking, this result is not quite correct. As noted by Kraichnan (1971), for an enstrophy spectrum $\Omega_k = k^2 E_k \sim k^{-1}$ resulting from (8.17) the transfer process is not local, i.e., restricted to, say, one octave, $k_n < k < 2k_n$, as in 3D turbulence, but instead any octave interval in the inertial range gives the same contribution. Taking this effect into account leads to an additional logarithmic factor $[\ln(k/k_{\rm in})]^{-1/3}$ in the spectrum (8.17), which makes the spectrum slightly steeper than k^{-3}.

Inserting the scaling $\delta v_l \sim \epsilon_\Omega^{1/3} l$ obtained from (8.17) into the balance relation between tranfer and dissipation $\delta v_l / l \sim \nu / l^2$ gives the enstrophy-dissipation scale

$$l_\Omega = k_\Omega^{-1} = (\nu^3/\epsilon_\Omega)^{1/6}. \tag{8.18}$$

The energy spectrum in the inverse cascade should again follow a Kolmogorov law, since the direction of the spectral tranfer is not crucial in the argument given in Section 5.3.2 and, of course, does not enter the dimensional analysis (5.37). Hence we have

$$E_k = C'_{\rm K}\epsilon^{2/3}k^{-5/3}, \quad k_E < k < k_{\rm in}. \tag{8.19}$$

The spectrum reaches a maximum at $k \sim k_E$, while for smaller k the energy spectrum is $E_k \sim k^3$, rather independently of the details of the turbulent dynamics. In fact, the small-k spectrum is mainly determined geometrically, $E_k \sim k^{D+1}$, where D is the spatial dimension, see (4.47), and we shall find the same k^3 energy spectrum in the inverse magnetic cascade discussed in Section 8.2.3. Since there is no simple physical dissipation process at small k, the inverse energy cascade will continue to smaller and smaller scales. For constant energy-input rate ϵ the time dependence of k_E can easily be derived. Since, up to numerical factors,

$$E = \int_{k_E(t)}^{k_{\rm in}} \epsilon^{2/3}k^{-5/3}\,dk \simeq \epsilon^{2/3}k_E^{-2/3},$$

we have

$$\frac{dE}{dt} = \epsilon = \epsilon^{2/3}\frac{dk_E^{-2/3}}{dt},$$

which can readily be integrated,

$$k_E = (\epsilon t^3)^{-1/2}. \tag{8.20}$$

While the Kolmogorov spectrum in the energy cascade is fully confirmed by numerical simulations, the energy spectrum in the enstrophy-cascade range depends more strongly on the individual properties of the turbulent state and is often considerably steeper than k^{-3}, or $k^{-3}(\ln k)^{-1/3}$ (e.g., Das *et al.*, 2001). Let us now consider the properties of 2D turbulence in the dual spectral range more closely, discussing also the higher-order statistics.

8.1.4 The enstrophy cascade

The most direct way to verify, or falsify, the phenomenological theory is by performing direct numerical simulations of turbulence, which in 2D can be performed at rather high Reynolds number. Numerous numerical studies of 2D turbulence performed since the early eighties now provide a rather reliable general picture, though results still vary in details. The connection with turbulence theory is clearest when we consider forced stationary conditions. As indicated in Fig. 8.1, stationarity in the presence of the dual cascade requires both an energy sink at large scales and an enstrophy sink at small scales, while energy (and enstrophy) is injected somewhere in between. For this purpose (8.6) is generalized by writing

$$\partial_t \omega + \boldsymbol{v}\,\nabla\omega = \mu_m(-1)^{m+1}\,\nabla^{-2m}\omega + \nu_n(-1)^{n+1}\,\nabla^{2n}\omega + f. \tag{8.21}$$

A high-order hyperviscosity n can be chosen, thus concentrating the dissipation of enstrophy at the high-k spectral boundary, since energy flows to small k and hence does not accumulate at the effective high-k cutoff. By contrast, the energy sink at small k should be gentle in order to avoid a pronounced bottleneck effect (see Section 5.3.6),[1] the safest choice being $m = 0$, i.e., a simple friction term $-\mu_0\omega$. The function f is usually chosen as a white-noise random force, $\langle f(\boldsymbol{k},t)f(\boldsymbol{k}',t')\rangle \propto \delta(\boldsymbol{k}+\boldsymbol{k}')\delta(t-t')$, applied isotropically in a narrow wavenumber band at $k \simeq k_{\text{in}}$.

Numerical simulations of forced 2D turbulence have, for instance, been conducted by Legras *et al.* (1988), Maltrud and Vallis (1991), and Borue (1993). Borue presented results at the highest effective Reynolds number. It is found that the energy spectrum is indeed close to k^{-3}, even the logarithmic correction could be identified;

$$E_k \simeq C_\Omega \epsilon_\Omega^{2/3} k^{-3}[\ln(k/k_{\text{in}})]^{-1/3}, \quad C_\Omega = 1.5-1.7. \tag{8.22}$$

The nonlocality of the transfer process in 2D turbulence, which results in the logarithmic factor in the spectrum, has also been confirmed directly by measuring the modal interactions. The value of the constant C_Ω had previously been derived theoretically by use of a renormalization-group approach by Olla (1991), who obtained $C_\Omega = 1.59$, in excellent agreement with the simulation result, whereas in an earlier paper Kraichnan (1971), considering a simple turbulence model, arrived at a somewhat larger value, $C_\Omega = 2.6$.

[1] In simulations of the inverse energy cascade by Borue (1994) using a high-order damping $m \gg 1$, a strong bottleneck hump is observed (though it was interpreted differently as a deviation from the expected Kolmogorov spectrum). The same assumption for m was made in a simulation study of the enstrophy cascade (Borue, 1993), in which the small-k bottleneck effect is also clearly visible but not detrimental for the scaling of the large-k spectrum.

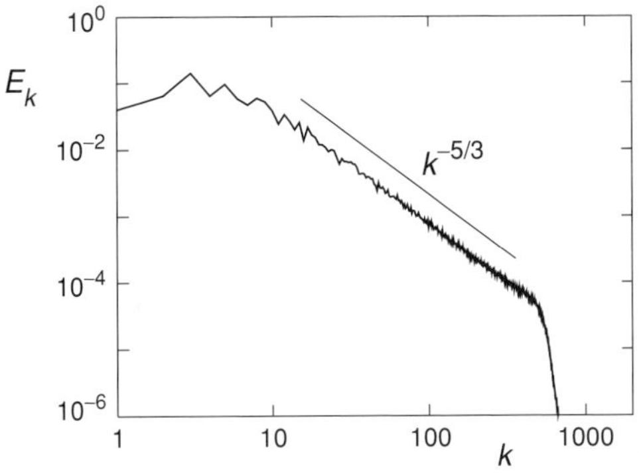

Figure 8.2: An energy spectrum from a numerical simulation of 2D turbulence forced at large wavenumber, $k \lesssim 600$. (Courtesy of Boffetta.)

In contrast to 3D turbulence, which decays essentially in a self-similar way, allowing the introduction of a normalized spectrum $\widehat{E}(\widehat{k})$, (5.57), decaying 2D turbulence is dominated by a self-organization process giving rise to coherent vortex structures. These were first observed numerically by McWilliams (1984), and thereafter in several studies at higher resolution, e.g., by Brachet *et al.* (1988) and Santangelo *et al.* (1989). Starting from a smooth initial state, first vorticity-gradient sheets develop, which are stretched and folded (this is similar to the development of steep gradients, the 'cliffs', in passive scalar turbulence), until the fluid is fully turbulent with a k^{-3} energy spectrum. As these random motions gradually decay, isolated coherent vortices appear. They lead to a change in the energy spectrum; only the high-k part of the spectrum preserves the k^{-3} shape, while at smaller k corresponding to the coherent structures the spectrum is significantly steeper. As time progresses these structures become increasingly more prominent, growing in size due to the occasional encounters and coalescence. In this state the system exhibits little resemblance to the fully developed turbulence of randomly folded vorticity-gradient sheets.

8.1.5 The inverse energy cascade

In the regime of the inverse energy cascade numerical simulations, e.g., by Maltrud and Vallis (1991), Smith and Yakhot (1993), and Boffetta *et al.* (2000) fully confirm the spectrum (8.19),

$$E_k = C'_{\mathrm{K}} \epsilon^{2/3} k^{-5/3}, \quad C'_{\mathrm{K}} \simeq 6, \tag{8.23}$$

which is illustrated in Fig. 8.2. The validity of the energy spectrum (8.23) in 2D indicates that the dimensional arguments on which the Kolmogorov spectrum is based are very robust, being independent of the direction of the flow of energy and insensitive to the degree of locality of the transfer process, as mode interactions in 2D are found to be much less well localized than are those in 3D. The value of the constant C'_K agrees with the renormalization-group analysis by Olla (1991) and also with Kraichnan's model (1971). It appears that the difference between the values of the constant in 2D ($C'_\mathrm{K} \simeq 6$) and 3D ($C_\mathrm{K} \simeq 1.6$, see (5.38)) is mainly due to geometrical effects connected with the different sizes of the k-space volume available for mode interactions, and is thus insensitive to the specific approximations in the various models of turbulence.

In contrast to the energy spectrum, higher-order statistics reveals significant differences between 2D turbulence in the energy-cascade regime and isotropic 3D turbulence. It has been noted (Smith and Yakhot, 1993) that, in the freely developing inverse cascade, intermittency is very weak; the probability distributions of the velocity increments δv_l are very close to Gaussian, resulting, in particular, in very small values of the odd-order moments (only when the cascade front reaches the largest scales in the system $k \simeq 1$, which leads to condensation of energy in this mode, do strong deviations from Gaussian statistics occur). However, even if the statistics is close to Gaussian, the statistics cannot be exactly Gaussian, since this would eliminate nonlinear interactions altogether and quench the turbulent transfer. In fact, even though they are small, deviations from Gaussian character are well defined, as analyzed by Boffetta *et al.* (2000). In particular, the third-order structure function satisfies an exact relation, namely the three-halves law,

$$S^{(3)}(l) = \langle(\delta v_{\parallel l})^3\rangle = \tfrac{3}{2}\epsilon l, \tag{8.24}$$

which can be derived in a similar way to Kolmogorov's four-fifths law (7.29) in 3D turbulence. The positive sign of the r.h.s. reflects the inverse direction of the energy transfer ϵ. This relation is indeed well reproduced numerically; see Fig. 8.3, which exhibits a scaling range of $S^{(3)}(l)$ of one decade. Though the relation looks very similar to the corresponding 3D expression, the relative magnitude of $S^{(3)}$ is small. A quantitative measure is provided by the skewness S,

$$S = S^{(3)}/(S^{(2)})^{3/2}.$$

Since the spectrum (8.23) implies the second-order structure function

$$S^{(2)} = C_2(\epsilon l)^{2/3}, \quad \text{with } C_2 = 2.2C'_\mathrm{K} \simeq 13,$$

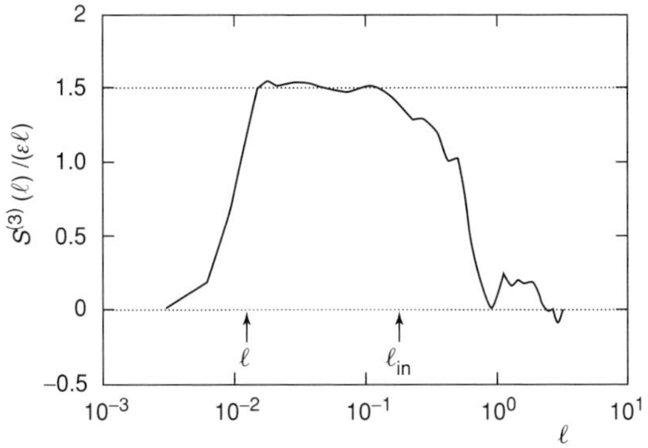

Figure 8.3: The compensated third-order longitudinal structure function $S^{(3)}(l)/(\epsilon l)$ in the inverse energy cascade in 2D turbulence, verifying (8.24). (From Boffetta *et al.*, 2000.)

the skewness is independent of the scale l and is numerically small,

$$S \simeq 0.03,$$

compared with $S \simeq 0.5$ in 3D turbulence. The origin of the small value of S in 2D is the large value of $C'_{\rm K}$, which means that only a small energy flux ϵ is required in order to generate the spectrum (8.23).

Also for higher orders $n > 3$, both even and odd, structure functions are found to follow very nearly the Kolmogorov scaling,

$$S^{(n)}(l) \simeq C_n(\epsilon l)^{n/3}, \tag{8.25}$$

i.e., the scaling exponents are

$$\zeta_n \simeq n/3. \tag{8.26}$$

Hence 2D turbulence in the energy-cascade range is nonintermittent, in distinct contrast to the intermittency of the 3D isotropic case, which is well described by the She–Lévêque formula (7.64).

8.2 Two-dimensional MHD turbulence

Let us now turn to the MHD case, for which, in the 2D approximation, $\partial_z = 0$, the reduced equations (8.4) and (8.5) for a strongly magnetized plasma read

$$\partial_t \omega + \boldsymbol{v} \cdot \nabla \omega - \boldsymbol{b} \cdot \nabla j = \nu \, \nabla^2 \omega, \tag{8.27}$$

$$\partial_t \psi + \boldsymbol{v} \cdot \nabla \psi = \eta \, \nabla^2 \psi. \tag{8.28}$$

Thus the induction equation reduces to a simple advection–diffusion equation for the scalar function ψ, which is, however, not a passive scalar, since it couples to the fluid motion through the Lorentz force. The addition of this term to the vorticity equation has a profound effect on the turbulence dynamics of the 2D system, as we shall see.

8.2.1 Properties of the ideal MHD sytem

As mentioned in Section 2.4, the 2D equations (8.27) and (8.28) have three quadratic ideal invariants, the total energy

$$E = \tfrac{1}{2}\int (v^2 + b^2)\, d^2x = \tfrac{1}{2}\sum_{\boldsymbol{k}} k^2(|\phi_{\boldsymbol{k}}|^2 + |\psi_{\boldsymbol{k}}|^2), \tag{8.29}$$

the cross-helicity

$$H^C = \int \boldsymbol{v}\cdot\boldsymbol{b}\, d^2x = \sum_{\boldsymbol{k}} k^2 \phi_{\boldsymbol{k}}\psi_{-\boldsymbol{k}}, \tag{8.30}$$

and the mean-square magnetic potential

$$A = \int \psi^2\, d^2x = \sum_{\boldsymbol{k}} |\psi_{\boldsymbol{k}}|^2. \tag{8.31}$$

The situation is hence very similar to that in 3D. While the first two invariants, E and H^C, are identical with the corresponding 3D expressions, the last, A, replaces the magnetic helicity H^M, (2.72), which vanishes in 2D. Following the treatment in Section 5.2.1, the absolute equilibrium spectra in the truncated k-space are obtained as moments of the Gibbs distribution

$$\rho_G = Z^{-1}\exp(-\alpha E - \beta A - \gamma H^C). \tag{8.32}$$

One finds (Fyfe and Montgomery, 1976)

$$E_k^K = \frac{2\pi k}{\alpha}\left(1 + \frac{k^2\tan^2\theta}{k^2+\kappa}\right), \quad E_k^M = \frac{2\pi k}{\alpha}\frac{k^2\sec^2\theta}{k^2+\kappa}, \tag{8.33}$$

$$E_k^R = E_k^K - E_k^M = \frac{2\pi k}{\alpha}\frac{\kappa}{k^2+\kappa}, \tag{8.34}$$

$$A_k = 2k^{-2}E_k^M, \tag{8.35}$$

$$H_k^C = -\frac{2\gamma}{\alpha}E_k^M, \tag{8.36}$$

which are written in forms similar to the corresponding 3D expressions in Section 5.2.1 with $\sin\theta = \gamma/(2\alpha)$ and $\kappa = (\beta/\alpha)\sec^2\theta$. While α is intrinsically positive, β and γ may also become negative. For $\beta < 0$ positivity of the denominator $k^2 + \kappa$ requires $|\beta| < k_{\min}^2 \cos^2\theta/\alpha$. In this case the expressions for E_k^K and E_k^M, and hence also H_k^C, are formally identical in 2D and 3D on setting $\kappa = -k_0^2$. In particular, the residual energy spectrum is $E_k^R < 0$, i.e., the magnetic modal energy exceeds the kinetic one, the difference being largest at the smallest k.

Therefore also the cascade properties are similar in 2D and 3D, quite in contrast to the hydrodynamic case; see Table 8.1. Since E_k and H_k^C are peaked at large k, they are expected to exhibit a direct cascade in the dissipative system, while A_k has an inverse cascade leading to self-organization, the formation of large-scale magnetic eddies.

8.2.2 Decay of 2D MHD turbulence

The decay laws of 3D MHD turbulence have been discussed at some length in Section 4.2.3, where it was specified that, in the case of finite magnetic helicity, the selective decay properties of the system govern the decay of turbulence. We found, however, certain deviations from the predictions of the basic theory, which result from the fact that the decay does not proceed in a self-similar way; in particular, the energy ratio E^K/E^M is not preserved but rather decreases on the same time scale as the energy itself, and hence the magnetic field decays more slowly than predicted, while the velocity decays more rapidly.

Also the decay of 2D MHD turbulence is ruled by a selective decay process, the conservation of the mean-square potential A on the time scale of the energy decay,

$$\frac{dA}{dt} = \eta \int d^2x\,(\nabla\psi)^2 = O(\eta), \tag{8.37}$$

while the energy-decay rate is found to be essentially independent of η. The situation is even somewhat simpler than that in 3D, for which we had to distinguish between the case of finite magnetic helicity and that of nearly vanishing magnetic helicity, while A and the magnetic energy E^M are intimately connected such that $E^M \neq 0$ implies $A \neq 0$. As in (4.52), we introduce a magnetic length scale L_M,

$$A = E^M L_M^2 \sim E L_M^2, \tag{8.38}$$

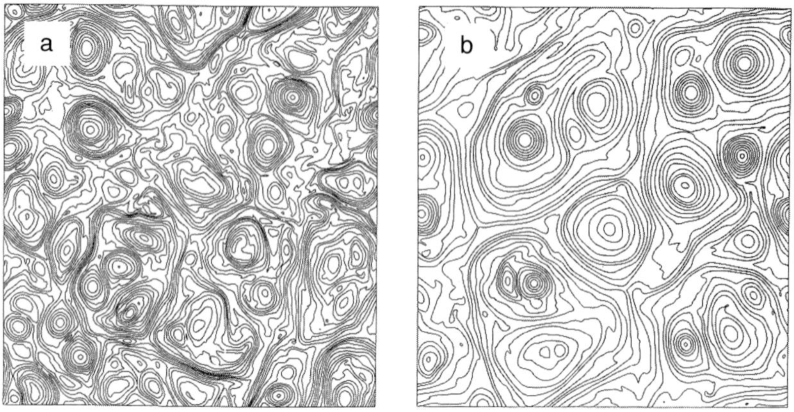

Figure 8.4: Contour plots of ψ at two different times during decay of turbulence in 2D MHD illustrating the increase of scale size of the magnetic eddies. (From Biskamp and Welter, 1989.)

where in the latter relation we assume that we have self-similarity, $E^M \sim E^K \sim E$. On the other hand, a dynamic macroscopic scale L is defined by (4.45),

$$-\frac{dE}{dt} \equiv \epsilon \sim \frac{E^{3/2}}{L}. \tag{8.39}$$

Identifying the two length scales, $L_M \sim L$, gives the relation

$$\frac{\epsilon}{E^2} \sim A^{-1/2} = \text{constant}, \tag{8.40}$$

which has the scaling solution (Hatori, 1984)

$$E \sim t^{-1}. \tag{8.41}$$

Physically, the decay of energy is caused by the inverse cascade of the magnetic potential, which proceeds as a sequence of coalescence processes leading to magnetic eddies of increasing size (see Fig. 8.4) and hence smaller gradients of ψ, $b^2 = (\nabla\psi)^2$. The numerical results, Fig. 8.5(a), agree qualitatively with this behavior. However, the log–log plot of $E(t)$ in Fig. 8.5(b) shows that the scaling (8.41) is reached only asymptotically after many turnover times, whereas during the intermediate period energy decays more slowly. This behavior can be attributed to the reorganization of the current density in the coherent magnetic eddies. The initially smoothly distributed current density contracts, becoming strongly peaked in the eddy centers, as illustrated in Fig. 8.6. This process occurs primarily during the dynamic events of coalescence of eddies, wherein the current density can be redistributed more efficiently than it can by resistive diffusion in a quasi-static eddy. Since the magnetic energy is higher

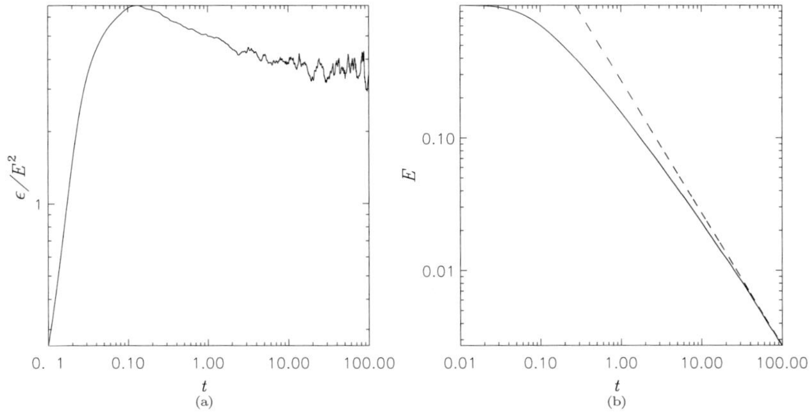

Figure 8.5: The decay of turbulence energy obtained from a numerical simulation followed for $10^3\tau_0$, where $\tau_0 \simeq 0.1$ is the initial eddy-turnover time. (a) The temporal evolution of ϵ/E^2, demonstrating the validity of relation (8.40); (b) $E(t)$ plotted on a log–log scale, with the dashed line indicating the power law t^{-1}. (From Biskamp and Schwarz, 2001.)

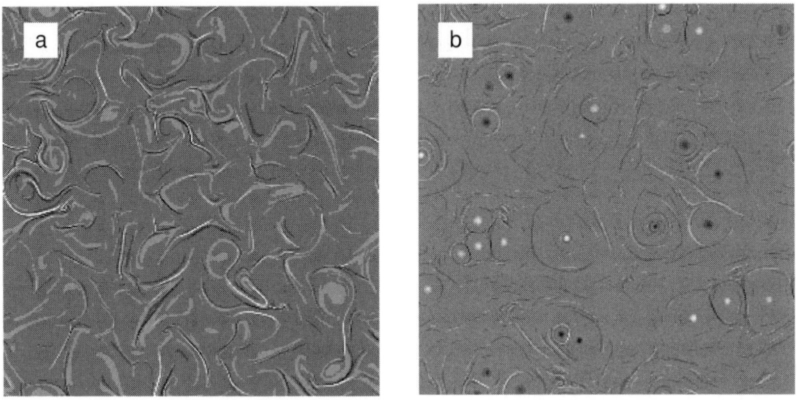

Figure 8.6: Grayscale plots of the current density in 2D MHD turbulence. (a) The current distribution at an early stage in the evolution of turbulence, during which large current densities are found only in current sheets at the boundaries of eddies, while the current distribution in the interiors of eddies is smooth. (b) The current density peaked in the centers of eddies during the late phase in the decay of turbulence. The state correponds to an ensemble of line currents. (From Biskamp and Schwarz, 2001.)

for a peaked current profile, the decrease of the ratio ϵ/E^2 observed in Fig. 8.5, which corresponds to slower decay of energy, is explained by invoking this condensation of current density, which compensates for part of the energy dissipated.

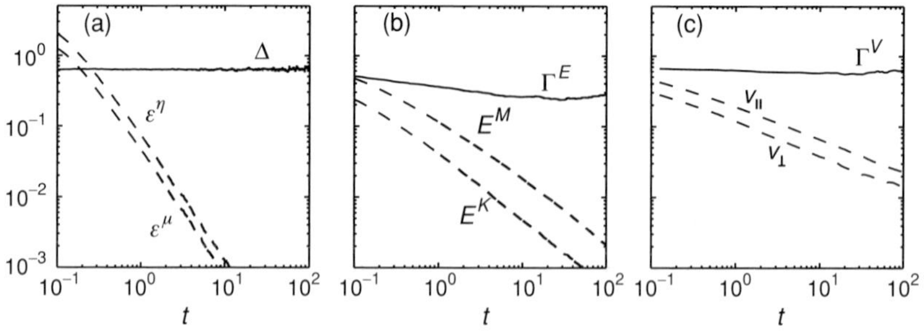

Figure 8.7: Log–log plots of the ratios of the dissipation rates, Δ (a), of the energies, Γ^E (b), and of the velocities, Γ^V (c) from the same simulation run of decaying 2D MHD turbulence as that shown in Fig. 8.5. Also plotted (dashed) are the individual quantities ϵ^μ, ϵ^η; E^K, E^M; $v_\perp$, $v_\parallel$. (From Biskamp and Schwarz, 2001.)

Apart from this effect, however, the decay proceeds in an almost self-similar way; in particular, the ratios of equivalent quantities remain essentially constant during decay of energy. This concerns the ratios of viscous and resistive dissipation rates, $\Delta = \epsilon^\mu/\epsilon^\eta$, of kinetic and magnetic energies, $\Gamma^E = E^K/E^M$, and of perpendicular and parallel velocities, $\Gamma^V = v_\perp/v_\parallel$. These velocities are defined by $v_\perp = \langle|\boldsymbol{v} \times \boldsymbol{b}|/b\rangle$ and $v_\parallel = \langle|\boldsymbol{v} \cdot \boldsymbol{b}|/b\rangle$, giving a measure of the local alignment of velocity along the magnetic field, in contrast to the cross-helicity $H^C = \langle\boldsymbol{v} \cdot \boldsymbol{b}\rangle$ which, because of cancellations, may remain small even in a locally completely aligned state.

The temporal evolution of these ratios is shown in Fig. 8.7, together with the decays of the individual quantities. The most strictly conserved ratio is Δ, as was first noted by Kinney *et al.* (1995), which seems to be a general property of a MHD system that is rather independent of the Reynolds number; the numerical value of Δ depends only on the magnetic Prandtl number and weakly on the Reynolds number. The energy ratio Γ^E is found to decrease slowly by about a factor of 2. This is connected with the condensation of current density discussed above, which increases the relative weight of the magnetic energy. The constancy of Γ^V is noteworthy, since it implies that, for weak initial velocity alignment, not only the correlation H^C/E but even the local alignment do *not* increase and the turbulence dynamics is not weakened. The dynamic alignment discussed in Section 4.2.2 seems to occur only if the initial alignment is already sufficiently large.

8.2.3 Spectra in 2D MHD turbulence

The analogy between the cascade properties in 2D and 3D MHD turbulence suggests that there should also be a similarity of the spectral laws. Regarding the energy spectrum the question of how strong the Alfvén effect is arises again. Is the transfer dynamics deeply affected by the nonlocal interaction with the large-scale magnetic field, resulting in an IK spectrum $E_k \sim k^{-3/2}$, or do the local interactions dominate, leading to a Kolmogorov spectrum $k^{-5/3}$? In the 3D case treated in Chapter 5 numerical simulations, though they are at still modest Reynolds numbers, indicate rather convincingly that the Alfvén effect does not control the spectral transfer, since the small-scale dynamics occurs mainly perpendicularly to the local field, such that the time scales for parallel and perpendicular mode interactions are of the same order, (5.47), which leads to an overall Kolmogorov spectrum. Since restriction to 2D allows numerical computations of turbulence at significantly higher Reynolds number, it should be even easier than it is in 3D to discriminate between the two alternatives. These expectations are, however, only partly fulfilled, as we shall see.

Results from the first numerical studies of 2D MHD turbulence of sufficiently high spatial resolution to allow the formation of an inertial range (Biskamp and Welter, 1989; Politano *et al.*, 1989) seemed to indicate the validity of the IK phenomenology. One finds the total energy spectrum

$$E_k = C_{\text{IK}}(\epsilon v_{\text{A}})^{1/2} k^{-3/2}, \quad \text{with } C_{\text{IK}} \simeq 1.8, \tag{8.42}$$

and the dissipation scale follows the IK prediction l_{IK}, (5.56), and so does the residual spectrum $E_k^R \sim k^{-2}$, (5.77) (Biskamp, 1995). These results are corroborated by the geometrical properties of turbulent eddies; in particular, the level surfaces are predicted to exhibit a fractal dimension depending on the scaling laws of the turbulence. On analyzing the level surfaces of the current density one finds the dimension $D = \frac{7}{4}$ (corresponding to $D = 11/4$ in 3D), which is consistent with the IK scaling $\delta z_l \sim l^{1/4}$, instead of $D = \frac{5}{3}$ ($D = \frac{8}{3}$ in 3D), which would be expected for the Kolmogorov scaling $\delta v_l \sim l^{1/3}$ (Biskamp, 1993b).

However, recent numerical studies at still higher Reynolds numbers, both for forced stationary turbulence (Politano *et al.*, 1998) and for freely decaying turbulence (Biskamp and Schwarz, 2001), have revealed certain anomalous features, which cast a slight slur on this seemingly clear picture. Figure 8.8 shows the normalized energy spectrum $\widehat{E}(\widehat{k})$, (5.60) multiplied by $\widehat{k}^{3/2}$ obtained from three simulation runs of decaying turbulence with different Reynolds numbers. The curves are plotted on a linear vertical scale to illustrate the anomalous features more distinctly. While the plot of the case with the lowest Reynolds number is horizontal, indicating that there is a pure $k^{-3/2}$ inertial-range spectrum extending over more than a decade, at higher Reynolds number the high-k

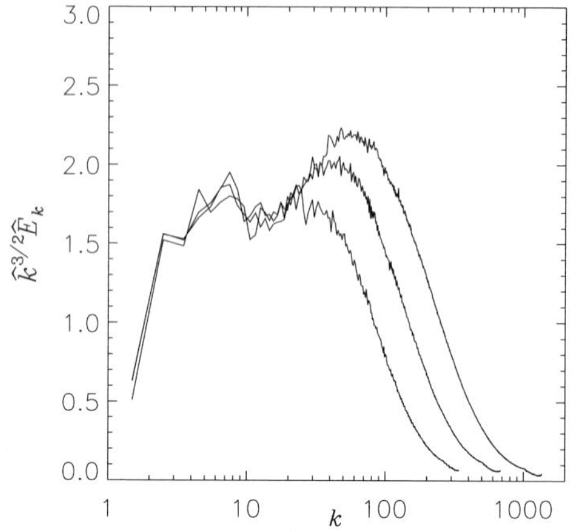

Figure 8.8: Compensated energy spectra from three simulation runs of decaying 2D MHD turbulence with different microscale Reynolds numbers $\mathrm{Rm}_\lambda = 150$, 280, and 450 (from the left). Note the linear vertical scale. (From Biskamp and Schwarz, 2001.)

part is modified by a hump with increasing amplitude and extending over an ever larger part of the inertial range, which appears to be associated with an anomalous, or nonlocal, bottleneck effect (see Section 5.3.6). (Note that the simulations shown in Fig. 8.8 were performed with normal diffusion, for which *a priori* no strong bottleneck effect is expected.) Though the physical origin of this phenomenon is not yet quite clear, it appears that it is connected with the nonlocalness of the mode interactions in 2D. In any case the spectrum is clearly flatter than $k^{-5/3}$, in contrast to the numerical results for 3D turbulence; see Figs. 5.2 and 5.3. We do not dwell on this point for the moment, postponing the interpretation of the difference between 2D and 3D MHD turbulence to Section 8.2.4, where we investigate the intermittency properties of 2D turbulence.

In the inverse cascade of the magnetic potential the large-scale magnetic field is rather weak, as we shall see below. Hence no Alfvén effect has to be considered in the spectral transfer and the spectrum of the mean-square magnetic potential A_k can be estimated by employing a simple Kolmogorov-type argument. Assuming that there is a constant transfer rate (i.e., input rate) of magnetic flux ϵ_ψ, where A_k and ϵ_ψ have the dimensions L^4T^{-3} and L^5T^{-2}, respectively, comparison of dimensions in the *Ansatz* $A_k \sim \epsilon_\psi^\alpha k^\beta$ gives

$$A_k \sim \epsilon_A^{2/3} k^{-7/3}. \tag{8.43}$$

This prediction has been verified in numerical solutions of the 2D closure equations (Pouquet, 1978; see Fig. 8.9), as well as in direct numerical simulations

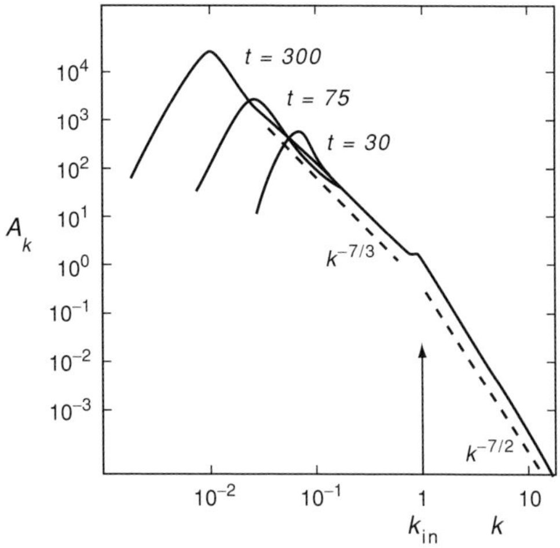

Figure 8.9: The inverse cascade of the mean-square potential A_k for $k < k_{in}$. The right-hand part of the spectrum ($k > k_{in}$) correponds to the direct energy cascade, for which $A_k = k^{-2}E_k \sim k^{-7/2}$. (From Pouquet, 1978.)

(Biskamp and Bremer, 1994). Results from the latter study show that the turbulent state in this regime consists of a random ensemble of current filaments, which tend to coalesce, driving the system to larger scales. As observed for the inverse energy cascade in 2D hydrodynamic turbulence, the turbulence is not intermittent but rather exhibits statistical properties that are very close to Gaussian, which hence seems to be a generic feature of an inverse turbulence cascade driven by some small-scale noise. Such behavior is, however, not unexpected, since the short correlation time of the source does not allow large-scale coherent structures with much longer correlation times to form.

8.2.4 Intermittency in 2D MHD turbulence

We have seen in the previous section that the energy spectrum of 2D MHD turbulence is less steep than the Kolmogorov spectrum found in isotropic 3D MHD turbulence, which is consistent with a $k^{-3/2}$ law (apart from the bottleneck effect). This agrees with the numerical observation discussed in Section 5.3.6, where an increase of the mean magnetic field in 3D turbulence leads to a flattening of the spectrum of the transverse modes, approaching the IK law for $B_0 \gg \langle \widetilde{b^2} \rangle^{1/2}$. The effect of the mean field became even more conspicuous on considering the scaling of the structure functions of the Elsässer fields $\langle (\delta z_l^{\pm})^n \rangle \sim l^{\zeta_n^{\pm}}$. We found in Section 7.5.2 that the values of the exponents $\zeta_n^{\pm}$ decreased with increasing strength of the mean field, approaching those observed in 2D MHD turbulence; see Fig. 7.10. Let us now analyze the 2D behavior more closely.

Though in 2D turbulence of much higher Reynolds number has been studied numerically, there is still some discrepancy of the detailed results published to date, the values of ζ_n being somewhat larger in the work of Politano *et al.* (1998) and Gomez *et al.* (1999) than was found by Biskamp and Schwarz (2001). The difference may be due to whether forced, roughly stationary turbulence is studied, as done by the former group, or decaying turbulence, as considered in the latter paper. This discrepancy may indicate that there is some nonuniversality in the behavior of 2D MHD turbulence. Nonetheless, the general tendency appears to be clear, namely that the spatial distribution of the turbulent fields is more intermittent in 2D than it is in 3D and is inconsistent with the Kolmogorov phenomenology. In particular, the results of both numerical studies show that the third-order exponent ζ_3 is clearly smaller than unity, which seems to contradict the exact relation (7.40), $\langle \delta z_\parallel^\mp \, \delta z_i^\pm \, \delta z_i^\pm \rangle = -2\epsilon^\pm l$. However, in contrast to Kolmogorov's four-fifths law in hydrodynamics, which involves only the third-order structure function for the velocity, the corresponding MHD relation involves both fields, δz_+ and δz_- (in addition, both longitudinal and transverse field increments occur). The scaling of the mixed structure functions $\langle |\delta z_\parallel^\mp \, \delta z_i^\pm \, \delta z_i^\pm|^{n/3} \rangle \sim l^{\alpha_n}$ does, indeed, exhibits a behavior different from that of the pure structure functions involving only one type of field. Gomez *et al.* (1999) found that the values of α_n agree rather well with the She–Lévêque formula, which was derived primarily for hydrodynamic turbulence. A similar behavior has been observed in thermal convection for the mixed structure functions $\langle |\delta v_l \, (\delta T_l)^2|^{n/3} \rangle$ (Section 9.3.1; see, in particular, Fig. 9.7).

9
Compressible turbulence and turbulent convection

In the previous chapters turbulence was assumed incompressible. As discussed in Section 2.3, this assumption is valid if either the sonic Mach number of the flow is small, $\mathrm{M_s} = v/c_s \ll 1$, or the Alfvén Mach number is small, $\mathrm{M_A} = v/v_A \ll 1$. The former condition applies to a weakly magnetized plasma, in which $v_A \ll c_s$, or to motions along the magnetic field, while the latter applies to motions perpendicular to the field. If the flow is turbulent, there is some arbitrariness in the definition of the Mach numbers, since one may choose (a) the mean flow velocity, (b) the r.m.s. velocity fluctuation $v = \langle \widetilde{v}^2 \rangle^{1/2} = (E^K)^{1/2}$, or (c) the local velocity. Following convention in turbulence theory, we refer to the Mach number in terms of the r.m.s. velocity, noting that local Mach numbers may be considerably higher.

Since laboratory plasmas are usually confined by a strong magnetic field, they can be considered incompressible, the dynamics consisting mainly of cross-field motions. Also the motions in the liquid core of the Earth, which drive the Earth's dynamo, are incompressible, since $\mathrm{M_s} \ll 1$ (here inertial effects are often neglected altogether, which is called the magnetostrophic approximation). By constrast, most astrophysical plasmas are compressible, for instance the interstellar medium, which is rather cold, such that, in the turbulent motions observed, $\mathrm{M_s}$, and possibly also $\mathrm{M_A}$, tend to be large (see Chapter 12), or the turbulence in the interplanetary plasma, which is riding on the supersonic and super-Alfvénic solar wind (Chapter 10). Also in stellar convection zones compressibility effects are important, such that the Boussinesq approximation, which is very appropriate for thermal convection in laboratory devices, does not apply.

For compressible systems one has to use the equation of motion in the velocity form (2.7),

$$(\partial_t + \boldsymbol{v} \cdot \nabla)\boldsymbol{v} = -\frac{1}{\rho} \nabla p + \frac{1}{c\rho} \boldsymbol{j} \times \boldsymbol{B} - \nabla \phi_g + \nu(\nabla^2 \boldsymbol{v} + \tfrac{1}{3} \nabla\nabla \cdot \boldsymbol{v}). \quad (9.1)$$

To distinguish between compression and vortex motions, it is useful to decompose the velocity,

$$\boldsymbol{v} = \boldsymbol{v}^c + \boldsymbol{v}^s, \tag{9.2}$$

where the compressive part $\boldsymbol{v}^c$ is irrotational, $\nabla \times \boldsymbol{v}^c = 0$, and the solenoidal part $\boldsymbol{v}^s$ is divergence-free, $\nabla \cdot \boldsymbol{v}^s = 0$. This representation corresponds to the Helmholtz decomposition, whereby a general vector field is written as the sum of a curl and a gradient,

$$\boldsymbol{v} = \nabla \times \boldsymbol{a} + \nabla \phi$$

where $\boldsymbol{a}$ and ϕ are solutions of Poisson's equation, $\nabla^2 \boldsymbol{a} = -\nabla \times \boldsymbol{v}$ (assuming that $\nabla \cdot \boldsymbol{a} = 0$ without loss of generality) and $\nabla^2 \phi = \nabla \cdot \boldsymbol{v}$. This decomposition leads us to introduce the parameter

$$\chi = \frac{E^c}{E^s}, \quad E^{c,s} = \tfrac{1}{2} \int d^3x \, (\boldsymbol{v}^{c,s})^2, \quad E^K = E^c + E^s. \tag{9.3}$$

Note that here E^K does not denote the kinetic energy $\frac{1}{2}\langle \rho \boldsymbol{v} \cdot \boldsymbol{v} \rangle$ but rather denotes the velocity variance $\frac{1}{2}\langle \boldsymbol{v} \cdot \boldsymbol{v} \rangle$, which is customary in compressible-turbulence theory. Also the velocity spectra E_k^c and E_k^s discussed below are defined in this way. Since the most important effect of compressibility in a supersonic flow is the formation of shock waves, the parameter χ measures the importance of shock-related motions compared with eddy motions, roughly speaking the velocity component perpendicular to a shock front compared with the component parallel to it. Note that χ is conceptually different from the Mach number M_s (or M_A), which compares the kinetic energy and thermal (or magnetic) pressure.

Taking the curl, or the divergence, of (9.1) gives the equations for the vorticity $\boldsymbol{\omega}$ and the velocity divergence $\nabla \cdot \boldsymbol{v}$, called dilatation,

$$\frac{d}{dt} \boldsymbol{\omega}_\rho = \boldsymbol{\omega}_\rho \cdot \nabla \boldsymbol{v} + \frac{1}{\rho^3} \nabla \rho \times \nabla p + \frac{1}{\rho} \nabla \times \frac{\boldsymbol{j} \times \boldsymbol{B}}{c\rho}, \tag{9.4}$$

$$\frac{d}{dt} \nabla \cdot \boldsymbol{v} = \omega^2 - \nabla \boldsymbol{v} : \nabla \boldsymbol{v} - \nabla \cdot \frac{\nabla p}{\rho} + \nabla \cdot \frac{\boldsymbol{j} \times \boldsymbol{B}}{c\rho} - \nabla^2 \phi_g, \tag{9.5}$$

omitting dissipation effects. Equation (9.4) has been written in terms of the potential vorticity $\boldsymbol{\omega}_\rho = \boldsymbol{\omega}/\rho$, resulting in a more compact form. The first term on the r.h.s. of (9.4) is the vortex-stretching effect, the well-known source of vorticity in Navier–Stokes turbulence. This term is absent in 2D, which implies that conservation of vorticity holds and leads to the inverse cascade of the energy discussed in Section 8.1. The second term, called the baroclinic effect, which vanishes for a polytropic equation of state $p \sim \rho^\gamma$, has been considered in problems of shock-wave interactions (Picone and Boris, 1988). It may also play a role in shear flows, even at low Mach number, by exerting a stabilizing effect on the Kelvin–Helmholtz instability (e.g., McMurtry *et al.*, 1989). In general, however, the baroclinic effect is found to be weak and we will therefore not discuss it any further, restricting our consideration to polytropic fluids. The third term in (9.4) arises from the Lorentz force, the main source of vorticity in 2D MHD turbulence. The physics of the generation of dilatation cannot easily be read off the corresponding equation (9.5) because of cancellation effects among the terms on the r.h.s., which is obvious in the incompressible limit $\nabla \cdot \boldsymbol{v} = 0$.

The properties of shock waves developing in a compressible MHD fluid are considered in Section 9.1. Section 9.2 discusses supersonic compressible isotropic turbulence, both in hydrodynamics and in MHD. Turbulent convection in a stratified fluid may be strongly affected by compressibility. In Section 9.3 we first summarize the main features of thermal convection in a neutral fluid, with particular emphasis on passive scalar turbulence, which has received much attention in recent years, and finally discuss the turbulent convection of magnetic fields.

9.1 MHD shock waves

Weak disturbances in a compressible fluid propagate in the form of linear waves, for instance sound waves in a neutral gas, while finite-amplitude disturbances steepen to form shock waves propagating faster than the linear group velocity. The basic agent making a signal steepen is the $\boldsymbol{v} \cdot \nabla \boldsymbol{v}$ term in the equation of motion. The prototypical model is Burgers' equation (see Section 3.1)[1]

$$\partial_t u + u\, \partial_x u = \nu\, \partial_x^2 u, \tag{9.6}$$

with $u = v_x$. If $\partial_x u < 0$, the fluid will be compressed, following fluid elements catching up with preceding ones, which leads to formation of quasi-discontinuities, where the steepening is limited only by viscosity and where hence strong dissipation occurs. An initially smooth velocity profile will develop into a series of sawteeth, as illustrated in Fig. 9.1. Burgers' turbulence, the solution of (9.6) developing from a broad-band initial velocity distribution, has attracted considerable interest as a model system of compressible turbulence,

[1] Turbulence developing in the (multidimensional) Burgers equation models high-Mach-number hydrodynamic turbulence. On normalizing the equation of motion, (9.1) with $\boldsymbol{B} = \nabla\phi_g = 0$, with respect to the mean velocity, the pressure force $\rho^{-1}\nabla p = M_s^{-1}\nabla \ln\rho$ (for a polytropic pressure law $p = a\rho^\gamma$) becomes negligible for $\mathsf{M}_s \to \infty$ (formally, it vanishes exactly for $\gamma = 0$), and the equation degenerates to Burgers' equation, while the density is advected and compressed passively.

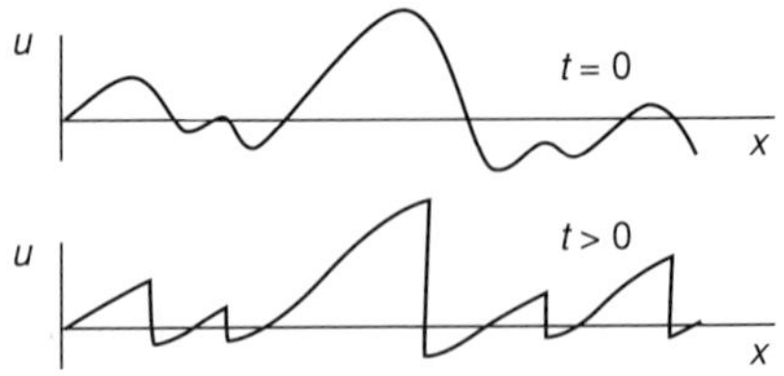

Figure 9.1: Shock formation in Burgers' turbulence. The negative-velocity-gradient parts of an initially smooth velocity profile steepen to form a series of shocks (shown schematically).

from which a number of exact results can be derived (e.g., Kraichnan, 1968; Kida, 1979; Bouchaud *et al.*, 1995), some of which will be discussed later in this chapter. One-dimensional MHD models generalizing Burgers' equation have been investigated by Yanase (1997) and Galtier and Pouquet (1998). In general, the presence of the $\boldsymbol{v} \cdot \nabla \boldsymbol{v}$ term is a necessary, but not sufficient, condition for shocks to form, since other terms in the equation counteract the steepening effect, which is obviously true in incompressible fluids, where the $\boldsymbol{v} \cdot \nabla \boldsymbol{v}$ term is compensated by the pressure gradient such that the flow becomes purely rotational.

Shock waves, or, briefly, shocks, are thin transition layers, where the dynamic variables of the fluid jump from their upstream values to their downstream values. In a local surrounding the shock front can be assumed plane and the conditions of the fluid in the adjacent upstream and downstream regions homogeneous. The thickness of the transition layer, which in a collision-dominated fluid is of the order of the mean free path,[2] can be neglected on a macroscopic scale. Moreover, the system is considered in a frame moving with the shock front, such that the configuration can be taken to be stationary. Here the conservation equations for mass, momentum, and energy assume the form $\nabla \cdot \boldsymbol{F} = 0$, from which the jump relations follow by integration. Let us derive these relations for MHD shock waves and discuss briefly their properties (for a more detailed treatment see, e.g., Jeffrey and Taniuti, 1964). Integration of the conservation laws $\nabla \cdot \rho \boldsymbol{v} = 0$, (2.16), $\nabla \cdot \mathcal{T} = 0$, (2.52), and $\nabla \cdot \boldsymbol{F}^E = 0$, (2.56), between the upstream and downstream states gives the jump conditions

$$\boldsymbol{n} \cdot [\rho \boldsymbol{v}] = 0, \tag{9.7}$$

[2] In dilute astrophysical plasmas collisions are often negligible, so shock waves are governed by collective processes, i.e., wave dispersion and turbulence. The best-known example is the bow shock in front of the Earth's magnetosphere. The shock front in these so-called collisionless shocks (for a recent review see, e.g., Russell, 1995) is usually highly oscillatory, much in contrast to the monotonic shock profile in a collision-dominated fluid, with waves propagating far into the upstream region, so that the shock thickness is often only poorly defined.

$$\boldsymbol{n} \cdot \left[\rho \boldsymbol{v}\boldsymbol{v} + \left(p + \frac{B^2}{8\pi} \right) \boldsymbol{I} - \frac{1}{4\pi} \boldsymbol{B}\boldsymbol{B} \right] = 0, \tag{9.8}$$

$$\boldsymbol{n} \cdot \left[\rho(\tfrac{1}{2}v^2 + h)\boldsymbol{v} + \frac{c}{4\pi} \boldsymbol{E} \times \boldsymbol{B} \right] = 0, \tag{9.9}$$

and the relations resulting from $\nabla \cdot \boldsymbol{B} = \nabla \times \boldsymbol{E} = 0$,

$$\boldsymbol{n} \cdot [\boldsymbol{B}] = 0, \tag{9.10}$$

$$\boldsymbol{n} \times [\boldsymbol{E}] = 0. \tag{9.11}$$

Here $h = u + p/\rho = [\gamma/(\gamma - 1)]p/\rho$ is the enthalpy, and $\boldsymbol{E} = -\boldsymbol{v} \times \boldsymbol{B}/c$. The unit vector $\boldsymbol{n}$ is the normal to the shock plane chosen such that $\boldsymbol{v} \cdot \boldsymbol{n} = v_n > 0$, and the brackets $[f] = f_2 - f_1$ denote the jump of f between the downstream state '2' and the upstream state '1'. Dissipation effects, which are important in the shock front, can be neglected outside. The jump conditions are usually called the *Rankine–Hugoniot relations*, though, strictly speaking, the name refers only to the hydrodynamic case ($\boldsymbol{B} = 0$), while in the MHD case they should be called de Hoffmann–Teller relations (de Hoffmann and Teller, 1950).

The formalism can be simplified by a Galilean transformation in the shock plane to a frame, where the tangential component $\boldsymbol{E}_t$, which is continuous across the shock, vanishes, such that $\boldsymbol{v}_t B_n = v_n \boldsymbol{B}_t$. This frame of reference is called the de Hoffmann–Teller frame. Since now $\boldsymbol{v}_t$ and $\boldsymbol{B}_t$ are parallel, it follows that also $\boldsymbol{E}_n = \boldsymbol{v}_t \times \boldsymbol{B}_t = 0$; hence $\boldsymbol{E} = 0$ and therefore $\boldsymbol{v} \parallel \boldsymbol{B}$ altogether. The jump conditions (9.7)–(9.11) can be written in the forms[3]

$$[\rho v_n] = 0, \tag{9.12}$$

$$m[v_n] + [p] + \frac{1}{8\pi}[B_t^2] = 0, \tag{9.13}$$

$$m[v_t] - \frac{1}{4\pi} B_n [B_t] = 0, \tag{9.14}$$

$$\tfrac{1}{2}[v_n^2 + v_t^2] + [h] = 0, \tag{9.15}$$

$$[B_n] = 0, \tag{9.16}$$

$$[E_t] = [v_t B_n - v_n B_t] = 0. \tag{9.17}$$

[3] The transformation to the de Hoffmann–Teller frame is possible only if $B_n \neq 0$. For $B_n = 0$ one can choose a reference frame, where $v_t = 0$. While the relations (9.12) and (9.13) remain valid, (9.15) is replaced by $\frac{1}{2}[v_n^2] + [h] + [B_t^2/\rho]/4\pi = 0$ and (9.16) by $[v_n B_t] = 0$, where the tangential electric field is $E_t = -v_n B_t/c = \text{constant}$.

Here m is the mass flow, $m = \rho_1 v_{n1} = \rho_2 v_{n2}$, and in (9.17) not only $[E_t]$ but also E_t itself is zero; hence $v_t B_n = v_n B_t$. Note that v_t and B_t are the tangential vector components, not absolute values. The six algebraic jump relations determine the downstream values ρ_2, v_{n2}, v_{t2}, B_{n2}, B_{t2}, and p_2 (or h_2) in terms of the upstream values. On eliminating, for instance, the velocities, one obtains, after some algebraic manipulations, the pressure ratio

$$\frac{p_2}{p_1} = \frac{\dfrac{\gamma+1}{\gamma-1}\dfrac{\rho_2}{\rho_1} - 1 + \dfrac{[B_t]^2}{8\pi p_1}\left(\dfrac{\rho_2}{\rho_1} - 1\right)}{\dfrac{\gamma+1}{\gamma-1} - \dfrac{\rho_2}{\rho_1}}. \tag{9.18}$$

Since a shock wave is dissipative, the second law of thermodynamics requires that the entropy must increase across the shock, which is satisfied if the shock wave is compressive, $\rho_2 \geq \rho_1$ and $p_2 \geq p_1$. Hence the numerator on the r.h.s. of (9.18) is positive and consequently the compression rate ρ_2/ρ_1 is limited,

$$\frac{\gamma+1}{\gamma-1} > \frac{\rho_2}{\rho_1} \geq 1, \tag{9.19}$$

$(\gamma+1)/(\gamma-1) = 4$ for $\gamma = \frac{5}{3}$, and the jumps of the variables satisfy the relations

$$[p] \geq 0, \quad [\rho] \geq 0, \quad [v_n] \leq 0. \tag{9.20}$$

Assuming, without loss of generality, that $B_n > 0$, (9.14) shows that

$$\text{either } [B_t] \geq 0,\ [v_t] \geq 0, \quad \text{or } [B_t] \leq 0,\ [v_t] \leq 0. \tag{9.21}$$

On inserting v_t from (9.17) into (9.14) one obtains

$$m[v_n B_t] = \frac{1}{4\pi} B_n^2 [B_t],$$

which can be written in the form

$$m[v_n]\langle B_t \rangle = -[B_t]\left(\langle \rho v_n^2 \rangle - \frac{B_n^2}{4\pi}\right), \tag{9.22}$$

where $\langle f \rangle = \frac{1}{2}(f_1 + f_2)$ and we used the relation

$$[fg] = [f]\langle g \rangle + [g]\langle f \rangle.$$

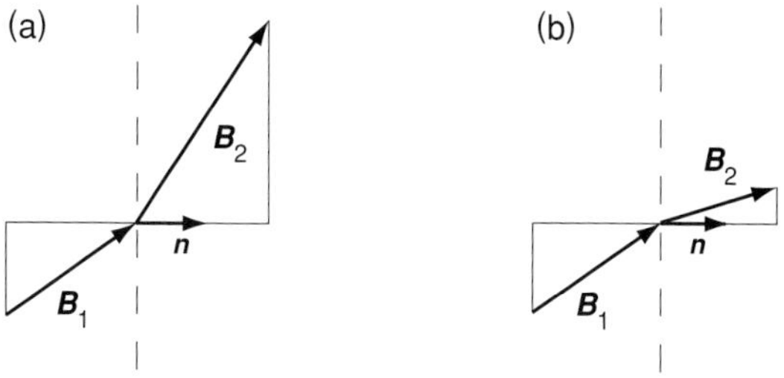

Figure 9.2: Changes of magnetic field in a fast shock (a) and a slow-mode shock (b). In the de Hoffmann–Teller frame, where $E_t = 0$, the velocity and magnetic field are parallel.

By rearranging terms we find the ratio

$$\frac{B_{t2}}{B_{t1}} = \frac{\rho_1 v_{n1}^2 - B_n^2}{\rho_2 v_{n2}^2 - B_n^2} = \frac{m v_{n1} - B_n^2}{m v_{n2} - B_n^2}. \tag{9.23}$$

To be definite, take v_{n1}, v_{t1}, B_n, and B_{t1} positive and assume also B_{t2} and hence v_{t2} to be positive.[4] It follows from (9.23) that either (a) $m v_{n1}, m v_{n2} > B_n^2$ or (b) $m v_{n1}, m v_{n2} < B_n^2$, whereas (9.22) tells us that $[B_t] > 0$ in case (a), i.e, the magnetic field in the downstream state is bent away from the normal, and $[B_t] < 0$ in case (b), i.e., the field is bent toward the normal, as illustrated in Fig. 9.2. In the weak-shock-limit cases (a) and (b) correspond to the two compressive MHD modes, the fast magnetosonic wave (2.93) and the slow mode (2.94), respectively. On linearizing the jump relations (9.12)–(9.17), $[f] = \delta f \ll f$, and noting that in the plasma restframe a weak shock moves at the linear group (i.e., phase) velocity v_g, $v_n = v_g + \delta v_n$, one obtains the equation

$$v_g^4 - v_g^2(c_s^2 + v_A^2) + c_s^2 v_{An}^2 = 0 \tag{9.24}$$

for v_g, with $v_{An}^2 = B_n^2/(4\pi\rho)$, which is, indeed, the second factor in (2.91), the dispersion relation for the compressive MHD modes, since $k^2 v_{An}^2 = (\boldsymbol{k} \cdot \boldsymbol{v}_A)^2 = k_\parallel^2 v_A^2$. One therefore speaks of fast, or magnetosonic, shocks in case (a) and slow, or slow-mode, shocks in case (b).[5]

[4] The opposite case is also possible; $B_{t2} = 0$ and $B_{t2} < 0$ are called the switch-off shock and intermediate shock, respectively, but these are not relevant in the present context.

[5] Two pairs of slow-mode shocks are the characteristic feature of Petschek's model of magnetic reconnection (Petschek, 1964). Since here the electric field E_t ($=E_z$ in the usual x, y-representation of the configuration) is finite, driving the reconnection flow, these shocks are not given in the de Hoffmann–Teller frame, and therefore velocities and magnetic fields are not parallel. While B_t is large in the upstream state and small in the downstream state, for v_t one has just the reverse behavior.

A general finite-amplitude perturbation of the fluid gives rise to formation both of fast-mode and of slow-mode shocks as well as transverse Alfvén waves and solenoidal eddy motions. For very low fluid velocities, the fast mode decouples and can be eliminated by averaging over the fast frequency, but the slow-mode shock survives in the incompressible limit, where $[\rho] = 0$ and the first two jump relations degenerate to $[v_n] = 0$ and $[p] + [B_t^2]/(8\pi) = 0$, i.e, quasi-static discontinuities, while the dynamics is determined by Alfvén waves and eddies. In the opposite limit of highly supersonic velocities the slow mode is negligible and the dynamics is determined by fast shocks and Alfvén waves.

Shock waves lead to efficient heating of the fluid, since the supersonic flow is rendered subsonic by decreasing the flow speed and increasing the pressure. Let us discuss the energy-dissipation rate, by which we mean the rate at which kinetic and magnetic energy is transformed irreversibly into heat when the fluid crosses the shock front. Conservation of the total energy flux (9.9) gives the dissipation rate, the energy per unit volume dissipated per second,

$$\epsilon = [\rho v_n h] = \tfrac{5}{2}[v_n p] \tag{9.25}$$

(with $\gamma = \frac{5}{3}$ to simplify the notation), which comprises the change of internal energy and the work done against the pressure force, since $\rho h = \rho u + p$.

To derive an expression for the dissipation rate in terms of the compression ratio ρ_2/ρ_1, we write $[v_n p] = [v_n]\langle p\rangle + [p]\langle v_n\rangle$, calculate $[p]$ and $\langle p\rangle$ from (9.18), and make the substitution $[v_n] = -\langle v_n\rangle[\rho]/\langle\rho\rangle$ to obtain

$$\epsilon = \frac{5}{2}\langle v_n\rangle\left(2p_1 + \frac{[B_t]^2}{8\pi}\frac{\rho_2 - \rho_1}{\rho_2 + \rho_1}\right)\frac{\rho_2/\rho_1 - 1}{4 - \rho_2/\rho_1}. \tag{9.26}$$

The efficiency of the dissipation of energy becomes obvious in the high-Mach-number limit. As has already been discussed, the increase in density is limited, (9.18). Equation (9.23) shows that, for fast shocks, the increase in magnetic field is restricted in the same way,

$$\frac{B_{t2}}{B_{t1}} \simeq \frac{\rho_1 v_{n1}^2}{\rho_2 v_{n2}^2} = \frac{\rho_2}{\rho_1} < 4,$$

while the downstream pressure may become arbitrarily large,

$$p_2 \propto \left(4 - \frac{\rho_2}{\rho_1}\right)^{-1}.$$

We thus find that, for $\mathsf{M} \gg 1$, most of the upstream flow energy is dissipated,

$$\epsilon \sim \mathsf{M}^3, \tag{9.27}$$

where $\mathsf{M} = \mathsf{M}_s$ or M_A. Hence dissipation by shock waves is expected to play a crucial role in compressible turbulence.

Let us finally mention a phenomenological approach, which has proven to be useful for describing fluid dynamics in molecular clouds in the interstellar medium; see Chapter 12. Since in these gases radiative cooling is more efficient the higher the density, the temperature in the compressed downstream state is rapidly decreased, becoming even lower than that in the upstream state. Hence energy is no longer conserved, invalidating the jump condition (9.15). Instead the pressure in (9.13) is described by a polytropic *Ansatz*, $p = a\rho^\gamma$, where γ satisfies only the condition $\gamma > 0$. Strong radiative cooling is accounted for by a value $\gamma < 1$. Here the increase in density in the downstream state is considerably larger than that for $\gamma > 1$.

9.2 Compressible homogeneous turbulence

As in the incompressible case it is convenient to study the intrinsic properties of supersonic flows in the framework of homogeneous turbulence. Finite compressiblity may affect both the macroscopic behavior, for instance the kinetic-energy decay rate, and the spectral properties reflecting the spatial distribution of the turbulent eddies; we will find, in particular, that both the magnetic field and the density become highly intermittent, in contrast to the smooth, if not uniform, distributions in incompressible turbulence. We first consider neutral fluids, before turning to the MHD properties of supersonic flows in plasmas.

9.2.1 Supersonic hydrodynamic turbulence

There is considerable interest in understanding supersonic turbulence, both in aerodynamics, for instance in the wake of a supersonic airplane, and in astrophysics, for instance in supersonic stellar winds. Highly supersonic velocities are also observed in the interstellar medium, where, in fact, these fast turbulent motions are believed to play an important role by retarding gravitational collapse and star formation, as we will see in Chapter 12. Though in these gases the interaction with magnetic fields is not negligible, hydrodynamic effects seem to dominate the dynamics. Hence we first discuss the properties of compressible hydrodynamic turbulence.

The energy-decay law

For incompressible turbulence a theory of the energy decay has been outlined in Section 4.2.3. From closure-theory arguments, the decay law $E \sim t^{-\lambda}$ is related to the energy spectrum at small k, (4.47), through the principle of permanence of large eddies, which predicts that $\lambda \simeq 1.4$ should be the most probable value in freely decaying turbulence. However, to date neither experiments nor numerical simulations could clearly verify this theory; instead, a lower value, $\lambda \sim 1$, has often been observed.

The dissipation process in supersonic flows is different from that in incompressible turbulence, taking place primarily in shock waves rather than in vorticity filaments. However, this does not necessarily imply the occurrence of a different macroscopic behavior, since the energy-decay rate is found to be essentially independent of the Reynolds number, i.e., of the magnitude of the dissipation coefficient, and also of the character of the dissipation process; use of higher-order diffusion operators seems to yield the same macroscopic properties and so do numerical solutions of the Euler equations, wherein dissipation arises only from numerical discreteness. The decay rate may, however, depend on macroscopic properties such as the initial Mach number and the polytropic index γ when one approximates the pressure by a polytropic law expressed by $p = a\rho^\gamma$.

The simplest model of compressible turbulence is Burgers' turbulence following (9.6), for which an exact decay law can be given. Kida (1979) showed that the exponent λ depends on the invariant integral of the velocity-correlation function $J = \int_0^\infty \langle u(x+l)u(x)\rangle \, dl$; in particular, one has $E \sim t^{-2/3}$ in the general case $J \neq 0$, whereas for $J = 0$ the decay is faster, $E \sim t^{-1}$.

For multidimensional compressible hydrodynamic turbulence no theory has yet been developed; hence the most convenient sources of information are numerical simulations. The most popular numerical methods for compressible hydrodynamics are briefly outlined in Section 9.2.3. The initial phase of decay of compressible turbulence has been studied numerically by Porter *et al.* (1992a, 1992b, 1994, 1998a, 1998b) using the piece-wise parabolic method (PPM) with the highest spatial resolutions attained to date (512^3 grid points; the run with 1024^3 points given in their 1998 papers is too short to allow full relaxation of the turbulence scales). The initial r.m.s. Mach number $\mathsf{M}_s = (E^K)^{1/2}/c_s \simeq 1$ makes the turbulence only weakly supersonic, or transonic, on the mean, though the local Mach number may still reach distinctly supersonic values even after M_s has dropped below unity. The polytropic index is $\gamma = 1.4$, the adiabatic index for a molecular gas. Starting from a smooth state, gradients steepen in the domains of negative velocity divergence to form extended shocks, a process that occurs in a fraction of an acoustic time $\tau_s = L/c_s$. Shock formation is accompanied by a rapid increase of the dissipation of compressive kinetic energy E^c. Vorticity, and hence enstrophy, which gives the dissipation of rotational kinetic energy, builds up more slowly

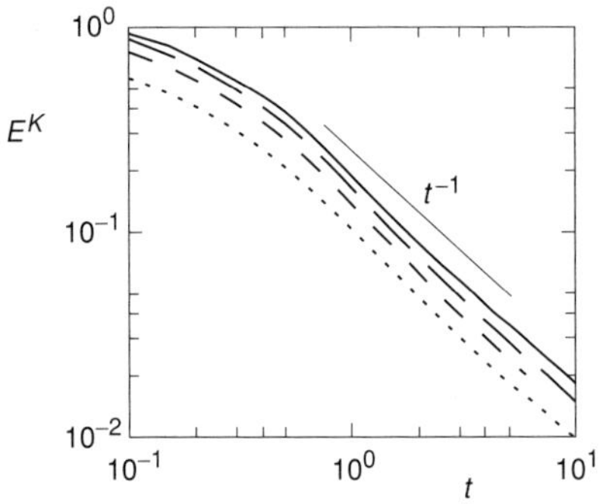

Figure 9.3: Decay of kinetic energy $E^K(t)$ in supersonic turbulence for various Reynolds numbers. (From Mac Low *et al.*, 1998.)

in several acoustic times, during which period the extended shock waves disintegrate into a large number of weaker shocks. The ratio of compressive and solenoidal energies remains about constant, $\chi \sim 0.1$, indicating that a certain equilibrium between compressive and rotational motions is attained. (As numerical simulations of driven, high-Mach-number turbulence show (Boldyrev *et al.*, 2001), this value appears to be rather universal for compressible turbulence.) By the time the enstrophy has reached a maximum, the turbulence is fully developed.

The subsequent decay process proceeds in a self-similar way, which has been studied numerically by Mac Low *et al.* (1998) using the ZEUS code with up to 256^3 grid points, isothermal pressure $\gamma = 1$, corresponding to strong radiative cooling, and initial Mach number $\mathrm{M_s} \simeq 5$. Since the temperature remains relatively low, the Mach number, $\mathrm{M_s} \propto T^{-1/2}$, decays more slowly than it would for an adiabatic pressure law. Figure 9.3 shows that, after the initial phase of several acoustic times, the decay of kinetic energy follows a power law $E^K \sim t^{-\lambda}$ with $\lambda \simeq 1.0$. The result is found to be independent of the Reynolds number, i.e., the numerical grid size, since in the ZEUS code dissipation occurs because of numerical-discreteness effects (the precise numbers for the decay exponent, called η in the paper, for the different runs should not be taken too literally). The decay rate is independent of the Mach number; a run with $\mathrm{M_s} \sim 0.1$ gives a similar decay law, in agreement with earlier results on incompressible turbulence. Hence compressibility effects seem to have little effect on the decay time scale of the turbulence. Only for highly supersonic, or hypersonic, conditions $\mathrm{M_s} \sim 50$, is the decay faster, $E^K \sim t^{-1.5}$ (Smith *et al.*, 2000), which is not surprising because of the very efficient dissipation by high-Mach-number shock waves. Note that the t^{-1} decay of the kinetic energy is also observed in incompressible MHD turbulence,

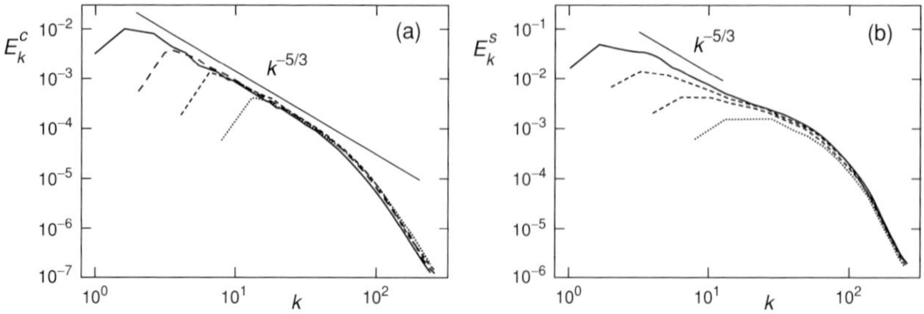

Figure 9.4: Kinetic-energy spectra E_k^c, (a), and E_k^s, (b), from four simulation runs with different Reynolds numbers. The spectra have been shifted horizontally to make the dissipative parts coincide. (From Porter *et al.*, 1994.)

both for finite and for vanishing magnetic helicity; see (4.59) and Fig. 4.3 in Section 4.2.3.

Spectral properties

Let us now consider the small-scale properties of compressible turbulence, *viz.*, Fourier spectra and spatial structures. For Burgers' turbulence the inertial-range energy spectrum $E_k = |u_k|^2$ is obtained analytically, $E_k \sim k^{-2}$ (Kida, 1979). This is just the Fourier spectrum for a single shock, i.e., a velocity discontinuity, which remains valid for a statistical ensemble of shocks. In hydrodynamic turbulence it is useful to distinguish between the compressive and the solenoidal velocity spectram, E_k^c and E_k^s. In the first phase of evolution of supersonic turbulence, which is dominated by extended shocks, one does indeed finds (Porter *et al.*, 1992a)

$$E_k^c \sim k^{-2}, \quad \mathsf{M_s} > 1,$$

which is similar to the spectrum in Burgers' turbulence. At later times, however, when the turbulence has become subsonic and these primary shocks have disintegrated into a myriad of weak shocks, the spectrum flattens, becoming Kolmogorov-like,

$$E_k^c \sim k^{-5/3}, \quad \mathsf{M_s} < 1,$$

as can be seen in Fig. 9.4(a).[6] A similar behavior is found for the solenoidal spectrum:

$$E_k^s \sim k^{-5/3}.$$

[6] This behavior is related to the spectrum of irrotational turbulence corresponding to an ensemble of sound waves, which has been studied by Zakharov and Sagdeev (1970), Kadomtsev and Petviashvili (1973), and Elsässer and Schamel (1976). Elsässer and Schamel found the spectral exponents $\lambda = 2, \frac{7}{4}$, and $\frac{3}{2}$ for spatial dimensions $D = 1, 2$, and 3, respectively.

The plausibility of a Kolmogorov spectrum in compressible turbulence has been discussed by Passot *et al.* (1988). However, the spectrum in Fig. 9.5(b) is modified by a pronounced hump, a bottleneck effect at the transition to the dissipation range (see Section 5.3.6), which is stronger than the corresponding effect found in Navier–Stokes turbulence. This effect seems to be caused by the abrupt onset of numerically induced dissipation at the smallest scales in the PPM method (see Section 9.2.3), rather than representing a special feature of compressible turbulence.[7] It is noteworthy that no bottleneck effect appears in the compressive-mode spectrum in Fig. 9.4(a), which may be due to the more efficient energy drain by shock dissipation. The spectrum of the total velocity, which would be measured experimentally, is dominated by the solenoidal part, since $E^s_k \gg E^c_k$.

A new feature arising in compressible turbulence is the occurrence of *density fluctuations*. While finite pressure fluctuations occur also in incompressible flows, the pressure being tied to the velocity by Poisson's equation $\nabla^2 p = -\rho_0 \nabla \cdot (\boldsymbol{v} \cdot \nabla \boldsymbol{v})$, density variations are negligible. For a polytropic law $p = a\rho^\gamma$ we have

$$\delta p = \frac{\gamma p}{\rho}\,\delta\rho = c_s^2\,\delta\rho. \tag{9.28}$$

Hence $\delta\rho \to 0$ in the incompressible limit $c_s \to \infty$. The spectrum of the pressure fluctuations in an incompressible fluid can readily be derived (Batchelor, 1951; Monin and Yaglom, 1975, Chapter 21.5). Since from Poisson's equation one has the scaling $\delta p \sim \rho\,\delta v^2$, dimensional analysis and use of Kolmogorov's spectrum $E^K_k \sim \epsilon^{2/3}k^{-5/3}$ give the pressure spectrum $E^P_k = \int d\Omega_{\boldsymbol{k}}\,|p_k|^2$ (dimension L^5T^{-4}),

$$\frac{1}{\rho_0^2}E^P_k \sim \epsilon^{4/3}k^{-7/3}. \tag{9.29}$$

Experimental results are in reasonably good agreement with this prediction (George *et al.*, 1984). The pressure spectrum should still be valid in a weakly compressible flow with, say, $\mathrm{M_s} \sim 0.1$. For a polytropic system (9.28) then gives the density spectrum $N^\rho_k = \int d\Omega_{\boldsymbol{k}}\,|\rho_k|^2$,

$$N^\rho_k \sim \epsilon^{4/3}c_s^{-4}k^{-7/3}. \tag{9.30}$$

For nearly incompressible MHD turbulence Montgomery *et al.* (1987) showed by considering a simple quasi-normal model that, for a Kolmogorov

[7] Previously, before the origin of the bottleneck effect had been realized, the spectral hump was interpreted phenomenologically as a local flattening of the spectrum from a $k^{-5/3}$ to a k^{-1} power law.

magnetic-energy spectrum, this spectrum can also be expected for the pressure fluctuations and the density fluctuations, $E_k^P \sim N_k^\rho \sim k^{-5/3}$, using again the linearized relation (9.28).

In supersonic flows, however, large values of the density contrast $(\rho_{\max} - \rho_{\min})/\langle\rho\rangle$ occur. Here the linearized relation (9.28) is no longer applicable, in that the density spectrum does not simply follow the pressure spectrum (whatever the latter is for highly compressible turbulence). Moreover, because of the strong asymmetry of these density fluctuations about the mean value, it is preferable to consider the statistical properties of the variable $\ln \rho$ instead of ρ. Porter *et al.* (1998a, 1998b), studying forced transonic turbulence, found a Kolmogorov density spectrum, but higher-order structure functions show that $\ln \rho$ is more strongly intermittent than the velocity. In highly supersonic turbulence the density exhibits a very spotty structure (see Fig. 9.6). In this case the p.d.f. gives a more intuitive description of the spatial structure of the density.

Spatial structures

While the Fourier spectrum provides a measure of the amplitude of the perturbation at a scale $l \sim k^{-1}$, the turbulence dynamics depends also on the spatial structure of the fluctuations. At small scales the compressive velocity gradient, the dilatation, is mainly localized in shock waves, which have a sheet-like structure. When shock fronts intersect, they also generate sheets of high-velocity shear, i.e., vorticity, which subsequently become Kelvin–Helmholtz unstable, breaking up into filaments. Hence the solenoidal component forms vorticity filaments much as in incompressible flows. This process of vorticity-sheet generation by intersecting shocks does not, however, mean that compressive motions directly generate solenoidal motions. The coupling between compressive and solenoidal components is, indeed, rather weak, the main reason being the disparity of time scales, which can be grasped intuitively by viewing an animated record of the turbulence dynamics in a simulation run (Porter *et al.*, 1992b), in which shock waves are seen to propagate rapidly across almost stationary vorticity eddies. The rotational motion present in the simulations results primarily from the initial conditions or, in driven turbulence, from the rotational part of the external force.

In spite of the presence of shock waves, the spatial distribution of the velocity structures in supersonic flows is rather similar to that in incompressible turbulence. The main reason is that, with $\chi \sim 0.1$, the velocity is still primarily solenoidal. For comparison with the known intermittency models the energy dissipation ϵ_l, (7.23), has been evaluated for the PPM runs discussed above (Porter *et al.*, 1999). The numerical values of the scaling exponents μ_n are found to be lower than those predicted by the She–Lévêque model (7.63). An alternative model, which may explain this discrepancy, has recently been proposed by Boldyrev (2001). The idea is that, while the inertial-range turbulence dynamics is Kolmogorov-like, implying that $x = \frac{2}{3}$ in the general formula of the log-Poisson model (7.61), the dissipation is dominated by shock waves, which have a sheet-like structure, instead of vorticity filaments; hence $C_0 = 1$ and therefore $\beta = \frac{1}{3}$,

$$\mu_n = -\frac{2}{3}n + 1 - \left(\frac{1}{3}\right)^n,$$

the same expression as that for μ_n^{MHD}, (7.65), which has been proposed for MHD turbulence. The scaling exponents of the velocity structure functions are then obtained by use of the refined similarity hypothesis (7.24),

$$\zeta_n = \frac{n}{9} + 1 - \left(\frac{1}{3}\right)^{n/3}.$$

Results from recent numerical studies of driven high-Mach-number ($\mathsf{M}_s \sim 10$) isothermal turbulence of rather high resolution confirm the validity of this model, at least for $n \lesssim 5$ (Boldyrev *et al.*, 2001). (It also seems that shocks, the dissipative elements in supersonic turbulence, which are more extended than the micro-current sheets in MHD turbulence, exhibit a rather complicated topological structure with a fractal dimension $D > 2$. Indeed, the choice of $D = 2.3$, and hence $C_0 = 0.7$, yields even better agreement with the numerical results.)

The p.d.f. of the vorticity components in supersonic turbulence is similar to that in incompressible turbulence, exhibiting symmetric 'stretched' exponential tails. The p.d.f. of the dilatation $\nabla \cdot \boldsymbol{v}$, however, is not symmetric but rather is skewed with an exponential tail extending to large negative values corresponding to shock waves. To obtain an idea of the spatial structures in supersonic turbulence, see the very impressive and instructive color frames of the vorticity and the dilatation, both in 2D and in 3D, presented in the papers by Porter *et al.* (1992b, 1994), who also illustrate in detail the dynamics of shock intersection and vorticity-sheet generation. The spatial distribution of shock waves in hypersonic turbulence, $\mathsf{M}_s \gg 1$, has been studied by Smith *et al.* (2000).

The density p.d.f.

The probability distribution of the density in supersonic turbulence, i.e., the fraction of the volume occupied by density of a particular value, is an interesting quantity, as it allows direct comparison with the distribution of the mass density observed, for instance, in the interstellar medium. It depends mainly on the equation of state, which for a polytropic pressure is determined by the polytropic index γ. Since the condition $\rho > 0$ makes the density fluctuations asymmetric about the mean value, it is natural to consider the statistics of $\ln \rho$. (In numerical studies the equations are usually written in terms of $\ln \rho$ in order to guarantee the positivity of ρ.) The isothermal case $\gamma = 1$ plays a special role, since in this case the hydrodynamic equations are invariant with respect to the transformation $\ln \rho \to \ln \rho + \ln c$, or $\rho \to c\rho$, where c is an arbitrary constant, i.e., do not depend on the mean value of the density. The pressure force in the momentum equation becomes $\rho^{-1} \nabla p = c_s^2 \nabla \ln \rho$ with $c_s =$ constant. (This is no longer true in an MHD fluid, since the Lorentz-force term destroys the exact invariance.) Consider the sequence $\rho_0 < \rho_1 < \cdots < \rho_n = \rho$ and write ρ/ρ_0 as the product of the fractions $\chi_j = \rho_j/\rho_{j-1}$,

$$\rho/\rho_0 = \widehat{\rho} = \chi_1 \chi_2 \cdots \chi_n,$$

which becomes a sum on taking the logarithm, $\ln \widehat{\rho} = s = \ln \chi_1 + \cdots + \ln \chi_n = s_1 + \cdots + s_n$. Because of the invariance property of the equations all individual summands s_j should be equivalent, exhibiting the same probability distribution. Hence, by application of the central-limit theorem, s, being the sum of a large number of the identically distributed variables s_j, is found to have a Gaussian, or normal, distribution, i.e., $\widehat{\rho}$ has a log-normal distribution (Scalo *et al.*, 1998; Vázquez-Semadini and Passot, 1999; Nordlund and Padoan, 1999),

$$p(\rho) = \frac{1}{\rho} \frac{1}{\sqrt{2\pi\sigma^2}} \exp\left(- \frac{(\ln \rho - s_0)^2}{2\sigma^2} \right), \tag{9.31}$$

using the relation $p(\rho)\, d\rho = p(\ln \rho)\, d \ln \rho$. The mean value s_0 is related to the variance, $s_0 = -\sigma^2/2$, which results from the condition $\langle \widehat{\rho} \rangle = 1$, which applies on choosing $\rho_0 = \langle \rho \rangle$. The derivation of (9.31) is similar to that of the log-normal intermittency model, Section 7.4.1, but, while the latter is based on physical assumptions and is, at most, approximately valid, the log-normal character of ρ, being based on an invariance property of the equations, should be exactly valid in isothermal turbulence. As we have seen in the discussion of the intermittency of the energy-dissipation rate in Section 7.4.1, the log-normal distribution implies that the density is sparsely distributed in space, occupying only a small fraction of the volume (a small 'volume-filling factor'). The log-normal character of the density distribution has been verified numerically by Scalo *et al.* in 2D simulations and by Nordlund and Padoan in 3D simulations, in which the width of the distribution function is found to increase with Mach number, $\sigma \propto \mathsf{M}_s$.

For $\gamma \neq 1$ deviations from the log-normal distribution are expected. The density distribution becomes skewed, with a tail developing at large argument

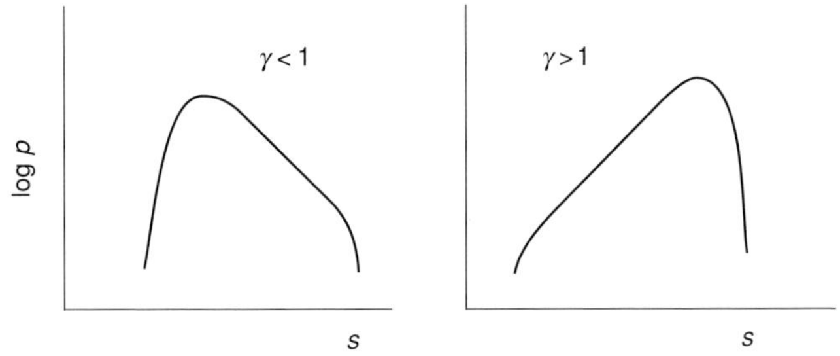

Figure 9.5: Logarithmic plots of the p.d.f. $p(s)$, $s = \ln \rho$, for $\gamma < 1$ and $\gamma > 1$ (shown schematically).

$s > s_0$ for $\gamma < 1$ and at small argument $s < s_0$ for $\gamma > 1$; see Fig. 9.5. The side to which the tail develops can readily be understood. For $\gamma > 1$ the temperature increases with increasing density; hence the pressure force in the high-density regions counteracts further compression, whereas for $\gamma < 1$, for which the temperature is reduced in compressed regions, the repulsive force is much weaker, allowing large values of the density to be reached. For high Reynolds numbers and Mach numbers the tail in the distribution function assumes a power-law character, such that the exponents depend on the degree of nonisothermality $\gamma - 1$ (Scalo *et al.*, 1998). In the limit $\gamma \to 0$, at which the pressure gradient vanishes, the hydrodynamic equations formally degenerate to Burgers' equation together with the density equation, which is now passive. The density p.d.f. obtained in 1D simulations of Burgers' equation by Gotoh and Kraichnan (1993) exhibits a clear power-law spectrum in the direction of increasing densities as expected for $\gamma < 1$ (Burgers' equation corresponds formally to $\gamma = 0$). Also results from numerical studies of a 1D hydrodynamic model of supersonic turbulence by Passot and Vázquez-Semadeni (1998) corroborate these properties of the density p.d.f., of which the large downstream values of the density at shocks for $\gamma < 1$ are noteworthy. Small values of γ are relevant for modeling the radiative cooling in dense molecular clouds in the interstellar medium (Chapter 12), where the high densities in star-forming cores are consistent with strong local density compression by supersonic flows.

9.2.2 Supersonic MHD turbulence

Since in a plasma the interaction with the magnetic field cannot be neglected, we have to consider also the induction equation (2.15),

$$(\partial_t + \boldsymbol{v} \cdot \nabla)\boldsymbol{B} = \boldsymbol{B} \cdot \nabla \boldsymbol{v} - \boldsymbol{B} \nabla \cdot \boldsymbol{v},$$

which shows that $\boldsymbol{B}$ changes due to field-line stretching and to compression. The equation can be written in a more compact form in terms of $\boldsymbol{B}_\rho = \boldsymbol{B}/\rho$, in analogy to (9.4) for the potential vorticity,

$$\frac{d}{dt}\boldsymbol{B}_\rho = \boldsymbol{B}_\rho \cdot \nabla \boldsymbol{v}, \tag{9.32}$$

incorporating the compression effect into the density factor, which can also be written in the form

$$\partial_t \boldsymbol{B}_\rho = \nabla \times (\boldsymbol{v}^c \times \boldsymbol{B}_\rho).$$

The strength of the magnetic field is measured in terms of the magnetic Mach number $\mathsf{M}_\mathrm{A} = v/v_\mathrm{A}$. If the field is relatively strong, $\mathsf{M}_\mathrm{A} \lesssim 1$, the perpendicular motions become essentially incompressible. In this case shock waves can be excited only along the field, but these constitute, nonetheless, the most important energy-dissipation process in supersonic flows, $\mathsf{M}_\mathrm{s} > 1$. In the super-Alfvénic case, $\mathsf{M}_\mathrm{A} > 1$, the tubulence can be isotropic.

Numerous studies of supersonic super-Alfvénic turbulence have been performed, mostly in the context of turbulence in molecular clouds, to which we shall return in Chapter 12. At this point we restrict the discussion to the basic properties of homogeneous supersonic MHD turbulence; magnetoconvection in a stratified configuration will be treated in Section 9.3.4. We consider, in particular, the three-dimensional simulations by Padoan and Nordlund (1999). The global properties of MHD turbulence such as the energy decay are found to be similar to those of supersonic hydrodynamic turbulence, especially in the super-Alfvénic case, in which the magnetic field is basically moved along with the plasma, since $E^K/E^M \sim M_\mathrm{A}^2 \gg 1$. Similarly to the density, the magnetic field is compressed in thin flux tubes, which leaves large fractions of space almost field-free, i.e., both the magnetic field and density are very intermittently distributed, with the p.d.f. exhibiting long exponential tails. As indicated by (9.32), density and magnetic field are strongly correlated, exhibiting the same compressive properties. Equation (9.32) might suggest that, due to the field-line-stretching term on the r.h.s., B_ρ can locally reach large values, implying that B is more strongly compressed than ρ is. However, that argument ignores the parallel dynamics. While B is compressed only in the perpendicular plane, density compression can occur in all three dimensions. Conservation of flux gives $Bl^2 =$ constant, while conservation of mass gives $\rho l^3 =$ constant; hence $B \sim \rho^{2/3}$. Numerical simulations yield a still smaller exponent,

$$B \sim \rho^{0.4}, \tag{9.33}$$

The value of the exponent is not far from the 0.5 observed for the interstellar medium; see Section 12.2.3. Strongly intermittent magnetic-field structures

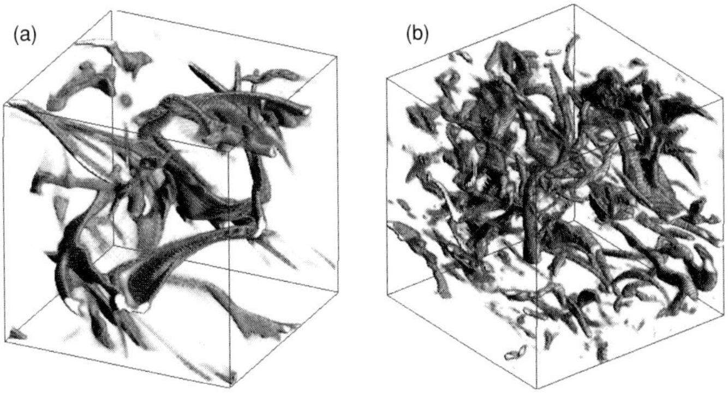

Figure 9.6: Density isosurfaces showing the highly intermittent distribution in supersonic turbulence with $\mathsf{M}_A \gg 1$, (a), and $\mathsf{M}_A \sim 1$, (b). In the latter the alignment of the density filaments with the mean magnetic field can be seen. (From Padoan and Nordlund, 1999.)

are a characteristic feature in magnetoconvection, to which we will return in Section 9.3.4.

9.2.3 Numerical methods in compressible hydrodynamics

In view of the importance of numerical simulations in turbulence theory, some remarks about numerical methods would seem to be relevant. In numerical studies of incompressible homogeneous turbulence spectral, or, more often, pseudospectral, methods are used (for a general introduction, see Canuto *et al.*, 1988). The advantages are clear: Poisson's equation for computing the pressure, (2.29), or the velocity in the vorticity formulation, (2.30), becomes algebraic in Fourier space, and so do the diffusion terms (of arbitrary order), which can even be advanced exactly. Periodic boundary conditions, the natural choice for homogeneous turbulence, are automatically satisfied. The most intricate numerical part is constituted by the convolutions arising from the nonlinear terms, which are solved by transformation to configuration space, multiplication and transformation back to Fourier space using the algorithm of the fast Fourier transform (FFT). Pseudospectral codes are remarkably simple, especially when one applies the two-thirds rule for removing aliasing errors. The central part of such a code is an efficient FFT solver, which for present-day massively parallel computers depends strongly on the specific machine architecture and is usually provided by the computer company.

In compressible hydrodynamics no elliptic equations have to be solved, but the quasi-discontinuities corresponding to shock waves are difficult to deal with in a finite Fourier representation because of the Gibbs phenomenon. While there have been some numerical studies applying spectral methods also to compressible MHD flows (Passot and Pouquet, 1987; Dahlburg and Picone, 1989; Kida and Orszag, 1992; for a review see Ghosh *et al.*, 1993), most numerical treatments use a finite-difference approach in configuration space. Since here the numerical algorithm is local, it is intrinsically better suited for parallel computers than are the nonlocal spectral methods.

There appear to be mainly two schools (for a general introduction to computational methods in astrophysical fluid dynamics, see, for instance, LeVeque *et al.*, 1998). The first uses relatively straightforward methods on an Eulerian grid, based, for instance, on the Lax–Wendroff or Runge–Kutta time-advancement schemes, which are simple to code and allow one to add further physical effects without great efforts. However, though they are formally of high-order accuracy, these schemes are rather dispersive. To prevent unphysical oscillations at steep gradients, some form of artificial viscosity is introduced, which, though broadening the thickness of the shock front, does not spoil the accuracy in the smooth parts of the solution. The second school is based on Godunov's method (see Richtmyer and Morton, 1967), which advances the solution along the characteristics of the system and is hence particularly well suited to deal with discontinuities of flow. This is basically a Lagrangian approach. To improve the accuracy of the Godunov method, which is only of first order, sophisticated interpolation schemes have been devised in order to smooth the solution between grid points and simultaneously satisfy the requirement for monotonicity, i.e., avoid introduction of oscillations at sharp gradients. These codes tend to be rather complex, especially in MHD with its host of different characteristics.

In general, the ideal-fluid equations (often called Euler's equations also in MHD) are solved. The idea is that, in astrophysical systems, collisional viscosities and resistivities are extremely small, corresponding to 'astronomical' values of the Reynolds numbers, so that it may be reasonable to omit dissipation terms altogether and rely only on numerically induced dissipation, for instance when magnetic reconnection is important. In contrast to incompressible turbulence, for which explicit sinks of kinetic and magnetic energy are required – the ideal system tends toward an absolute equilibrium distribution –, compressible hydrodynamics automatically accounts for heating by solving the energy equation. Numerically one has to make sure that the total energy is conserved and that the Rankine–Hugoniot conditions are satisfied at discontinuities. The tacit assumption is that macroscopic flow properties are independent of the dissipation processes, for instance that the magnetic reconnection rate does not depend on the value of the resistivity, which might not be strictly true. Also the spectra can be affected by the abrupt onset of grid-induced dissipation, exhibiting a pronounced bottleneck effect; see Fig. 9.4(b).

Let me briefly mention some codes that are being applied in numerical studies of compressive turbulence. The best-known code of the first type is the ZEUS code developed by Stone and Norman and documented in a series of papers (Stone and Norman, 1992a–c), which is widely used for fluid-dynamic problems in astrophysics. The code can deal with general MHD flows including radiation transport. Its basic algorithm has also been used in a number of different code developments (e.g., the MOCCT method devised by Hawley and Stone, 1995) with special emphasis on treating the condition $\nabla \cdot \boldsymbol{B} = 0$. Of a similar character to ZEUS is the code developed by Nordlund *et al.* (1992) for astrophysical MHD and radiation transport, primarily to study solar convection.

The prototypical code using a higher-order extension of the Godunov method is the piecewise parabolic method (PPM) introduced by Colella and Woodward (1984), which was mainly applied to compressible hydrodynamic turbulence, both in 2D and in 3D. Subsequently an MHD version has been developed (Dai and Woodward, 1994) and applied to several 2D systems (Dai and Woodward, 1998). More recently several different numerical approaches to the MHD problem based on the Godunov method have been written (e.g., Balsara, 2001).

9.3 Turbulent convection

Convection in a stratified medium is a very common source of turbulence. In most cases convection is driven thermally by heating a layer from below and cooling it from above, though there may also be different sources such as the gradient of chemical composition, which is believed to be responsible for the convection in the Earth's liquid core. The mechanism driving convection, namely the Rayleigh–Taylor instability discussed in Section 3.2.2, sets in when the temperature gradient becomes superadiabatic, (3.50). In this section we now consider the nonlinear dynamics resulting from this instability which, for sufficiently high Rayleigh number, leads to fully developed turbulence.

9.3.1 Turbulence in a Boussinesq fluid

For terrestrial conditions the typical turbulent eddy size l is small compared with the equilibrium scale height $L_g = p/(g\rho)$, (2.85), both in laboratory experiments, due to the limited size of the system L, $l \leq L \ll L_g$, and in the atmosphere because of the finite vertical extent of the unstable region. Moreover, convective motions are slow compared with the speed of sound, since the maximum energy a fluid element may gain from buoyancy in the gravitational field is

$$v^2 \sim gL \sim (L/L_g)c_s^2 \ll c_s^2,$$

i.e., the motion is incompressible, and we can apply the Boussinesq equations (3.54)–(3.56), where the instability condition is simply $dT_0/dz < 0$ or $d\rho_0/dz > 0$ (heavy fluid on top of light fluid). In this section we consider the simplest case of a nonrotating neutral fluid, leaving the discussion of magnetic effects to Section 9.3.4 and the discussion of rotation to Section 11.4. We are hence dealing with two dynamic variables, the velocity $\boldsymbol{v}$ and the temperature fluctuation T, obeying equations (2.37) and (2.43),

$$\partial_t \boldsymbol{v} + \boldsymbol{v} \cdot \nabla \boldsymbol{v} = -\frac{1}{\rho_0} \nabla p + \alpha_\rho g T \boldsymbol{e}_z + \nu \nabla^2 \boldsymbol{v}, \quad \nabla \cdot \boldsymbol{v} = 0, \tag{9.34}$$

$$\partial_t T + \boldsymbol{v} \cdot \nabla T = -v_z T_0' + \kappa \nabla^2 T, \tag{9.35}$$

where we have used (2.42) and assumed that the force due to gravity is exerted in the $-z$-direction, $\boldsymbol{e}_g = -\boldsymbol{e}_z$. The first term on the r.h.s. of (9.35), namely $-v_z T_0'$, represents the effect of the mean temperature gradient. It acts like an external force on the temperature fluctuations T, which would otherwise decay whatever the velocity. The dynamics, in turn, is driven by the buoyancy force, the α_ρ-term on the r.h.s. of (9.34). Since in the Boussinesq approximation the equations do not contain an explicit space dependence, we can assume the turbulence to be homogeneous, at least for open boundary conditions, though it is, in general, not isotropic. There is a net heat flux $\langle v_z T \rangle$ opposite to the temperature gradient, which is usually measured in terms of the Nusselt number Nu, the ratio of convective and diffusive heat fluxes, $\mathsf{Nu} = |\langle v_z T \rangle / (\kappa T_0')|$. From (9.34) one obtains the equation for the kinetic energy:

$$\frac{1}{2} \frac{d}{dt} \langle v^2 \rangle = \alpha_\rho g \langle v_z T \rangle - \nu \langle \omega^2 \rangle, \tag{9.36}$$

while (9.35) gives the equation for the mean-square temperature fluctuation,

$$\frac{1}{2} \frac{d}{dt} \langle T^2 \rangle = -T_0' \langle v_z T \rangle - \kappa \langle (\nabla T)^2 \rangle. \tag{9.37}$$

Hence, to maintain a stationary level of turbulence, the heat flux must be opposite to the mean temperature gradient which, in turn, must be opposite to the gravitational force.

Let us now consider the scaling properties of convective turbulence described by (9.34) and (9.35). We first derive the velocity and temperature spectra using a Kolmogorov-type argument. Since in the inertial range the spectra are independent of the dissipation coefficients, they can depend only on the spectral transfer rates ϵ_V and ϵ_T of the kinetic energy and the mean-square temperature fluctuation, respectively, and on the buoyancy parameter $\alpha_\rho g$. The presence of the latter introduces the length scale $L_{\rm B}$,

$$L_{\rm B} = \frac{\epsilon_V^{5/4}}{\epsilon_T^{3/4}(\alpha_\rho g)^{3/2}}, \tag{9.38}$$

called the *Bolgiano length* (Bolgiano, 1959), which is the only combination of the three quantities with the dimension of a length. Dimensional arguments lead to the following expressions for the spectra E_k^K and $E_k^T = \int d\Omega_{\boldsymbol{k}}\, |T_k|^2$:

$$E_k^K \sim \epsilon_V^{2/3} k^{-5/3} \phi(kL_{\rm B}), \quad E_k^T \sim \epsilon_T \epsilon_V^{-1/3} k^{-5/3} \psi(kL_{\rm B}), \tag{9.39}$$

where ϕ and ψ are arbitrary dimensionless functions. The form of the temperature spectrum results from the requirement that E_k^T should be proportional to the temperature transfer rate ϵ_T, since the basic equation (9.37) is linear in T.

Since the buoyancy force contains no spatial derivatives, it acts primarily on large scales, being negligible at small scales, such that, for high wavenumbers $kL_{\rm B} > 1$, the temperature fluctuation becomes effectively a passive scalar. In this range the turbulence therefore does not depend on the buoyancy parameter $\alpha_\rho g$, i.e., on $L_{\rm B}$, such that E_k^K reduces to the Kolmogorov spectrum and E_k^T follows the same power law,

$$E_k^K = C_{\rm K} \epsilon_V^{2/3} k^{-5/3}, \quad E_k^T = C_T \epsilon_T \epsilon_V^{-1/3} k^{-5/3}, \quad kL_{\rm B} \gg 1, \tag{9.40}$$

i.e., we have $\phi(\infty) = \psi(\infty) = 1$. At large scales, by contrast, the buoyancy effect determines the dynamics. Here the energy received locally from this force exceeds that received through the spectral tranfer ϵ_V. Hence, for $kL_{\rm B} \ll 1$, the spectrum should be independent of ϵ_V, resulting in the *Bolgiano spectrum* (Bolgiano 1959 and 1962)[8]

$$E_k^K = C\epsilon_T^{2/5}(\alpha_\rho g)^{4/5} k^{-11/5}, \quad E_k^T = C' \epsilon_T^{4/5}(\alpha_\rho g)^{-2/5} k^{-7/5}, \quad kL_{\rm B} \ll 1, \tag{9.41}$$

i.e., in this regime one has $\phi(kL_{\rm B}) \sim (kL_{\rm B})^{-8/15}$ and $\psi(kL_{\rm B}) \sim (kL_{\rm B})^{4/15}$. These spectral power laws can also be derived from the scaling relations between the increments δv_l and δT_l obtained from the balance of the advection term and the buoyancy force in (9.34), $(\delta v_l)^2/l \sim \alpha_\rho g\, \delta T_l$, and from Yaglom's four-thirds law (7.37), $\delta v_l\, (\delta T_l)^2 \sim \epsilon_T l$, which is valid for any scalar field, passive or not. Combining these two relations gives

$$\delta v_l \sim \epsilon_T^{1/5}(\alpha_\rho g)^{2/5} l^{3/5}, \quad \delta T_l \sim \epsilon_T^{2/5}(\alpha_\rho g)^{-1/5} l^{1/5}, \tag{9.42}$$

called the *Bolgiano scaling*, from which the spectra (9.41) follow immediately.

[8] Bolgiano assumed that the medium is stably stratified. The result is, however, valid also for unstable stratification; see L'vov (1991).

Most laboratory experiments on thermal convection are performed in a closed vessel. Here turbulent transport is dominated by the mean flow along the strongly inhomogeneous boundary layer (for a general review see Siggia, 1994). Some distance away from the walls, however, one finds conditions under which mean gradients are small and turbulence can be considered homogeneous. Observations show that both regimes, the Kolmogorov spectrum for $kL_{\mathrm{B}} > 1$ and the Bolgiano spectrum for $kL_{\mathrm{B}} < 1$, may be present (Wu *et al.*, 1990; Chilla *et al.*, 1993; Ching, 2000). The Bolgiano length L_{B}, which in the form (9.38) is difficult to measure experimentally, can also be expressed in terms of integral quantities (Chilla *et al.*, 1993),

$$L_{\mathrm{B}} = \frac{\mathsf{Nu}^{1/2}}{(\mathsf{RaPr})^{1/4}} L, \tag{9.43}$$

where L is the height of the vessel. Since for large Rayleigh number one finds the scaling $\mathsf{Nu} \sim \mathsf{Ra}^{2/7}$ (Castaing *et al.*, 1989), the Bolgiano length should decrease with increasing Ra, $L_{\mathrm{B}} \sim \mathsf{Ra}^{3/28}$; hence the spectral range of Bolgiano scaling should increase. On the other hand, the dependence of L_{B} on the Prandtl number indicates that, in a low-Prandtl-number fluid, the Bolgiano scaling range should be reduced in favor of the Kolmogorov scaling. This is indeed observed in experiments using mercury with $\mathsf{Pr} = 0.02$ (Cioni *et al.*, 1995).

Temperature fluctuations in the atmosphere tend to exhibit a $k^{-5/3}$ rather than a $k^{-7/5}$ spectrum, since the turbulence is often driven by wind, i.e., shear flow, rather than by buoyancy, so that the temperature behaves as a passive scalar.

Two-dimensional convection arises in the presence of a strong magnetic field or for fast rotation, where the dynamics is restricted to the plane perpendicular to $\boldsymbol{B}$ or the rotation vector $\boldsymbol{\Omega}$. Since in 2D no Kolmogorov-type direct energy cascade exists, ϵ_V is small, and so is L_{B}, so a Bolgiano scaling is expected across the entire inertial range. This is indeed seen in numerical simulations of 2D Boussinesq convection with open boundary conditions (Biskamp *et al.*, 2001). The temperature fluctuations, however, are highly intermittent, as can be seen in Fig. 9.7, which shows the scaling exponents ζ_n^T, ζ_n^V, and z_n of the structure functions $\langle|\delta T_l|^n\rangle \sim l^{\zeta_n^T}$, $\langle|\delta v_l|^n\rangle \sim l^{\zeta_n^V}$, and $\langle|\delta v_l\, \delta T_l^2|^{n/3}\rangle \sim l^{z_n}$. While the velocity is only weakly intermittent, ζ_n^V being almost linear, ζ_n^T exhibits a highly anomalous scaling indicating strong intermittency. This reflects the ramp-and-cliff structure of the turbulent temperature field typical of a passive scalar, which will be discussed in the subsequent section; see Fig. 9.8. It is noteworthy that the scaling exponents z_n of the mixed structure functions are found to follow the same intermittency law as do the structure functions of the velocity in 3D hydrodynamic turbulence, which is well described by the She–Lévêque

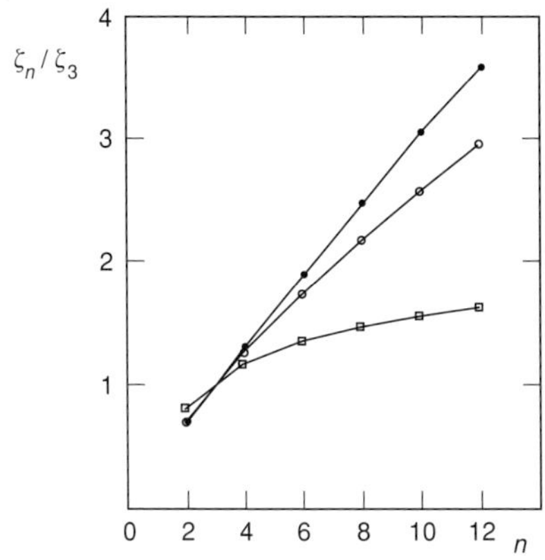

Figure 9.7: Relative scaling exponents ζ_n^T/ζ_3^T (squares), ζ_n^V/ζ_3^V (full dots), and z_n/z_3 (open dots) of the structure functions $\langle|\delta T_l|^n\rangle$, $\langle|\delta v_l|^n\rangle$, and $\langle|\delta v_l\, \delta T_l^2|^{n/3}\rangle$, respectively, obtained from 2D numerical simulations of Boussinesq turbulent convection. (From Biskamp *et al.*, 2001.)

formula (7.64). Remember that, in both cases, the third-order exponents equal unity exactly by virtue of Kolmogorov's four-fifths law and Yaglom's four-thirds law.

9.3.2 Passive scalar turbulence

We have seen that, for $kL_B > 1$, the buoyancy force is negligible, such that the temperature fluctuation no longer reacts on the dynamics but rather is swept along passively. Hence the temperature equation (9.35) is linear with a prescribed velocity field $\boldsymbol{v}(\boldsymbol{x}, t)$,

$$\partial_t \theta + \boldsymbol{v} \cdot \nabla \theta = \kappa \, \nabla^2 \theta + f, \tag{9.44}$$

where in nondimensional form κ is the inverse Péclet number (3.37), and f is an external force. We write θ instead of T to indicate that this may be any passive scalar field, not only the temperature, but also, for instance, the concentration of a pollutant emitted into the atmosphere, or a dye added to a turbulent liquid such as a drop of milk in a cup of coffee stirred with a spoon. Because of the relative simplicity and the various applications passive scalar turbulence has attracted much interest (for recent reviews, see Warhaft, 2000; Shraiman and Siggia, 2000).

The formal linearity of the equation is, however, deceptive, since our interest is in the *statistics* of the scalar field in terms of the statistics of the velocity field, which is again a nonlinear problem, leading, in particular, to a similar hierarchy of moments to that as in nonlinear fluid turbulence. The problem is, nevertheless, made more easily accessible to a theoretical treatment by choosing the velocity suitably, which has led to exactly, or quasi-exactly, solvable systems, the most prominent being Kraichnan's model (Kraichnan, 1994). A detailed discussion of these model systems would, however, lead too far. Here we concentrate on passive scalar turbulence driven by the velocity in an incompressible turbulent neutral fluid.

The scalar power spectrum can be obtained by simple phenomonological arguments. The external forcing should be restricted to the longest scales; think, for instance, of an externally imposed mean temperature gradient, such that at smaller scales the scalar follows the homogeneous equation

$$\partial_t \theta + \boldsymbol{v} \cdot \nabla \theta = \kappa \, \nabla^2 \theta. \tag{9.45}$$

The dissipation rate of the scalar 'energy' $\langle \theta^2 \rangle$,

$$-\frac{d}{dt} \langle \theta^2 \rangle = \kappa \langle (\nabla \theta)^2 \rangle := \epsilon_\theta, \tag{9.46}$$

is also the spectral transfer rate. The linearity of (9.45) requires that the scalar spectrum $E_k^\theta = \int d\Omega_{\boldsymbol{k}} \, |\theta_{\boldsymbol{k}}|^2$ is proportional to ϵ_θ. Then in the *inertial–convective subrange*, where neither viscosity nor diffusion is important, dimensional analysis gives the spectrum

$$E_k^\theta = C_\theta \epsilon_\theta \epsilon_V^{-1/3} k^{-5/3}, \quad k < \min\{l_{\mathrm{K}}^{-1}, l_\theta^{-1}\}, \tag{9.47}$$

which is, for instance, the temperature spectrum (9.39) in the regime where buoyancy is negligible. The $k^{-5/3}$ temperature spectrum has been observed experimentally in turbulence of sufficiently high Reynolds number, mainly in grid-generated wind-tunnel turbulence (e.g., Sreenivasan, 1996) and in atmospheric turbulence (e.g., Monin and Yaglom, 1975). The parameter C_θ, called the Obukhov–Corrsin constant, is found in the range $C_\theta \simeq 0.45$–0.55 (Mydlarski and Warhaft, 1998). It should, however, be mentioned that, in such real-world passive-scalar turbulence, which is always driven in an anisotropic way, the anisotropy is not wiped out by turbulent mixing but persists down to the smallest scales (see, e.g., Warhaft, 2000), which must be considered when one is discussing the spectrum in detail. This anisotropy is also found in numerical simulations when turbulence is driven by an average temperature gradient (Celani *et al.*, 2000).

The diffusion scale l_θ is determined by the balance of advection and diffusion in (9.45), $\delta v_l \, \delta\theta_l / l \sim \kappa \, \delta\theta_l / l^2$. If $\mathrm{Pr} = \nu/\kappa < 1$,[9] we have $l_\theta > l_{\mathrm{K}}$, where

[9] This ratio is, strictly speaking, called the Schmidt number for a general scalar field, while the name Prandtl number is reserved for the case of temperature, when κ is the thermal diffusivity.

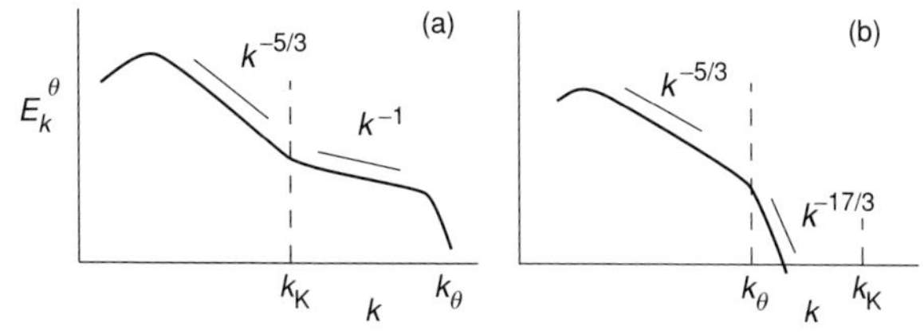

Figure 9.8: Energy spectra of passive scalar turbulence, for (a) $\mathsf{Pr} > 1$ and (b) $\mathsf{Pr} < 1$ (shown schematically).

$l_K = (\nu^3/\epsilon_V)^{1/4} = k_K^{-1}$ is the Kolmogorov microscale, such that l_θ lies in the inertial range of the velocity field, where $\delta v_l \sim \epsilon_V^{1/3} l^{1/3}$; hence we find

$$l_\theta = (\kappa^3/\epsilon_V)^{1/4} = l_K \mathsf{Pr}^{-3/4}, \quad \mathsf{Pr} < 1. \tag{9.48}$$

For $l_\theta^{-1} < k < l_K^{-1}$, the *inertial–diffusive subrange*, the scalar energy spectrum falls off steeply but still follows a power law. In this subrange, where the scalar modes are no longer governed by spectral transfer, E_k^θ is driven locally by the velocity; hence E_k^θ should be proportional to the velocity spectrum E_k^K, in particular to $\epsilon_V^{2/3}$, and independent of the viscosity. Since in general $E_k^\theta \sim \epsilon_\theta \epsilon_V^{-1/3} k^{-5/3} f(kl_\theta, \mathsf{Pr})$, the scaling *Ansatz* $f = (kl_\theta)^\alpha \mathsf{Pr}^\beta$ gives $\alpha = -4$ and $\beta = 0$; hence

$$E_k^\theta \sim \epsilon_\theta \epsilon_V^{2/3} \kappa^{-3} k^{-17/3}, \quad l_\theta^{-1} < k < l_K^{-1}, \tag{9.49}$$

the inertial–diffusive-range spectrum first proposed by Batchelor *et al.* (1959).

In the opposite case, $\mathsf{Pr} > 1$, we have the *viscous–convective subrange* $l_K^{-1} < k < l_\theta^{-1}$, where the scalar energy spectrum follows a power law, which is flatter than $k^{-5/3}$. To determine this spectrum, visualize the dynamics in this regime. The smallest velocity structures have the scale l_K, producing the vorticity $\delta v_{l_K}/l_K = (\epsilon_V/\nu)^{1/2}$. A fluid element of smaller scale is therefore strained by a uniform shear flow $(\epsilon_V/\nu)^{1/2}$, such that the scalar transfer rate is proportional to this strain,

$$\epsilon_\theta \sim k E_k^\theta \left(\frac{\epsilon_V}{\nu}\right)^{1/2}, \tag{9.50}$$

whence

$$E_k^\theta \sim \epsilon_\theta \left(\frac{\nu}{\epsilon_V}\right)^{1/2} k^{-1}, \quad l_K^{-1} < k < l_\theta^{-1}, \tag{9.51}$$

the viscous–convective subrange spectrum (Batchelor, 1959). In this case the diffusion length l_θ is, however, larger than would be obtained by using the

Figure 9.9: Grayscale plots of passive scalar turbulence from a 2D numerical simulation, showing the typical cliff-and-ramp structure. (From Celani *et al.*, 2000.)

argument given above. Instead, it is determined by the consistency condition

$$\epsilon_\theta \sim \kappa \int_0^{k_\theta} k^2 E_k^\theta \, dk \sim \kappa \int_0^{k_\theta} \epsilon_\theta \left(\frac{\nu}{\epsilon_V}\right)^{1/2} k \, dk,$$

where $k_\theta = l_\theta^{-1}$; hence

$$l_\theta = (\nu\kappa^2/\epsilon_V)^{1/4} = l_{\mathrm{K}} \mathsf{Pr}^{-1/2}, \quad \mathsf{Pr} > 1. \tag{9.52}$$

The scalar spectrum, which we just derived from simple statistical principles, provides little information about the spatial distribution of the scalar field. Its characteristic feature is the ramp-and-cliff structure shown in Fig. 9.9. Regions of finite scalar gradient are sheared perpendicularly to the gradient, while simultaneously shrinking along the gradient because of incompressibility, which leads to a local steepening of the gradient, ultimately forming gradient sheets (cliffs) bounding regions of relatively smooth behavior (ramps).

Obviously, this turbulence pattern is highly intermittent, which is reflected in a strongly nonlinear behavior of the scaling exponents ζ_n^θ of the structure functions $\langle \delta\theta_l^n \rangle \sim l^{\zeta_n^\theta}$ and in corresponding non-Gaussian p.d.f.s. Kraichnan (1997) has shown that, in the case of a rapidly varying velocity field ('white-noise

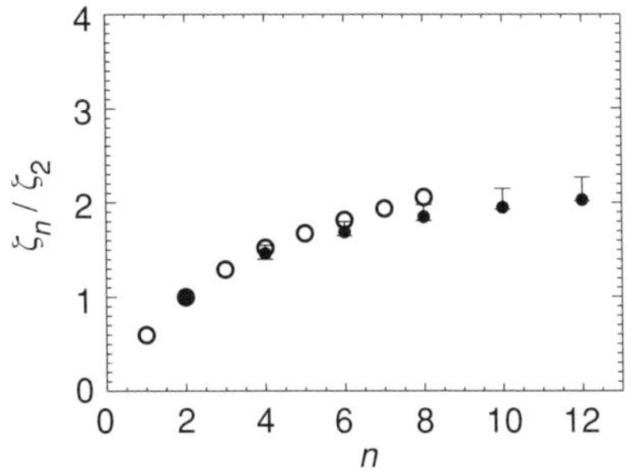

Figure 9.10: Relative scaling exponents of scalar turbulence ζ_n/ζ_2: 2D numerical simulations of passive scalar turbulence (Celani *et al.*, 2000) (error bars); 2D simulation of thermal convection (Biskamp *et al.*, 2001) (full dots); and temperature measurements in wind-tunnel turbulence (Ruiz-Chavarria *et al.*, 1996) (open dots).

turbulence'), the scaling exponents increase as $\zeta_n^\theta \sim \sqrt{n}$, i.e., there is no asymptotic linear behavior, in contrast to the behavior of the exponents of the velocity structure functions, for instance $\zeta_n^V \to n/9$ in the She–Lévêque formula (7.64). Numerical simulations and experimental observations of scalar turbulence indicate that ζ_n^θ may even saturate, though the asymptotic behavior is difficult to assess. There seems to be a certain universality of the relative exponents ζ_n/ζ_2 of scalar turbulence (the absolute values depend on the velocity spectrum, as can be seen from Yaglom's relation $\langle \delta v_l \, \delta\theta_l^2 \rangle \sim l$). The relative exponents for a passive scalar from a 2D simulation driven by the turbulent velocity of the inverse $k^{-5/3}$ cascade (Celani *et al.*, 2000) agree with those of the temperature from a 2D simulation of turbulent convection (Biskamp *et al.*, 2001), and the relative scaling exponents from observations of the temperature fluctuations in wind-tunnel turbulence, which are of course fully 3D, are only insignificantly larger; see Fig. 9.10.

The p.d.f. of the scalar increments $\delta\theta_l$ exhibits two characteristic features, a cusp-like form for small values, representing the extended ramps, where the increment is small, and extended tails for large values, representing the cliffs. A remarkable feature observed in experiments on thermal convection at high Rayleigh number is the exponential shape of the p.d.f. of the temperature itself, which should be Gaussian for homogeneous turbulence according to the central-limit theorem, as discussed in Section 7.1 (e.g., Gollub *et al.*, 1991). Indeed, these exponential temperature distributions have become the hallmark of so-called hard turbulence in Rayleigh–Bénard convection (Castaing *et al.*, 1989). The phenomenon results from the large convective eddies which are generated occasionally, which move fluid parcels across distances exceeding the integral scale and thus explore the full mean gradient of the scalar field, leading to

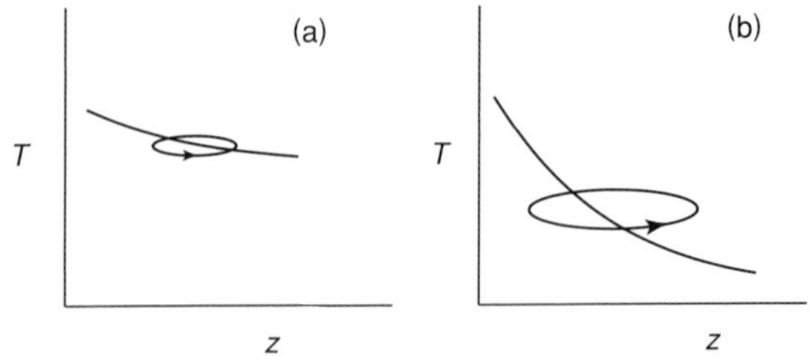

Figure 9.11: Illustrations of convection in (a) a Boussinesq fluid and (b) a general compressible fluid.

'large events' with much higher probabilities than their Gaussian values (Pumir *et al.*, 1991). Hence a mean gradient, even if it is very weak, not only causes a persistent small-scale anisotropy of the scalar turbulence but also affects the probability distribution.

9.3.3 Compressible turbulent convection

Convection zones in astrophysical systems, for instance in stars, are strongly stratified, extending over many pressure- and density-scale heights, $h = -p/(dp/dz)$, and typical convective eddies have scales $l \gtrsim h$; see Fig. 9.11. As a consequence, velocities are high, reaching, and even exceeding, the speed of sound, $\mathsf{M}_s \sim 1$. One must therefore give up the convenient Boussinesq approximation and consider the full compressible system instead. The condition for convective instability is given by (3.50); hence the turbulence, in particular the heat flux $\langle v_z \widetilde{T} \rangle$, will be driven by the superadiabatic temperature gradient

$$\beta_s = -\left(\frac{dT}{dz} - (\gamma - 1)\frac{T}{\rho}\frac{d\rho}{dz} \right) = -\left(\frac{dT}{dz} - \left.\frac{dT}{dz}\right|_{\text{ad}} \right). \tag{9.53}$$

Let us estimate the heat flux using a simple heuristic approach. The vertical velocity of a fluid element shifted upward by a distance l is given by the kinetic energy the element gains by acceleration through the buoyancy force,

$$\rho v_z^2 \simeq \widetilde{\rho} g l \sim \frac{\widetilde{T}}{T} \rho g l,$$

using (2.42), which should still be approximately valid. The level of turbulence is determined by the mixing length l_m, (4.13), connecting the fluctuation amplitude with the mean gradient driving the turbulence,

$$\widetilde{T} \sim l_m \beta_s.$$

One now relates the mixing length to the only available length, the pressure scale height h, by introducing the dimensionless parameter α,

$$l_m = \alpha h, \tag{9.54}$$

such that the heat flux can be written in the form

$$\langle v_z \widetilde{T} \rangle \sim \alpha^2 (\mathsf{RaPr})^{1/2} \kappa \beta_s, \tag{9.55}$$

where here the Rayleigh number is defined in terms of h and β_s, $\mathsf{Ra} = gh^4 \beta_s/(T\nu\kappa)$. The phenomenological parameter α, the normalized mixing length, which we shall encounter again in Section 11.2.2, is connected to the Bolgiano length, (9.43), $\alpha \sim L_\mathrm{B}/L$, since $\mathsf{Nu} = \langle v_z \widetilde{T} \rangle/(\kappa\beta_s)$. α contains the specific properties of the turbulence and must be determined by comparison with observations or numerical simulations.

The numerical method staying closest to the Boussinesq approximation is the anelastic approximation (e.g., Gilman and Glatzmaier, 1981), which deals with systems extending over several scale heights but still eliminates the sound wave. The essential modification consists of replacing the incompressiblity condition $\nabla \cdot \boldsymbol{v} = 0$ by

$$\nabla \cdot (\rho \boldsymbol{v}) = 0. \tag{9.56}$$

Numerical simulations using the anelastic approximation have been performed primarily for studying convection in planetary cores, whereas, in the case of stellar convection zones, where supersonic motions may occur, a fully compressible treatment appears indispensable. Because of the limitations in Rayleigh number and Prandtl number ($\mathsf{Ra} \lesssim 10^6$ and $\mathsf{Pr} \gtrsim 10^{-1}$ compared with 10^{20} and 10^{-9}, respectively, on the Sun), either only a small section can be computed or the parameters have to be drastically rescaled. The characteristic feature of compressible convection is a pronounced up–down asymmetry of the flow with relatively slowly expanding upflows and a network of rapidly converging sheet-like or filamentary downdrafts (Stein and Nordlund, 1989; Toomre *et al.*, 1990). This asymmetry, which is absent from Boussinesq flows, is caused by the effect of horizontal pressure fluctuations. Since $\widetilde{p} \propto \widetilde{T}$, upflows with $\widetilde{T} > 0$ are slowed down by the gradient of the local pressure enhancement, a process called *buoyancy braking*, while downflows are accelerated. As a consequence there is a mean downward flux of kinetic energy, while the heat, or more precisely enthalpy, flux is, of course, upward.

Since convective transport of heat is very efficient, the system stays close to the marginal state. In laboratory experiments this means that the temperature gradient is concentrated in thin layers at the upper and lower walls, while the temperature profile is practically flat in the central part. In stellar convection zones the temperature profile is very close to adiabatic, except at the outer edge, where strong radiative cooling can compete with convection.

9.3.4 Magnetoconvection

Turbulent convection of magnetic fields in a stratified layer is intimately connected with the dynamo problem, in particular the generation of magnetic fields in planets and stars. As mentioned in Section 4.1.4, a natural source of kinetic helicity, the basic agent for dynamo action, is the combination of buoyancy and rotation; see (4.31). The generation of large-scale fields is related nonlinearly to the inverse cascade of the magnetic helicity; see Section 6.2.1 and Fig. 6.4. However, dynamo theory with all its ramifications and applications has developed into a major topic in its own right, a detailed discussion of which would lead us beyond the framework of the book. I therefore refer the interested reader to several reviews of the field, e.g., the classical treatise by Moffatt (1978), the review paper by Roberts and Soward (1992), and Chapter 5 of my book on magnetic reconnection (Biskamp, 2000). Here we are concerned primarily with the structure of the magnetic field in turbulent convection.

Whereas in a closed configuration growth of the large-scale field by dynamo action will ultimately lead to $E^M \gg E^K$, in many astrophysical systems field energy is continuously lost by expulsion from the convection zone, for instance by emergence of flux tubes into the solar corona, so the average field energy in the unstable region remains limited. The case in which the magnetic energy is smaller than the kinetic energy, $E^M < E^K$, is particularly interesting. Here the spatial structure of the turbulent magnetic field is different from the smooth space-filling distribution in a system with $E^M \sim E^K$ assumed before, confined in thin flux tubes. This behavior can already be understood from the form of the equations. If the field is sufficiently weak, the Lorentz force can be neglected and the vorticity and magnetic field follow similar equations:

$$\partial_t \boldsymbol{B} + \boldsymbol{v} \cdot \nabla \boldsymbol{B} = \boldsymbol{B} \cdot \nabla \boldsymbol{v} - \boldsymbol{B} \nabla \cdot \boldsymbol{v} + \eta \nabla^2 \boldsymbol{B},$$

$$\partial_t \boldsymbol{\omega} + \boldsymbol{v} \cdot \nabla \boldsymbol{\omega} = \boldsymbol{\omega} \cdot \nabla \boldsymbol{v} - \boldsymbol{\omega} \nabla \cdot \boldsymbol{v} + \nu \nabla^2 \boldsymbol{\omega},$$

which for $\mathsf{Pr}_m = 1$ are formally identical. Hence the solutions $\boldsymbol{B}$ and $\boldsymbol{\omega}$ are expected to be very similar, if $\boldsymbol{B}$ is driven by the same velocity, since the initial correlation between $\boldsymbol{v}$ and $\boldsymbol{\omega}$ is rapidly wiped out by the random character of the motion. The only difference is the amplitude, $\boldsymbol{B} \ll \boldsymbol{v}$ being assumed (Batchelor, 1950), as $\boldsymbol{B}$ is passively convected. For $\mathsf{Pr}_m \neq 1$, the strict analogy between the magnetic field and the vorticity no longer holds, but they should still be similar in the inertial range. For a Kolmogorov kinetic-energy spectrum

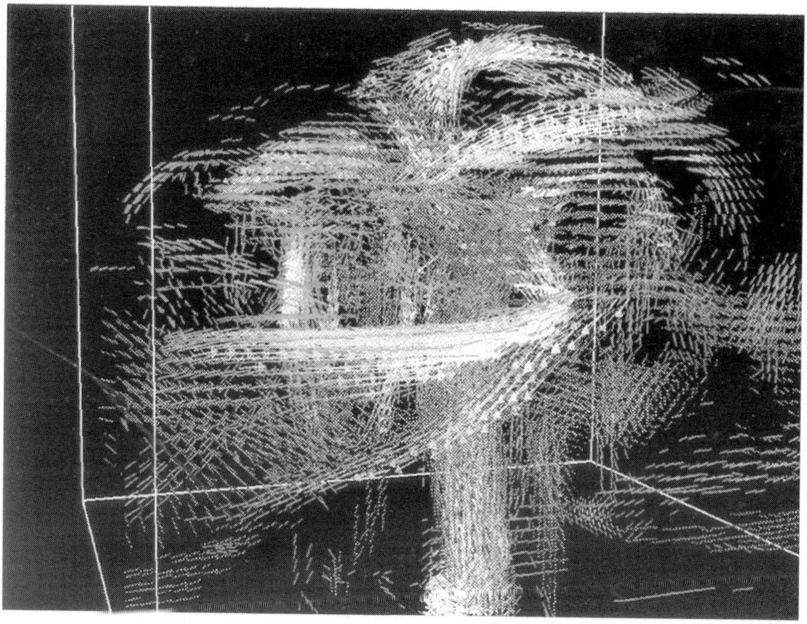

Figure 9.12: Magnetic flux tubes in magnetoconvection. (Courtesy of A. Brandenburg.)

the analogy implies that

$$E_k^M \sim \Omega_k = k^2 E_k^K \sim k^{1/3}. \tag{9.57}$$

Since the vorticity in Navier–Stokes turbulence is concentrated in thin tubes, or filaments, the magnetic field is expected to exhibit a similar behavior, which is indeed observed in simulations (Meneguzzi *et al.*, 1981; Nordlund *et al.*, 1992); see Fig. 9.12. (The argument leading to the spectrum (9.57) should, however, be considered with some caution. Since the magnetic energy is not ideally invariant, the spectrum cannot be inferred from a Kolmogorov-type cascade for the magnetic energy, but is solely based on the analogy with the vorticity.)

Any smoothly distributed 'seed' magnetic field will rapidly be stretched, compressed, and twisted into tubes with diameter limited only by resistive diffusion. At sufficiently large Reynolds number even a very weak initial field will thus reach the kinetic-energy level $E_k^M \sim E_k^K$ at large k, such that, at these scales, the Lorentz force can no longer be neglected. Even when the total magnetic energy is still small, $E^M \ll E^K$, one expects a certain spectral range which, due to the Alfvén effect, the two spectra essentially coincide, $E_k^M \gtrsim E_k^K$, as illustrated in Fig. 9.13. For $E^M \gg E^K$ the spectrum looks similar as sketched in Fig. 6.5.

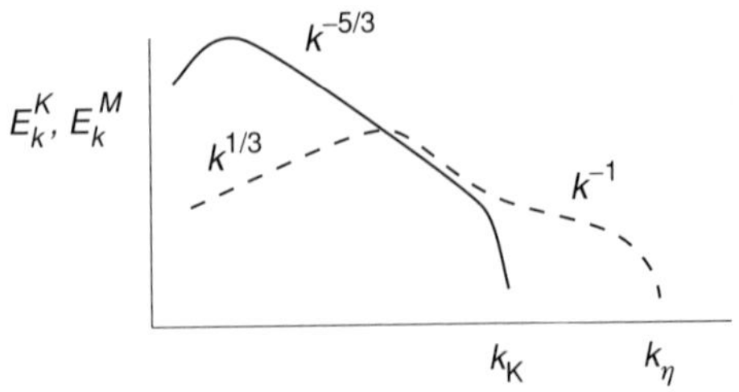

Figure 9.13: Kinetic (continuous) and magnetic (dashed) energy spectra in kinetically dominated turbulent convection for $\mathsf{Pr}_m \gg 1$ (shown schematically).

The spectral behavior at small scales depends on the magnetic Prandtl number. Since $\mathsf{Pr}_m \sim T_e^4/n$, we have $\mathsf{Pr}_m \gg 1$ in dilute plasmas, such that resistive dissipation occurs at much smaller scales than does viscous dissipation and the magnetic field exhibits a further inertial range $l_{\mathrm{K}}^{-1} < k < l_\eta^{-1}$, within which the spectrum is flatter, $E_k^M \sim k^{-1}$, in analogy to the viscous–convective subrange of the passive scalar (9.51). Three-dimensional simulations have, for instance, been performed by Kida *et al.* (1991), who studied the statistical properties of kinetically driven incompressible MHD turbulence.

Numerical studies of compressible magnetoconvective turbulence have been performed by a number of groups; by Hurlburt *et al.* (1994), Brandenburg *et al.* (1990, 1996), Nordlund *et al.* (1992), to name a few. For low magnetic-field energy the characteristic features of the convection are similar to those of the hydrodynamic case. Since convection wants to reduce the driving force, the superadiabatic gradient β_s, (9.53), the general tendency is to flatten the temperature gradient by an upward heat flux and steepen the density gradient by a net downward flow of mass. (The actual values of these fluxes depend on the boundary conditions.) Hence also the magnetic field tends to be convected downward, mainly in the form of flux tubes wrapped around vortex tubes, accumulating at the bottom of the convection region (and in the adjacent overshoot region). Here the field is essentially in static equilibrium; the density is reduced by draining the plasma along field lines. Thus individual strands of flux, which may be ripped off this magnetic sea, become buoyant, drifting upward and finally emerging across the stellar surface into the corona. Simulations of turbulent magnetoconvection are in general consistent with this picture, but numerical resolution is still far too low to allow one to determine any spatial scaling properties.

10
Turbulence in the solar wind

The solar wind provides an almost ideal laboratory for studying high-Reynolds-number MHD turbulence. Turbulence is free to evolve unconstrained and unperturbed by *in situ* diagnostics, satellite-mounted magnetometers, probes and particle detectors. We will see that many features of homogeneous MHD turbulence discussed in the previous chapters are discovered in solar-wind turbulence, but there are also unexpected and still unexplained features. Since the turbulence varies significantly in the different regions of interplanetary space depending on the local solar-wind conditions, it is useful to first give at least a rough picture of the mean solar-wind properties, before discussing the properties of the turbulent fluctuations about the mean state.

10.1 Mean properties of the solar wind

Stars lose not only energy by radiation but also mass (and angular momentum) by a, more or less, continuous radial flow called the stellar wind, the solar wind in the case of the Sun. The origin of this flow is the high temperature in the corona, which means that the coronal plasma is not gravitationally bound and, if it is not confined by magnetic loops, expands into interplanetary space, giving rise to the supersonic solar wind. The flow extends radially out to a distance beyond the planetary system, before it is slowed down by the termination shock expected at roughly 100 AU (1 AU $\simeq 1.5 \times 10^8$ km is the Earth's orbital radius). It thus forms a bubble in interstellar space, the heliosphere. The solar-wind plasma consisting of fully ionized hydrogen with a small admixture of helium soon reaches supersonic and super-Alfvénic speeds. Beyond the corona this plasma is practically collisionless; hence dissipation processes are solely due to collective effects.

Detailed information about the solar-wind properties in the different regions of the heliosphere has been collected by several satellite missions. The inner heliosphere from 0.3 AU to 1 AU has been scanned by the two spacecraft of the Helios mission. Though they were launched in the 1970s, the data from this mission are still being evaluated, forming the basis of a large part of the literature on solar-wind turbulence, because of the high quality of the instrumentation and because they are still the only source of information about the interesting region close to the Sun. The outer heliosphere beyond 1 AU has been sampled by several spacecraft, namely Pioneers 10 and 11 and Voyagers 1 and 2. While their orbits remain close to the ecliptic, the Ulysses spacecraft traveling on a highly inclined orbit is also sampling data from high solar latitudes, the polar regions, at distances between about 1 AU and 5 AU. Combining these data now gives a rather complete picture of the solar wind with its different facets. In this section we first present the classical hydrodynamic model of the solar wind and then discuss the modifications caused by the magnetic field.

10.1.1 Hydrodynamic model of the solar wind

The basic physics of the solar wind is described by a one-fluid hydrodynamic model in which it is assumed that there is a stationary radial flow driven only by pressure and gravity and we neglect, in particular, the solar magnetic field. In view of the collisionless character of the plasma, a fluid approach might appear inappropriate, but, on the large global scales considered here, kinetic effects are, indeed, negligible. The medium is treated as an ideal gas with thermodynamics characterized by a polytropic equation of state, $p = a\rho^\gamma$. The basic equations for the mass density ρ and the radial velocity v are

$$\frac{d}{dr}(r^2\rho v) = 0, \tag{10.1}$$

$$v\frac{dv}{dr} = -\frac{GM_\odot}{r^2} - \frac{1}{\rho}\frac{dp}{dr} = \frac{d}{dr}\left(\frac{GM_\odot}{r} - \frac{c_s^2}{\gamma - 1}\right), \tag{10.2}$$

where $c_s = \sqrt{\gamma p/\rho}$ is the local sound speed and $M_\odot$ is the solar mass. By inserting (10.1) we may also write (10.2) in the useful form

$$\frac{1}{2}\left(1 - \frac{c_s^2}{v^2}\right)\frac{dv}{dr} = \frac{2c_s^2}{r} - \frac{GM_\odot}{r^2}. \tag{10.3}$$

In this equation there occurs a critical point, where dv/dr is not defined, for $v = c_s$ at $GM_\odot/(2c_s^2) = r_s$, called the sonic radius.

Equations (10.1) and (10.2) can readily be integrated, yielding

$$r^2\rho v = \Gamma, \quad \frac{1}{2}v^2 + \frac{c_s^2}{\gamma - 1} - \frac{GM_\odot}{r} = W, \tag{10.4}$$

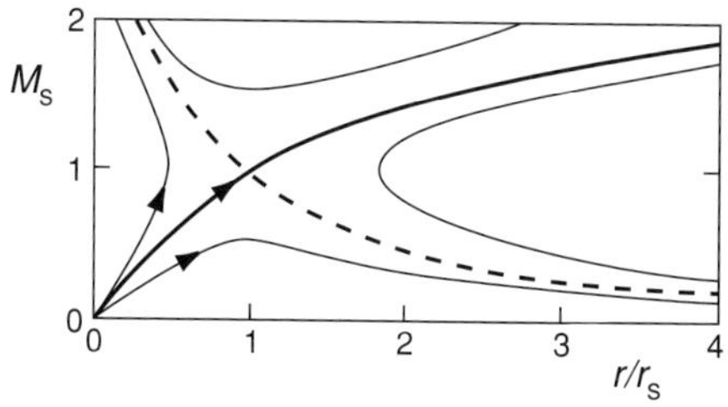

Figure 10.1: The manifold of solutions of (10.4).

Table 10.1: *Solar-wind parameters: since the velocity of the wind is usually compared with the local Alfvén speed, the sonic Mach number is not given; one has typically* $\mathsf{M}_s > \mathsf{M}_A$

	n (cm^{-3})	T (K)	v (km s^{-1})	B (G)	M_A
Base of corona	10^9	2×10^6	<1	10	$\ll 1$
1 AU	3–20	10^4–10^6	300–800	$(3–20) \times 10^{-5}$	4–12

which, on substitution of ρ, gives the equation for $v(r)$. The integration constants Γ and W depend on the boundary conditions at the base of the solar corona. The different classes of solutions are illustrated in the M_s, r-diagram in Fig. 10.1. Observations indicate that the velocity increases monotonically from small subsonic to highly supersonic values; hence the heavy line in Fig. 10.1 is the relevant solution. One thus finds that the gravitational field acts in a manner analogous to a Laval nozzle.[1] From (10.4) one obtains the asymptotic behavior

$$v \simeq \sqrt{2W}, \quad \rho \sim \frac{1}{r^2}, \quad T \sim \frac{1}{r^{2(\gamma-1)}}, \quad \mathsf{M}_s = \frac{v}{c_s} \sim r^{\gamma-1}. \tag{10.5}$$

While the velocity remains constant, the Mach number increases because of continuous cooling of the gas. For a temperature at the base of the corona of 2×10^6 K the sound speed is $c_s \simeq 1.3 \times 10^2$ km s^{-1}; hence $r_s \sim (5–10)r_\odot$ depending on γ. For the Earth's orbit ($r = 1\,\text{AU} \simeq 215 r_\odot$) the velocity is $v \sim 500$ km s^{-1} and the Mach number reaches $\mathsf{M}_s \sim 10$. Typical solar-wind parameters are summarized in Table 10.1.

The one-fluid approximation actually gives a more realistic description than does a simple two-fluid model for ions and electrons, since anomalous heating and transport processes keep electron and ion temperatures similar. (In two-fluid theory one would have $T_e \gg T_i$ because of high parallel conduction of heat

[1] A Laval nozzle is a convergent–divergent nozzle producing a supersonic jet.

by electrons.) However, the effects of these processes vary with distance from the Sun, which would correspond to an r-dependent polytropic coefficient. In fact, for distances that are not too large, the isothermal pressure assumed by Parker (1958) in his celebrated solar-wind paper is often closer to reality than is an adiabatic pressure law. For $\gamma = 1$ the pressure force in (10.2) becomes $-c_s^2(d/dr)\ln\rho$ with constant c_s, whence, after integration and fixing the constant of integration, the equation for the velocity assumes the simple form

$$\mathrm{M}_s^2 - \ln \mathrm{M}_s^2 = 4\left[\ln\left(\frac{r}{r_s}\right) + \frac{r_s}{r}\right] - 3, \qquad (10.6)$$

with the asymptotic behavior $\mathrm{M}_s \simeq 2\sqrt{\ln(r/r_s)}$.

10.1.2 Effects of the solar magnetic field

In the corona we are dealing with a low-β plasma; hence the magnetic field plays a dominant role, which restricts the outflow to regions of open field lines, the coronal holes covering mainly the polar regions. A further effect arises from the rotation of the Sun. As long as the field is strong and the flow is slow, i.e., sub-Alfvénic, $\mathrm{M}_\mathrm{A} < 1$, the plasma is dragged along with the field frozen into the photosphere and is thus forced to corotate with the Sun. However, since the flow speed increases with radius, while the field decreases, $B \sim r^{-2}$ due to flux conservation, the flow becomes super-Alfvénic beyond a certain radius, the Alfvén radius $r = r_\mathrm{A}$, and the field is now dragged along with the plasma. With the footpoints anchored in the rotating photosphere, field lines assume the form of an Archimedean spiral (a 'Parker spiral'), for which the angle ψ with respect to the Sun–Earth line increases with distance, $\tan\psi \simeq \Omega r/v$, $\psi \simeq 45°$ for the Earth's orbit.

Including the Lorentz force and rotation, the simplest configuration consists of a radial and an azimuthal flow component, $\boldsymbol{v} = \{v_r(r), 0, v_\phi(r)\}$ in spherical coordinates, coupled to the magnetic field with similar components, $\boldsymbol{B} = \{B_r(r), 0, B_\phi(r)\}$. The corresponding equations have been solved by Weber and Davis (1967). There are now three critical points, where the flow speed reaches the phase velocities of the three MHD modes, namely the slow mode, the Alfvén mode, and the fast mode, v_slow, v_A, and v_fast (Section 2.5.1). In the low-β coronal plasma $v_\mathrm{slow} \simeq c_s \ll v_\mathrm{A} \simeq v_\mathrm{fast}$; hence the sonic radius r_s is smaller than the Alfvén radius r_A, such that the latter plays essentially the role of the former in Parker's hydrodynamic solution and the asymptotic behavior is similar.

Since the solar magnetic field is, of course, not unipolar, but dipolar (with multipolar contributions), the flow depends also on the latitude. Pneuman and

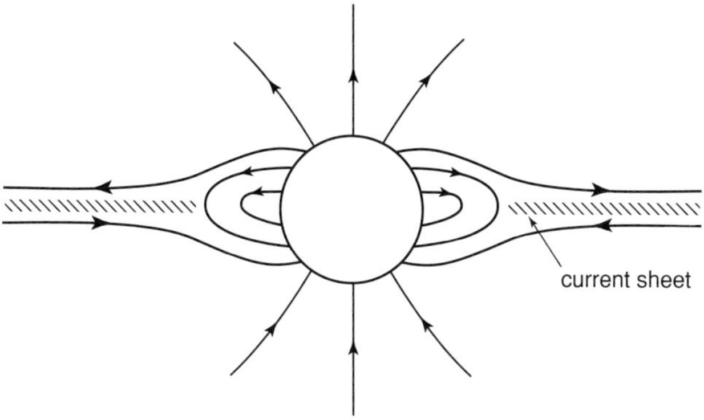

Figure 10.2: A schematic drawing of the solar dipole field modified by the solar wind.

Kopp (1971) solved the MHD equations for a steady axisymmetric expansion assuming a dipolar field at the base of the corona. The resulting configuration, illustrated in Fig. 10.2, has the shape of a helmet streamer. Only the magnetic loops near the equator preserve the dipolar character confining coronal plasma. Field lines with footpoints at higher latitudes are dragged along with, and opened up by, the solar wind, forming a neutral sheet further out, where the field reverses direction, which is called the heliospheric current sheet. On field lines originating from the polar regions, where the field is relatively strong, the flow becomes first supersonic and then super-Alfvénic, while in the region close to the current sheet, where the field is weak, this process occurs in the reverse order. Ideally the current sheet is located in the ecliptic; in reality, however, it is found to be more or less strongly wrapped about this plane because of deviations of the coronal field distribution from axisymmetry, such that, because of solar rotation, the field seen at the Earth comes alternatingly from the northern and the southern polar holes. Hence the field configuration in the ecliptic consists of different sectors of opposite field directions separated by rather sharp boundaries, tangential discontinuities, which correspond to the lines where the current sheet crosses the plane. There are typically four sectors, as illustrated in Fig. 10.3. While at times close to sunspot minimum, for which the field is mainly dipolar, conditions are qualitatively as shown in Fig. 10.2, at sunspot maximum, when the solar field is strong and irregularly distributed, the polar holes are small, but further holes are present at low altitudes, so that the current sheet is more strongly perturbed, reaching out to high latitudes, and may even be split into several such sheets, corresponding to a complex field structure. In any case, conditions in the ecliptic, in particular at the radius of the Earth's orbit, are much more variable than are conditions at higher latitudes.

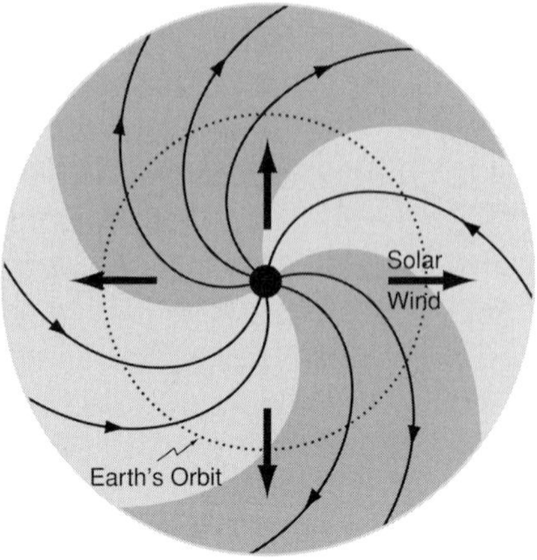

Figure 10.3: The sector structure of the interplanetary magnetic field in the equatorial plane (shown schematically).

10.1.3 Fast and slow winds

Because of the structure of the magnetic field, the solar wind cannot be radially symmetric, even far away from the complex coronal loops. While the flow originating from the central parts of the coronal holes is fast, $v \lesssim 800\,\mathrm{km\,s^{-1}}$, it is much slower in the region close to the neutral sheet, $v \gtrsim 300\,\mathrm{km\,s^{-1}}$. (However, since the density in the slow wind is considerably higher than that in the fast wind, the mass flow ρv tends to be even higher in the former.) The boundaries between the different flows are rather sharp. While the fast wind is fairly uniform, the velocity of the slow wind varies considerably, as can be seen in Fig. 10.4. Because of the solar rotation, the fast wind catches up with the slow wind, forming a compression region of high density and magnetic field, called the corotating interaction region (CIR), which steepens into a shock front at larger heliospheric distances. To an observer on the Earth the flow speed varies quasi-periodically, with a steep rise to high velocity on entering the CIR, which is followed by a gradual decrease to the slow-wind level in the rarefaction region, where the slow wind flows behind the fast one.

Fast and slow winds exhibit also different thermodynamic properties. The proton temperature in the slow wind drops more rapidly with radius than does that in the fast flow; typically $T \sim r^{-1.3}$ in the former, which corresponds to an adiabatic expansion, (10.5), for $\gamma = \frac{5}{3}$, and $T \sim r^{-0.8}$ in the latter, corresponding to $\gamma \lesssim 1.4$, which is closer to isothermal expansion, indicating that there is substantial local input of heat by conduction and wave dissipation.

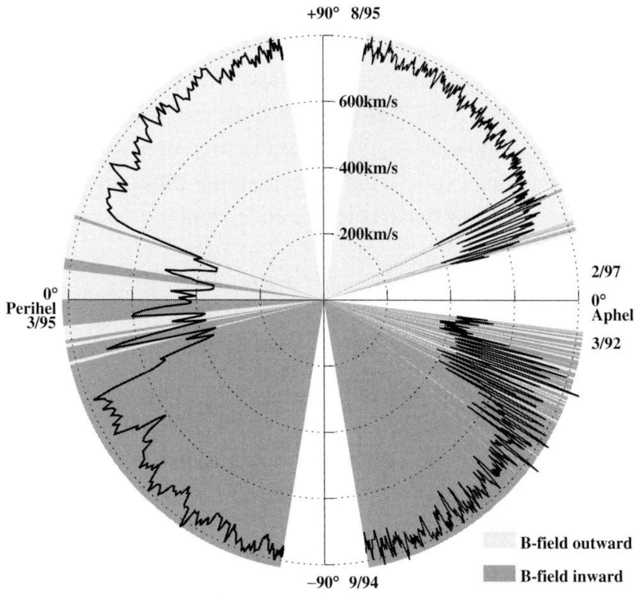

Figure 10.4: Solar-wind-speed and magnetic-field-polarity observations from the Ulysses spacecraft during the passage across both polar regions. (Courtesy of J. Woch.).

The acceleration process of the fast wind is not fully understood. It is true that, in the coronal holes, plasma may expand freely along field lines, such that flow conditions are similar to those assumed in Parker's hydrodynamic model. However, while the latter gives the correct *mean* wind properties, $v \sim 500\,\mathrm{km\,s^{-1}}$ for a coronal temperature of about 2×10^6 K, it cannot explain the fast wind out of the coronal holes, which has a substantially higher velocity, even though the coronal temperature in this region is lower, $T \lesssim 10^6$K, as is directly visible in the X-ray images, on which the holes appear dark. That the temperature in coronal holes is lower is easily understood, since the plasma is not confined and hence cannot be heated as efficiently by whatever the coronal heating mechanism is; but why does it generate higher wind speeds? It has been suggested that the dissipation of intense high-frequency Alfvén waves originating from the convection zone leads to a high-energy ion population – observations of strongly broadened spectral lines indicate that effective temperatures of up to 10^8K occur –, which gives rise to an efficient acceleration process in addition to the effect of the thermal pressure gradient.

Neither is the origin of the slow wind clear. As can be seen in Fig. 10.4, the slow wind emanates from the magnetically complex equatorial belt, where there are small areas of open field lines between closed-loop structures, which, in some sense, brake the free Parker-like outflow and thus give rise to lower velocities. The sharp boundary between fast and slow winds indicates that the two types originate from regions with different magnetic structures. As we shall see in the following section, also the properties of the MHD turbulence riding on the mean solar-wind flow are quite different for the two flow speeds. They may give additional clues for understanding the origin of the slow flows.

10.2 MHD fluctuations in the solar wind

Early satellite observations, such as those made by Mariner 2, already showed that the solar wind is far from steady but carries fluctuations of substantial amplitudes. Such fluctuations are seen in all variables, namely velocity, magnetic field, density, and temperature. Since wavelengths are typically much larger than the intrinsic plasma scales, in particular the ion Larmor radius, they can be described in the MHD framework, although, because of the collisionless character of the plasma, wave dissipation is caused by kinetic processes, most probably cyclotron-resonance effects.

The presence of fluctuations is not surprising. The wind originates in the corona, which is the site of intense magnetic activity occurring on practically all scales, driven by the turbulent dynamics in the solar convection zone. Hence the flow emanating from the corona should be modulated by broadband fluctuations or 'waves'. In flight these waves, being of 'nonlinear' amplitude, will interact, or 'cascade', and thus randomize further. Moreover, instabilities in the solar wind driven by sheared flows, shocks, beams, or anisotropries give rise to a continous resuscitation of the turbulence, which would otherwise decay following one or the other of the selective decay routes.

Solar-wind turbulence comes mainly in two forms, either with a high velocity–magnetic-field correlation, corresponding to outgoing Alfvén waves, or as essentially uncorrelated fully developed turbulence exhibiting a Kolmogorov spectrum. These observations had previously divided the solar-wind community into essentially two opposite schools, the wave school emphasizing the coronal origin of the fluctuations and the turbulence school a local excitation. During the last decade a unifying picture has evolved, explaining the gross properties of the fluctuations, though many questions concerning particular features remain to be answered. Here I give a brief overview of the present status of understanding while referring the interested reader to several reviews of the field, notably Roberts and Goldstein (1991), Marsch (1991), Goldstein *et al.* (1995), and Tu and Marsch (1995).

10.2.1 Wave types and turbulence spectra

Waves in the fast wind

Let us first discuss the fluctuations observed in the fast flow. For distances that are not too large, say $r < 0.5$ AU, these are characterized by large-amplitude oscillations of magnetic field and velocity transverse to the mean field which, even though they are completely irregular, exhibit strong velocity–magnetic-field correlation, $\delta \boldsymbol{v}_\perp \simeq \pm \delta \boldsymbol{b}_\perp$. Hence we are dealing with almost unidirectional Alfvén waves (Section 2.6), for which the observed sign is such that wave propagation is outward, i.e., a minus sign for $\boldsymbol{B}_0$ pointing outward and a plus sign for the opposite case. To conform with the solar-wind literature, we adopt in this chapter the notation that z^+ and the energy $E^+ = \frac{1}{4}(z^+)^2$ refer to outgoing waves, and z^- and the energy $E^- = \frac{1}{4}(z^-)^2$ to ingoing waves, regardless of the polarity of the field. The first, and most natural, explanation of this finding is due to Belcher and Davis (1971). The fluctuations are assumed to be generated by magnetic processes in the corona and advected with the wind into the interplanetary space. The fact that only outgoing waves are observed implies that these processes, which in general give rise to waves propagating in both directions, are not active beyond the Alfvén radius. Since compressive waves, which should also have been excited, are more strongly damped by dissipation in shocks, only incompressible modes, Alfvén waves, survive. In fact, it is found that $\delta B_\perp \gg \delta B_\parallel$ and that local fluctuations in density are weak, $\delta n/n \lesssim 0.1$. Hence solar-wind turbulence, at wavelengths exceeding the ion Larmor radius, can be treated in the framework of incompressible MHD.

These MHD fluctuations extend over a broad range of scales, which are usually measured as frequencies, $10^{-5}\,\mathrm{Hz} < f < 10^{-1}\,\mathrm{Hz}$; a wavetrain advected with the solar wind can be regarded as 'frozen in', since the mean velocity is much higher than the phase velocity and the amplitude of the fluctuations. Hence one can relate the wavenumber $k = k_r$ parallel to the flow v_0 to the frequency f, $kv_0 = \omega = 2\pi f$, which is called the Taylor hypothesis, such that the frequency range corresponds to the wavenumber range $10^{-8}\,\mathrm{km}^{-1} < k/(2\pi) < 10^{-4}\,\mathrm{km}^{-1}$.[2] The largest scales are of the order of the distance r, while the smallest are still well above the proton Larmor radius r_p. This scale range can also be used to define an effective Reynolds number, since the collisional diffusivities ν and η are negligible. From (5.62) we obtain $\mathrm{Re}_{\mathrm{eff}} \lesssim (r/r_p)^{4/3} \sim 10^7$.

[2] In reality, the inertial range is shorter, $k/(2\pi) > 10^{-7}\,\mathrm{km}^{-1}$ or $f > 10^{-4}\,\mathrm{Hz}$, since perturbations with larger scales should be regarded as large-scale inhomogeneities advected by the solar wind from the corona.

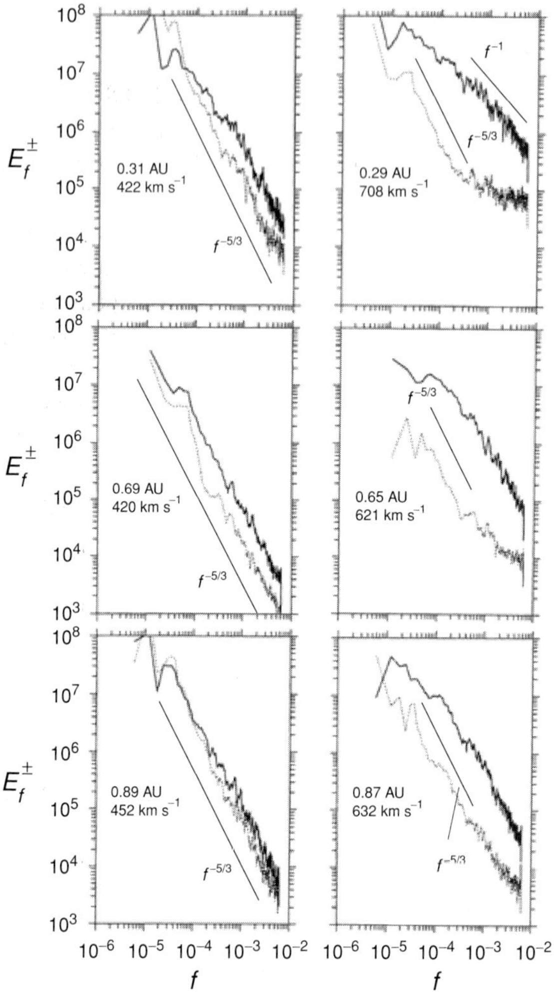

Figure 10.5: Evolution with distance of the power spectra of outgoing waves E_f^+ (heavy) and ingoing waves E_f^- (dotted) for slow wind (left-hand column) and fast wind (right-hand column). The straight lines indicate the power laws $f^{-5/3}$ and f^{-1}. (From Marsch and Tu, 1990a.)

In the region close to the Sun, $r \lesssim 0.3$ AU, the outward-propagating wave spectrum in the fast wind follows roughly a k^{-1} power law; see the top chart on the right in Fig. 10.5. A possible interpretation has been given by Velli *et al.* (1989), who showed that the linear coupling of the outgoing waves to the large-scale inhomogeneity of the solar wind generates a weak ingoing wave

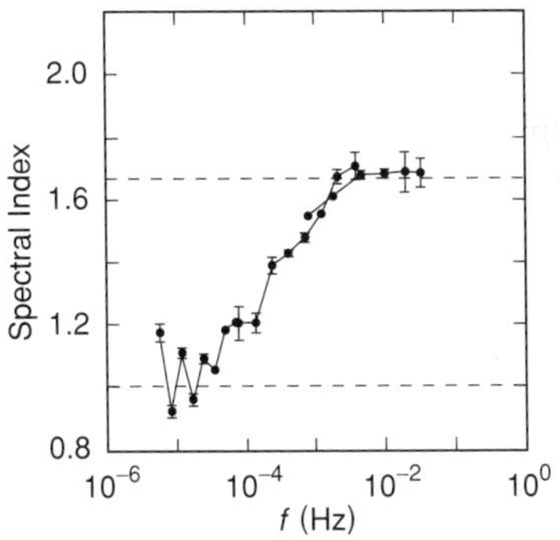

Figure 10.6: The variation with frequency of the spectral index of the magnetic energy, $-\log E_k^M$, at $r \simeq 2.4$ AU in the high-speed solar wind. (Adapted from Horbury, 1999.)

component; the nonlinear interaction of outgoing and ingoing waves then leads to a turbulent cascade with a k^{-1} spectrum. The shape of the observed spectrum is, however, not invariant but rather evolves with increasing distance from the Sun, as illustrated in the right-hand column of Fig. 10.5. The interaction of the dominant outward-propagating wave spectrum E_k^+ with the low-intensity inward-propagating wave spectrum E_k^- is such that first the spectra approach each other at high k, at which E_k^- is pulled up, thus flattening the spectrum, while E_k^+ becomes steeper. At large distances $r \gg 1$ AU (not shown in Fig. 10.5) the two spectra finally merge into a Kolmogorov spectrum, $E_k^+ \simeq E_k^- \sim k^{-5/3}$. Figure 10.6 gives the spectral index of the magnetic energy taken at some intermediate distance, $r \simeq 2.4$ AU, which demonstrates the transition from the k^{-1} behavior at small k (actually frequency f) to the Kolmogorov behavior at large k. Also the cross-helicity $H^C = E^+ - E^-$ decreases with distance and the Alfvén ratio E_k^K / E_k^M decreases from 1 to about 0.5 over most of the spectral range.

Why does this evolutionary process happen? Sections 4.3.2 and 5.3.5 tell us that, in a freely decaying system, once one component exceeds the other, it will grow further at the expense of the other until the turbulence is fully aligned, i.e., carries only unidirectional Alfvén waves, which will then be dynamically inert. In the solar wind just the opposite process seems to occur. One can argue that the turbulence in the fast wind cannot be regarded as freely decaying, since there are local processes reinjecting energy, as is clearly indicated by the slower than adiabatic drop of the temperature mentioned before. To the extent to which the spiraling mean magnetic field becomes weaker and more perpendicular to the flow with increasing distance from the Sun, its stabilizing effect on a shear flow is reduced, such that local tangential variations of the flow profile can render Kelvin–Helmholtz modes unstable (Section 3.2.1).

The turbulence is generally found to be anisotropic with respect to the local field direction, which may vary substantially compared with the mean field direction, since the amplitude of the low-frequency waves is large. The small-scale turbulence is more intense perpendicular to the local field than it is parallel to the local field, as would be expected from the properties of homogeneous 3D turbulence (Sections 5.3.3 and 7.5.2). It is, however, difficult to obtain a fully 3D picture of solar-wind turbulence from data produced by some isolated spacecraft such as the Helios mission, because they sample data in one direction only. More detailed information will be obtained from the multiple-satellite programs such as the four spacecraft of the Cluster mission (though the latter, being mainly confined to the Earth's magnetosphere, can only marginally sample the interplanetary plasma).

It is also worth noting that the residual spectrum $E_k^R = \frac{1}{2} z_k^+ \cdot z_k^-$, (5.66), is also observed to be close to Kolmogorov (Marsch and Tu, 1990a), in contrast to the theoretical spectrum (5.77) in the IK framework and to the result from 3D turbulence simulations, Fig. 5.5, both of which predict steeper power laws.

Turbulence in the slow wind

In the slow wind fluctuation amplitudes are smaller than they are in the fast wind. The lower level of turbulence implies that less wave energy can be dissipated locally, so the radial variation of the wind temperature is close to adiabatic, as mentioned above. Fluctuations in density are somewhat higher than those in the fast wind; hence compression effects are stronger but still small enough for the assumption of incompressibility to remain approximately valid. The energy spectra approach a Kolmogorov law, with E_k^+ only slightly larger than E_k^-, as illustrated in the left-hand column in Fig. 10.5. The shape of the spectrum is nearly invariant radially outward from 0.3 AU, the smallest distance from the Sun at which fluctuations have been measured, i.e., the turbulence is fully developed, or 'dynamically old'; there are some classical, often cited, observations of Kolmogorov-like magnetic-energy spectra, such as from Voyager 2 (e.g., Matthaeus *et al.*, 1982) and from Mariner 10 (e.g., Goldstein *et al.*, 1995). The decrease in the level of turbulence observed is due to expansion of the solar wind and does not imply that there is a free decay of the turbulence.[3] There appears to be a continuous input of energy, mainly by local shear flows, shocks, and compression effects in the CIRs. The main reason for the turbulence being further developed than it is in the fast wind seems to be the absence of a strong

[3] The variation of the small-scale fluctuation amplitude in the WKB approximation for large Mach number and $v \simeq$ constant (10.5) is $\langle \delta B^2 \rangle / B^2 \sim B^{-1/2}$ (Hollweg, 1974). While the absolute amplitude decreases, the relative amplitude increases, since $B \sim r^{-2}$.

guide field, since the slow wind blows in the region near the heliospheric current sheet, where the field reverses sign, which favors 3D randomization, leading naturally to a more Kolmogorov-like spectrum.

10.2.2 Intermittency in solar-wind turbulence

It has been, and still is, very challenging to understand the structure of the solar-wind turbulence by studying, in particular, the scaling properties of the structure functions; see, for instance, Burlaga (1991, 1993), Marsch and Tu (1993), Carbone (1994), Ruzmaikin *et al.* (1995), and Horbury and Balogh (1997). A comprehensive review of the subject has been given by Marsch and Tu (1997). Because of the high effective Reynolds number, an extended scaling range could be expected, resulting in rather precise values of the scaling exponents. In reality conditions are, however, far less favorable. In the fast wind originating from the polar regions conditions are rather steady, giving rise to quasi-stationary turbulence, but, at the solar distances sampled by the Ulysses spacecraft (<5 AU), the turbulence is still in a transitional phase, as the energy spectrum clearly shows; see Fig. 10.6. The relaxed, dynamically old, $k^{-5/3}$ range covers less than one decade at the high-frequency end of the spectrum. In the slow wind in the equatorial plane, on the other hand, the turbulence is further evolved, often exhibiting a Kolmogorov spectrum over more than two decades, but flow conditions vary rapidly, so reliable averages are difficult to obtain. For comparison with theory, additional problems arise from the lack of homogeneity and isotropy. Only for, say, $f > 10^4\,\mathrm{Hz}$ may the inhomogeneity due to expansion of the solar wind be considered small. Thus the solar-wind turbulence can hardly be expected to obey one of the intermittency models discussed in Chapter 7, which may therefore only serve as general guidelines.

One considers the functions

$$\langle |f(t+\tau) - f(t)|^n \rangle \sim \tau^{\zeta_n},$$

where $f(t)$ is either the radial velocity, which is the case for most data from the Helios and Voyager spacecraft in the equatorial plane, or a component of the magnetic field when data from the Ulysses spacecraft sampling the fast wind from the polar regions are evaluated. Scaling properties obtained in the different studies vary substantially. Consider, for instance, the work by Marsch and Tu (1993), who analyzed the velocity structure functions for fast- and slow-wind conditions at different radii in the inner heliosphere. The turbulence is obviously very intermittent in the sense that the scaling exponents ζ_n deviate strongly from the linear law determined by ζ_2, i.e., from the velocity spectrum. More importantly, the exponents differ from data set to data set without exhibiting a systematic dependence on wind speed and radius, such that, at this stage, comparison with theory is not possible. Most of these data seem to correspond to states of still-evolving, dynamically young, turbulence, in spite of the fact that some exhibit a surprisingly broad scaling range of the structure functions. In a later paper, Tu and Marsch (1996) tried to account for the dynamic evolution of the turbulence by generalizing the p-model of Meneveau and Sreenivasan (1987) by introducing a second free parameter into the model. Though good fits of the observed scaling exponents were obtained, no clear physical trend in the parameter values could be found.

Horbury and Balogh (1997) analyzed the magnetic-field structure functions in fast-wind turbulence of polar origin. Because of the short scaling range only the low-order exponents $\zeta_n, n \leq 4$, could be determined with sufficient accuracy. The primary interest is to decide whether the basic scaling of the turbulence is Kolmogorov-like, with $\zeta_3 = 1$ in particular, or more IK-like, with $\zeta_4 = 1$. Data clearly indicate that there is a Kolmogorov-like scaling and thus rule out intermittency models based on IK scaling such as (7.69). By comparing the results for ζ_n with several intermittency models based on approximated Kolmogorov scaling the authors could refute the random-β model and the She–Lévêque model (7.64) for hydrodynamic turbulence, while good agreement with the MHD log-Poisson model ζ_n^{MHD}, (7.66), was found. Hence this simple model, which also fits the results from simulations of homogeneous turbulence, as discussed in Section 7.5.1, appears to catch essential properties of MHD turbulence.

10.2.3 Dissipation of turbulence

We have mentioned before that the temperature in the solar wind decreases more slowly than adiabatically, which implies that there is a local input of energy. While heating by shock waves is probably the most efficient mechanism, also the dissipation of the Alfvénic fluctuations, the main, solenoidal, component of the turbulence, should give a major contribution, especially in the fast polar wind, where amplitudes of fluctuations are particularly large. Because of the absence of interparticle collisions, dissipation occurs by collective effects, notably Landau and cyclotron damping of waves (see, for instance, Stix, 1962). While Landau damping dominates for small-scale longitudinal waves, the transverse fluctuations of the solar-wind turbulence are most strongly affected by cyclotron damping. For a wave that is circularly polarized in the direction of gyration of particles about the local magnetic field a resonance occurs if the

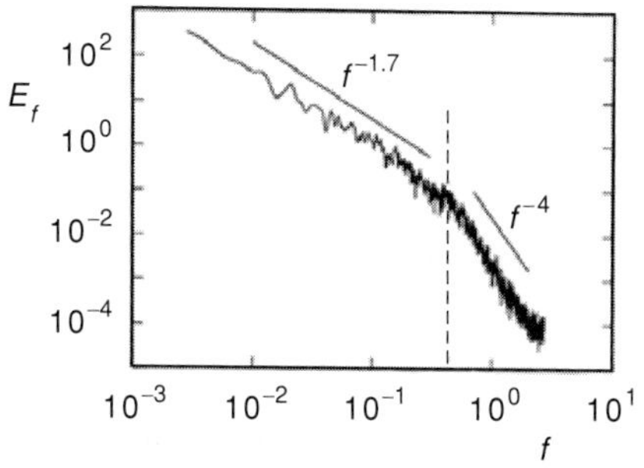

Figure 10.7: A typical energy spectrum for turbulence, showing the dissipative falloff at $f > 0.4$ Hz corresponding to cyclotron-resonance damping of Alfvén waves. (From Leamon *et al.*, 1998.)

Doppler-shifted frequency equals the cyclotron frequency,

$$\varpi - k_\parallel v_\parallel = \Omega_i, \tag{10.7}$$

where $\Omega_i = eB/(m_i c)$ is the ion cyclotron frequency (we consider only the ion cyclotron resonance, since the electron cyclotron frequency is too high for resonance with the fluctuations in the solar wind). In the plasma frame $\omega \simeq k_\parallel v_\mathrm{A}$ and significant damping can be expected if $v_\parallel \sim v_{thi}$, the thermal spread of the velocity distribution. Hence dissipation of the turbulence should occur at wavenumbers

$$k_\parallel = k_d \simeq \frac{\Omega_i}{v_\mathrm{A} + v_{thi}} = \frac{\Omega_i}{v_\mathrm{A}(1 + \beta^{1/2})}, \tag{10.8}$$

such that the observed frequency spectrum should fall off steeply at $f \simeq k_d v/(2\pi) \simeq \frac{1}{2}[\Omega_i/(2\pi)]\mathrm{M_A} \simeq 0.5$ Hz for typical values $B = 5 \times 10^{-5}$ G and $\mathrm{M_A} = 10$. This is, indeed, observed in the energy spectrum shown in Fig. 10.7. The observation of the power-law spectrum with the falloff in the spectral range where strong damping is expected gives further evidence for the presence of a turbulent cascade.

10.2.4 Compressive fluctuations

We know already that globally the solar wind, because of the expansion, is a highly compressible medium the mean density decreasing as $n \sim r^{-2}$, (10.5), while, on the other hand, small-scale fluctuations are essentially incompressible. Let us now consider effects of compression more closely. On large scales strong compression occurs in the corotating interaction region mentioned before, where fast wind catches up with slow wind. Also in the transitional scale range, the so-called mesoscales, say, $10^{-8}\,\mathrm{km}^{-1} < k/(2\pi) \lesssim 10^{-7}\,\mathrm{km}^{-1}$, substantial variations in density, $\delta n/\overline{n} \sim 1$, are observed. They do not, however, represent fast shock waves, but rather tangential discontinuities with approximate pressure balance $\delta[B^2/(8\pi) + p] \simeq 0$. At smaller scales, $k/(2\pi) > 10^{-7}\,\mathrm{km}^{-1}$, effects of compression are weak; field lines are mainly bent, not compressed, $\delta \boldsymbol{b} \simeq \delta \boldsymbol{b}_\perp$, and $\delta n \ll \overline{n}$. Nonetheless, definite spectral laws for the fluctuations of density and field strength have been obtained from observations (e.g., Marsch and Tu, 1990b). Systematic differences are found for low-speed and high-speed wind conditions, following mainly the properties of the velocity and magnetic-field spectra discussed before. Not surprisingly, fluctuations are more compressive in the slow wind. Here the density spectrum is consistent with a $k^{-5/3}$ law. It appears to be fully evolved and does not vary radially except for the amplitude, which follows the general expansion. It is difficult to associate the density fluctuations with a specific mode characteristic such as the fast mode, since the compressive part of the velocity cannot be isolated by a single spacecraft. In interpreting compressive effects in the solar wind, one should consider that solar-wind turbulence is, at most, transonic, since the velocity fluctuations are usually subsonic, only the mean flow being highly supersonic. Hence the theory developed by Montgomery *et al.* (1987), which predicts that, in a weakly compressible plasma, the density spectrum should follow a Kolmogorov law if the magnetic-field spectrum does, seems to be most adequate.

In the fast wind, where fluctuations are essentially Alfvénic, fluctuations in density are very small. As the turbulence evolves radially, “aging dynamically”, the evolution of the density spectrum is found to be similar to that of the subdominant inward-propagating Elsässer field z^-, shown by the dotted lines in the right-hand column in Fig. 10.5.

11

Turbulence in accretion disks

In this chapter we talk about accretion disks, a widespread phenomenon in astrophysics, wherein magnetic turbulence is present not just as a byproduct but rather is essential for its very existence, as is now generally believed. Accretion, the accumulation of mass onto a central object due to gravitational attraction, naturally leads to the formation of disk-like structures, since the infalling matter, due to conservation of angular momentum, tends to rotate about the center of gravity. The system is in approximate equilibrium, in that the radial component of the gravitational force is balanced by the centrifugal force and the axial component by the pressure gradient. Since the disk material moves on Keplerian orbits with angular velocity $\Omega(r) \propto r^{-3/2}$, the angular momentum $\propto r^2\Omega$ decreases with decreasing radius. Hence, when the matter moves inward, conservation of angular momentum requires that the excess is transferred outward. Thus the rate of accretion of mass is determined by the transport of angular momentum, which therefore becomes the crucial issue for understanding the dynamics of these systems. If a transport mechanism is provided, material is spiraling in toward the central object (just as conservation of vorticity leads to a spiraling flow of the water from a bathtub). Half of the gravitational energy set free by the inward motion goes into kinetic energy, speeding up the flow as it spirals more rapidly at smaller radii, while the rest is dissipated into heat, which makes the disk luminous. The intense energetic radiation is in fact the hallmark of accretion disks in observations.

In Section 11.1 we briefly outline the basic physics of accretion disks as well as the observational evidence, concentrating on close binary systems. Section 11.2 gives the steady-state equations for a Keplerian disk, describing the configuration in terms of the stress tensor, which accounts for the transport of angular momentum, the basic motor of the accretion process. Since molecular viscosity can be neglected, it is generally assumed that the transport is caused by turbulence in the disk. Shakura and Sunyaev (1973) introduced a simple turbulent viscosity in terms of a parameter α, resulting in the classical α-disk model, whose scaling laws are very useful for a general understanding of accreting systems and are essentially confirmed by observations. The crucial problem for a first-principles theory is the origin of the presumed turbulence since, contrary to intuitive reasoning, a Keplerian shear flow is found to be hydrodynamically stable, both linearly and nonlinearly (Section 11.3). The clue for solving this mystery is provided by the magnetorotational instability introduced by Balbus and Hawley (1991), which demonstrates the fundamental importance of even weak magnetic fields for the disk dynamics. We first consider the basic properties of the linear instability and then discuss the various aspects of the nonlinear behavior. Though the saturation level is not fully universal but may depend on the initial state, the effective magnetoviscosity is of the correct order of magnitude required by the observed rates of accretion (Section 11.4).

It is obvious that this chapter can give only a rather cursory introduction to this rich and multifacetted topic, which has been a central theme in astrophysics for many years. The interested reader is therefore referred to several reviews that have appeared in the literature, in particular those by Pringle (1981) and, more recently, by Papaloizou and Lin (1995), who discuss the theoretical aspects, while a sequel to the latter (Lin and Papaloizou, 1996) presents an overview of the observations. A collection of contributions giving a broad coverage of the field is contained in an ASP Conference Series (Wickramasinghe *et al.*, 1997) and in a volume of lecture notes (Meyer-Hofmeister and Spruit, 1997). Concerning more specifically the topic treated in this book, instability theory and MHD turbulence, an excellent review is offered by the article by Balbus and Hawley (1998). On a more elementary level, the textbook by Frank *et al.* (1992) gives a general introduction to accretion phenomena.

11.1 Basic properties of accretion disks

Accretion disks are numerous in the universe and occur with very different sizes and physical parameters. Roughly speaking, there are three groups of accreting systems, (i) protostellar disks, (ii) close binaries, and (iii) active galactive nuclei (AGNs). Protostellar disks, which lead to formation of stars and planets, consist of relatively cool and dense gas, with a major fraction of dust, and thus constitute rather intricate systems, in which complex chemical processes play an important role. However, since in these very weakly ionized gases with their low electrical conductivity effects of magnetic fields can be neglected, this book will have little to say about them. In binary stars and AGNs, on the other hand, the degree of ionization is high, so plasma properties, and hence the coupling to the magnetic field, are essential. In AGNs magnetic fields are very important; in particular, the collimated flows ejected on both sides along the axis, the jets, require the presence of a strong field. Also the disk-like structure can be observed directly and many fascinating features of these distant objects have been disclosed.

Most observational information about accretion disks is, however, obtained from binary-star systems, because more details are known for them than for other astronomical objects, notably their masses and dimensions, though the spatial structure cannot be directly imaged. In addition, there are many of these, even rather close to our solar system. In fact, the majority of stars form binaries, and conditions for accretion disks to occur in binaries are relatively common. Detailed observations of the temporal variation and spectral composition of the radiation have provided a consistent picture of their structure and dynamics, which, in essence, agrees with the theoretical predictions. In the following, when we are thinking of applications, mainly accretion in binaries is meant.

Let us briefly discuss how mass transfer and accretion occur in such a system. The gravitational field in the neighborhood of two stars is best illustrated by the equipotentials in the orbital plane. These consist, broadly, of two classes, lines encircling both stars at distances much larger than their separation and those encircling each star in its neighborhood. Both classes are separated by a figure-eight line, called the Roche lobes of the two stars. There is a saddle point of the potential in between at the line crossing, called the inner Lagrange point L_1,[1] which is, figuratively speaking, the pass between the potential troughs around each star. As long as both stars are detached, i.e., much smaller than their lobes, transfer of mass can occur only through the stellar winds. If, however, during its evolution, one star swells, eventually filling its Roche lobe, then material flows into the Roche lobe of the compact star, a process called Roche-lobe overflow (illustrated in Fig. 11.1), which occurs across L_1 and its vicinity. As the region around M_1 fills up with gas, this flow is no longer freely propagating but gradually becomes thermalized, radiating energy until it reaches the lowest energy state compatible with its angular momentum, a circular orbit, from which it slowly spirals inward as angular momentum is transported outward. The accretion process is also illustrated stereographically in Fig. 11.2.

[1] Actually, the situation is somewhat more complicated. The binary system is conveniently considered in the rotating center-of-mass frame of reference. The effective potential has the form

$$\phi_g = -\frac{GM_1}{|\boldsymbol{r}-\boldsymbol{r}_1|} - \frac{GM_2}{|\boldsymbol{r}-\boldsymbol{r}_2|} - \tfrac{1}{2}(\boldsymbol{r}\times\boldsymbol{\Omega})^2,$$

where $\boldsymbol{r}_{1,2}$ are the radius vectors of the stars from the center of mass and the last term stems from the centrifugal force; see (2.27). The latter gives rise to four additional potential extrema outside the Roche lobes, the outer Lagrange points, which are, however, not important in the present context.

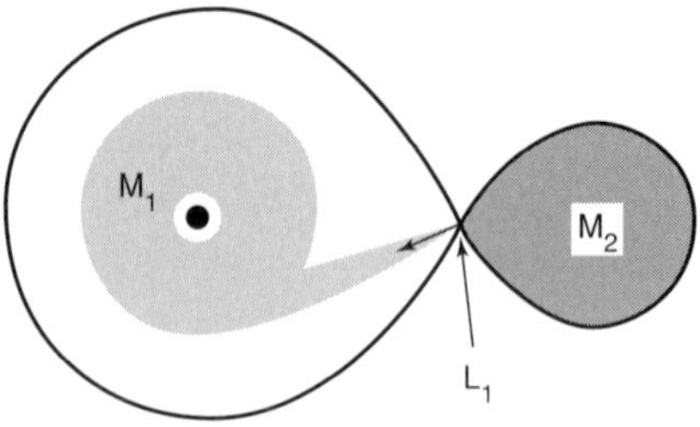

Figure 11.1: A schematic illustration of the transfer of mass and disk formation in a binary system M_1, M_2 with the donor star M_2 filling its Roche lobe.

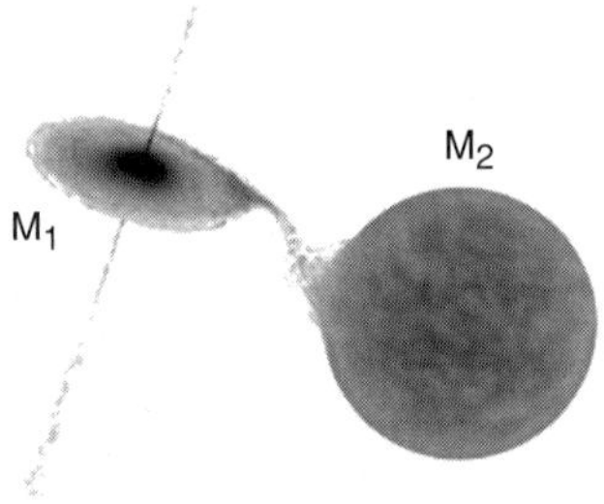

Figure 11.2: A stereographic illustration of the accretion onto a compact star M_1 from a donor star M_2. The figure shows also the jet emitted along the axis of rotation of the accretion disk.

The luminosity, the total radiated energy flux, of an accreting object is limited by the Eddington luminosity. This limit is reached when the radiation pressure balances the gravitational force, since at greater luminosities the outward radiation pressure would exceed the gravitational attraction, so accretion could no longer occur. The Eddington luminosity is given by the expression[2]

$$L_{\mathrm{E}} = 1.3 \times 10^{38}(M/M_{\odot})\,\mathrm{erg\ s^{-1}}, \tag{11.1}$$

[2] The radiation acts primarily on the electrons. If S is the energy flux per unit surface, the force on the electrons is $S\sigma_{\mathrm{T}}/c$, where σ_{T} is the Thompson cross-section. Charge neutrality (i.e., quasi-neutrality, see Section 2.1) requires that the same force also acts on the ions, which are mainly protons with mass m_p. The gravitational force on each proton is GMm_p/r^2. Expressing the radiation force in terms of the luminosity $L = 4\pi r^2 S$, the net inward force is

$$\frac{GMm_p}{r^2} - \frac{L\sigma_{\mathrm{T}}}{4\pi c r^2}.$$

Hence the luminosity for which this force vanishes is

$$L = L_{\mathrm{E}} = \frac{4\pi GMm_p c}{\sigma_{\mathrm{T}}} = 1.3 \times 10^{38}\frac{M}{M_{\odot}}\,\mathrm{erg\ s^{-1}},$$

where $M_{\odot}$ is the solar mass.

which depends only on the mass of the central object and fundamental constants and is, in particular, independent of the radius. Typical luminosities of binary X-ray sources are $L \sim 10^{37}\,\mathrm{erg\,s^{-1}}$, which is not very far off the limit. On the other hand, the luminosity is determined by the accretion rate $\dot{M}$, the amount of matter falling onto the stellar surface per second. If its kinetic energy is completely transformed into radiation at the stellar surface, the maximum luminosity of the accreting object is

$$L_{\mathrm{acc}} = GM\dot{M}/r_*, \tag{11.2}$$

where r_* is the stellar radius. Thus the condition $L_{\mathrm{acc}} < L_E$ implies that there is an upper limit on the rate of accretion.

Accretion disks form primarily around compact objects, i.e., white dwarfs,[3] neutron stars, and black holes. From (11.2) one obtains a estimate of the spectral range of the radiation emitted from such accreting objects. Assuming the gas to be optically thick, such that radiation occurs in the form of blackbody radiation of temperature T, the mean photon energy is $h\nu \sim k_{\mathrm{B}}T$, which is related to the luminosity L_{acc} of the object, $L_{\mathrm{acc}} = 4\pi r_*^2 \sigma T^4$ with the Stefan–Boltzmann constant $\sigma = 5.67 \times 10^{-5}\,\mathrm{erg\,s^{-1}}$,

$$T = (L_{\mathrm{acc}}/4\pi r_*^2 \sigma)^{1/4}. \tag{11.3}$$

Hence T depends only weakly on the luminosity. For neutron stars we may therefore take the Eddington luminosity in (11.3), which gives

$$k_{\mathrm{B}}T \sim 10^{10} (M/M_{\odot})^{1/4} r_*^{-1/2}\ \mathrm{eV}, \tag{11.4}$$

where r_* is in centimeters. For white dwarfs, however, the luminosity is considerably smaller, $L_{\mathrm{acc}} \sim 10^{33}\,\mathrm{erg\,s^{-3}}$; hence $k_{\mathrm{B}}T$ is smaller by an order of magnitude. With typical radii $r_* \sim 10^9\,\mathrm{cm}$ and $10^6\,\mathrm{cm}$ for white dwarfs and neutron stars, respectively, and masses of roughly one solar mass, the thermal photon energies emitted from the central part of the disk should be of the order of 1 keV for the latter and 10 eV for the former.[4] Hence accretion disks around

[3] Of particular interest are cataclysmic variables consisting of a low-mass main-sequence star as secondary partner of the binary system filling its Roche lobe transferring matter onto a white dwarf as primary. In these systems, however, accretion does not occur in a continuous way but rather as a sequence of bursts.

[4] Nonthermal photons of much higher energy may be emitted directly from the stellar surface, where an upper limit on the photon energy is given by the energy of infalling particles GMm_p/r_* converted into radiation,

$$h\nu \lesssim 10^{14} (M/M_{\odot}) r_*^{-1}\ \mathrm{eV}. \tag{11.5}$$

This corresponds to photon energies up to 10^2 MeV for neutron stars and 10^2 keV for white dwarfs.

neutron stars are expected to be X-ray sources, whereas those around white dwarfs radiate mainly in the ultraviolet.

Accretion around a black hole is somewhat different, since at the Schwarzschild radius $r_S = 2GM/c^2 = 3(M/M_\odot)$ km, the horizon of the black hole, all infalling matter is simply swallowed. Only at distances significantly larger than r_S, for which general-relativity effects are still weak, does the virial theorem (see footnote in Section 11.2.1) apply by virtue of which half of the energy set free is thermalized and radiated. As a result the total luminosities and the photon energies are similar to those of accretion disks around neutron stars. The main difference is the absence of a magnetic field in a black hole; hence there is no magnetosphere, which strongly affects the accretion process in the innermost region around a neutron star.

It is also useful to have a rough idea of the density of the matter in accretion disks. Characteristic number densities are $n \sim 10^{15}\,\text{cm}^{-3}$, which is much higher than the number density in molecular cloud cores with typically $10^6\,\text{cm}^{-3}$ (see Chapter 12), but much lower than that in stars with $\sim 10^{25}\,\text{cm}^{-3}$. Surface temperatures decrease as $r^{-3/4}$, see Section 11.2.2, as one moves away from the central region; hence the outer parts of the disk are not very hot but still sufficiently ionized to be treated as a plasma.

11.2 The standard disk model

From these basic properties of accretion disks we now derive a more quantitative model. A major simplification arises through the thin-disk approximation, which is usually well satisfied in the main part of the disk outside the region close to the stellar surface, where physical conditions become rather complicated. The transport of angular momentum, which determines the accretion rate, is described in a phenomenological way.

11.2.1 Keplerian disks

We consider an accretion disk around a point mass M located at the center $r = z = 0$ of a cylindrical coordinate system (r, θ, z). The gravitational force is entirely due to the central mass, neglecting the self-gravitation of the matter in the disk. This is in general a very good approximation because of the low density of the disk material. In the steady state the disk is characterized by a flow with a dominant azimuthal component v_θ, the Keplerian motion, and a small radial component v_r, the accretion motion, $v_r \ll v_\theta$. In the r-direction gravity is balanced by the centrifugal force,

$$v_\theta^2/r = \Omega^2 r = GM/r^2, \tag{11.6}$$

whence Kepler's law $\Omega \propto r^{-3/2}$. In the axial direction the gravitational force can be balanced only by the pressure gradient, which is much weaker than the centrifugal force, since in the main part of the disk the Keplerian motion is strongly supersonic, $r\Omega \gg c_s$. We thus have

$$\partial_z p = -\frac{GM\rho}{r^3} z = -\rho\,\Omega^2 z, \tag{11.7}$$

which, for a constant temperature across the disk, has the solution

$$\rho = \rho_c \mathrm{e}^{-z^2/(2h^2)}, \quad h = c_s/\Omega \tag{11.8}$$

with the isothermal sound speed $c_\mathrm{s} = \sqrt{p/\rho}$. The pressure consists of the gas and the radiation pressures,

$$p = \frac{2\rho}{m_p} k_\mathrm{B} T + \frac{4}{3c}\sigma T^4.$$

In general the temperature is not constant across the disk, but the change from the midplane value T_c to the value at the surface T is much smaller than the change of the density. One usually defines the sound speed in terms of T_c. T_c and T are connected by radial transport, which is described by the integrated optical depth τ,

$$\sigma T_c^4 = \tau \sigma T^4,$$

where τ can be calculated from the standard free–free-absorption model of the opacity (see, e.g., Frank *et al.*, 1992).

The pressure scale height h is small compared with the radius, $h/r = c_s/v_\theta \ll 1$; thus h/r can be regarded as the expansion parameter in this thin-disk approximation. The equilibrium force balance does not determine the slow radial inward flow v_r. To obtain a relation for this quantity, we consider the radial component of the angular-momentum flux, (2.54),

$$F_r = r\left(\rho v_\theta v_r - \frac{B_\theta B_r}{4\pi}\right), \tag{11.9}$$

which is divergence-free in the steady state, $\partial_r r\, F_r = 0$. Viscosity is neglected in (11.9) since, because of the low gas density, collisional effects on the transport of angular momentum are negligible. Instead it is generally believed that the transport is due to turbulence, the origin of which will be the subject of the subsequent sections. We therefore write $\boldsymbol{v} = \langle \boldsymbol{v} \rangle + \widetilde{\boldsymbol{v}}$ and $\boldsymbol{B} = \widetilde{\boldsymbol{b}}$, assuming that the mean magnetic field is weak and can be neglected in the equilibrium. In this section the average is taken not only over the fluctuations, which can be interpreted as a time average, but also over the vertical disk structure,

$$\langle f \rangle = \frac{1}{\rho_\Sigma} \int_{-\infty}^{\infty} dz \, \overline{\rho f}, \tag{11.10}$$

where the overbar indicates the time average and ρ_Σ is the surface density,

$$\rho_\Sigma = \int_{-\infty}^{\infty} dz \, \rho = \sqrt{2\pi} h \rho_c \tag{11.11}$$

with $\rho_c(r)$ the density in the midplane of the disk. With $\langle v_\theta \rangle = r\Omega$ the average of the angular-momentum flux (11.9) assumes the form

$$\langle F_r \rangle = \rho_\Sigma r \Big(r\Omega \langle v_r \rangle + R_{r\theta} \Big) = \frac{\text{constant}}{r}, \tag{11.12}$$

where

$$R_{r\theta} = \langle \widetilde{v}_r \widetilde{v}_\theta - \widetilde{v}_{\mathrm{A}r} \widetilde{v}_{\mathrm{A}\theta} \rangle, \quad \widetilde{\boldsymbol{v}}_\mathrm{A} = \frac{\widetilde{\boldsymbol{b}}}{\sqrt{4\pi\rho}} \tag{11.13}$$

is the r, θ component of the turbulent stress tensor (4.4) (up to a sign factor). The 'fluctuating Alfvén speed' $\widetilde{\boldsymbol{v}}_\mathrm{A}$ should be considered only as the suitably normalized field fluctuation and has no physical meaning as a velocity fluctuation, since the Alfvén speed is defined as the speed of propagation of magnetic perturbations and hence refers to the average field. The first term in the expression (11.12) is the part of the angular-momentum flux which is coupled to the accretion flow, and hence is directed inward, while the second term is the angular-momentum flux directed outward. The constant is fixed at the inner boundary of the disk, which we take as the stellar surface r_*, where the stress tensor vanishes. On substituting the radial flow by the accretion rate $\dot{M}$,

$$\dot{M} = -2\pi r \rho_\Sigma \langle v_r \rangle, \tag{11.14}$$

(11.12) can be written in the form

$$-\frac{\dot{M} r \Omega}{2\pi} + \rho_\Sigma r R_{r\theta} = -\frac{\dot{M}}{2\pi r} r_*^2 \Omega_*,$$

which can be solved for $R_{r\theta}$,

$$R_{r\theta} = \frac{\dot{M}\Omega}{2\pi\rho_\Sigma}\left[1 - \left(\frac{r_*}{r}\right)^{1/2}\right]. \tag{11.15}$$

Note that, for $r \gg r_*$, the essential r-dependence of $R_{r\theta}$ arises through $\Omega \sim r^{-3/2}$ and $\rho_\Sigma(r)$.

We now derive the radial energy flux, which is connected with the luminosity of the disk. From (2.57) we obtain, neglecting the pressure which is of order $(h/r)^2$,

$$\begin{aligned}\langle F_r^E \rangle &= \left(\tfrac{1}{2}\rho_\Sigma r^2\Omega^2 + \rho_\Sigma \phi_g\right)\langle v_r \rangle + \rho_\Sigma r \Omega R_{r\theta} \\ &= \frac{\dot{M} r \Omega^2}{4\pi} + \rho_\Sigma r \Omega R_{r\theta},\end{aligned} \tag{11.16}$$

where in the second line we made use of the virial theorem (see Section 12.3.1)[5] $\phi_g = -r^2\Omega^2$ and of (11.14), or, inserting (11.15),

$$\langle F_r^E \rangle = \frac{3GM\dot{M}}{4\pi r^2}\left[1 - \frac{2}{3}\left(\frac{r_*}{r}\right)^{1/2}\right]. \tag{11.18}$$

The divergence of the energy flux gives the surface emissivity S,

$$S = \frac{1}{r}\,\partial_r r\,\langle F_r^E \rangle = \frac{3GM\dot{M}}{8\pi r^3}\left[1 - \left(\frac{r_*}{r}\right)^{1/2}\right]. \tag{11.19}$$

The luminosity of the disk is the integral of S over the surface of the disk,

$$L = 2\int_{r_*}^{\infty} S(r) 2\pi r\, dr = 2\pi r_* \langle F_r^E(r_*) \rangle = \frac{GM\dot{M}}{2r_*}, \tag{11.20}$$

where the factor of 2 arises from the two sides of the disk. Expression (11.20) is just half the accretion luminosity $L_{\rm acc}$, (11.2), the second half coming from the kinetic energy, which eventually will also be transformed into radiation when the material impinges on the stellar surface.

It is thus clear that most of the luminosity of the accreting system originates from the emission at, or close to, the stellar surface with $r \gtrsim r_*$, since the emissivity of the disk declines rapidly on moving away from the center, $S \propto r^{-3}$ for $r \gg r_*$. From $S = \sigma T^4$ we obtain also the radial dependence of the temperature, $T \propto r^{-3/4}$. While the radiation spectrum in the neighborhood of the accretion center of a neutron star or black hole, $r \sim 10^6$ cm, is in the X-ray range $h\nu \sim 1$ keV, the photon energy drops to $h\nu \sim 0.1$ eV, i.e., the infrared range, at a distance of one solar radius, $r_\odot \sim 10^{11}$ cm.

[5] The virial theorem for a particle moving under a central force $\boldsymbol{f} = -\nabla\phi(r)$ reads

$$\tfrac{1}{2}m\overline{v^2} = -\tfrac{1}{2}\overline{\boldsymbol{f}\cdot\boldsymbol{r}} = \tfrac{1}{2}\overline{r\,\partial_r\phi},$$

where the overbar indicates the time average. For a power law $\phi \propto r^n$ the virial theorem connects kinetic and potential energy,

$$\tfrac{1}{2}m\overline{v^2} = \tfrac{1}{2}n\overline{\phi}. \tag{11.17}$$

In the case of the gravitational field $n = -1$, and one obtains $\frac{1}{2}\overline{v^2} = -\frac{1}{2}\overline{\phi_g}$.

11.2.2 The α-disk model

In the preceding section we derived a number of relations that determine the main variables of the disk, namely the temperature $T(r)$, the surface density $\rho_\Sigma(r)$, and the thickness of the disk $h(r)$, as functions of the stress tensor $R_{r\theta}$. This is as far as we can go without some knowledge about the transport of angular momentum. Since the latter is not due to classical viscosity, the most probable cause is the action of some kind of turbulence. In the absence of a strong magnetic field, i.e., for a high-β plasma, there are mainly two sources of free energy in the disk to drive turbulence, namely the shear flow connected with the Keplerian motion and the pressure gradient in the axial direction. While we shall consider the actual mechanism of the generation of turbulence in the subsequent sections, at this point it is sufficient to know that there are good reasons for the presence of a sufficiently high level of fluid turbulence, which can be accounted for by a phenomenological model. As outlined in Section 4.1.2, the simplest approach is to introduce an eddy viscosity ν_t, (4.12),

$$\nu_t = \widetilde{v} l_m = \alpha c_s h. \tag{11.21}$$

Here the mixing length l_m is taken to be the pressure scale height h, which is the only intrinsic length scale and therefore the natural choice for buoyancy-driven convection, and the assumption regarding the velocity fluctuation, namely that $\widetilde{v} \lesssim c_s$, is reasonable since higher velocities would give rise to shocks and hence be rapidly damped by dissipation. The proportionality constant α is a free parameter, which contains all the missing information on the properties of the turbulence.

The parameter α was first introduced, in a slightly different way, by Shakura and Sunyaev (1973) using essentially dimensional arguments. They made the *Ansatz* $R_{r\theta} = \alpha c_s^2$, since in a high-$\beta$ plasma the speed of sound is the only intrinsic quantity with the dimension of a velocity. Indeed, this *Ansatz* is equivalent to (11.21) if one when assumes that the shear flow is the free-energy source, see (4.10),

$$R_{r\theta} = -\nu_t \frac{dv_\theta}{dr} \sim \alpha c_s h \Omega = \alpha c_s^2, \tag{11.22}$$

replacing Ω from (11.8).

Convincing as the *Ansatz* (11.21) might look, some caution is appropriate. The turbulence may well be strongly anisotropic with radial scales of the order of r, since in the case of a shear-flow instability the maximum growth rate $\gamma_{\max} \sim dv_\theta/dr$ occurs for $kr \sim 1$; see (3.33). Even if the Keplerian flow is not directly unstable, as discussed in Section 11.3, it is the main source of free energy and may well influence the turbulence in the fully developed state. In addition, turbulent motions amplify the magnetic field by dynamo action, which could invalidate the assumption of a primarily nonmagnetic disk, introducing the Alfvén speed beside the sound speed in (11.21).

Nonetheless the simple *Ansatz* (11.22) has proven very useful for constructing a simple reference model of an accretion disk, which yields properties that are also accessible to, and corroborated by, observations. The model consists of scaling laws for the physical quantities of the disk in terms of the two parameters $\dot{M}$ and α. These are the midplane temperature T_c, the midplane density ρ_c, and the thickness h, which are connected by the relations

$$\sigma T_c^4 = \tau S, \tag{11.23}$$

$$S = \frac{3GM\dot{M}}{8\pi r^3}, \tag{11.24}$$

$$\rho_c = \frac{\rho_\Sigma}{\sqrt{2\pi} h}, \tag{11.25}$$

$$\rho_\Sigma = \frac{\dot{M}\Omega}{2\pi R_{r\phi}}, \tag{11.26}$$

$$h = c_{\mathrm{s}}/\Omega, \tag{11.27}$$

$$R_{r\theta} = \alpha c_{\mathrm{s}}^2. \tag{11.28}$$

In these equations we have ignored for simplicity factors of $1 - (r/r_*)^{1/2}$, thus limiting their applicability to regions $r \gg r_*$ well away from the central object, which is not a crucial omission, since in any case the rather intricate conditions in the stellar atmosphere and the interaction of the stellar magnetosphere with processes in the accretion disk (see. e.g., Miller and Stone, 1997) are only very crudely described by including these factors. It is worth pointing out that α and $\dot{M}$ are free parameters that are independent of each other though, of course, conceptually related as only a finite α allows accretion to occur.

From (11.23)–(11.28) one obtains, after some straightforward manipulations, the scaling laws of the standard α-disk:

$$T_c \sim M^{1/4} \dot{M}^{3/10} \alpha^{-1/5} r^{-3/4}, \tag{11.29}$$

$$\rho_c \sim M^{5/8} \dot{M}^{11/20} \alpha^{-7/10} r^{-15/8}, \tag{11.30}$$

$$h \sim M^{-3/8} \dot{M}^{3/20} \alpha^{-1/10} r^{9/8}. \tag{11.31}$$

Figure 11.3: The cross-section of an accretion disk (shown schematically).

The exact equations including proportionality constants can be found, for instance, in the book by Frank *et al.* (1992). In the qualitative discussion in this section we want only to point out the weak dependence, in particular of T_c and h, on the parameters $\dot{M}$ and α, such that the configuration of the disk is rather insensitive to the actual numerical values chosen. Values of $\dot{M}$ are obtained from the observed luminosity L_{acc}, (11.2), by inserting the mass and the radius of the accreting object. One finds that $\dot{M} \sim 10^{16}$ g s^{-1} both for neutron stars and for white dwarfs, since the luminosity is roughly inversely proportional to the radius. Typical values of α assumed in modeling cataclysmic variables are $\alpha \sim$ 0.01–0.1 (Cannizo *et al.*, 1988). It is also interesting to note that the density decreases much more rapidly with radius than does the temperature and that the thickness of the disk increases nearly linearly, as illustrated in Fig. 11.3.

11.3 Hydrodynamic stability of accretion disks

Since only turbulence allows angular-momentum transport of sufficient magnitude to explain accretion rates inferred from observations, one would like to know the mechanism for the generation of turbulence and the main properties in the fully developed nonlinear state. The phenomenological *Ansatz* (11.21), or (11.22), implies not just a certain level of turbulence energy but also a sufficiently strong correlation between the components of the velocity fluctuations $\widetilde{v}_r$ and $\widetilde{v}_\theta$ with the correct signs. This problem has been studied intensively during the last two decades, and it was found that those types of turbulence, which seemed to be rather naturally excited, either do not exist at all in an accretion disk, namely the classical shear-flow instability, or, in the case of convective turbulence, lead to very inefficient transport of angular momentum, which, in addition, has the wrong sign, as we shall discuss. Only rather recently has the crucial role of the magnetic field been disclosed. It now appears that the magnetorotational instability discussed by Balbus and Hawley (1991) and the resulting turbulence do indeed satisfy most of the requirements for an adequate transport mechanism. In this section we consider the hydrodynamic case in order to find out why the Keplerian flow is stable, even nonlinearly, and why convective turbulence is ineffective. In the subsequent section we then add a (weak) magnetic field, which gives rise to instability, making the disk fully turbulent.

11.3.1 Shear-flow stability of a Keplerian disk

Hydrodynamic shear-flow stability of a rotating incompressible fluid is determined by Rayleigh's criterion (3.44), which we have derived in Section 3.2.1. The fluid is stable if the specific angular momentum increases outward,

$$\frac{d(r^2\Omega)^2}{dr} > 0,$$

which is valid for sufficiently smooth profiles $\Omega(r)$. This is satisfied for the Keplerian flow $\Omega \sim r^{-3/2}$, which is hence shear-flow stable. In speaking of instability, one usually implies linear instability, i.e., the growth of an infinitesimal perturbation. In cases of linear stability the question which arises is that of what happens if perturbations of finite, though still small, amplitude are applied, since it is well known that nonrotating flows are often nonlinearly unstable, for instance the Couette flow between two parallel plates moving relative to each other, and the Poiseuille flow in a pipe, which are linearly stable but become fully turbulent if the perturbation exceeds some threshold (see, e.g., Orszag and Kells, 1980). This property of nonrotating flows led to the idea that velocity shear will always give rise to turbulence if only the Reynolds number is sufficiently high, and thus will also give rise to the Keplerian shear flow in an accretion disk. The latter is, however, not true, as numerical simulations have demonstrated (Balbus *et al.*, 1996); simulations show that even large initial disturbances decay. To understand this behavior, we consider the hydrodynamic equation (2.7) for $\boldsymbol{B} = 0$. We split the velocity into the Keplerian flow and the deviations therefrom,

$$\boldsymbol{v} = r\Omega \boldsymbol{e}_\theta + \widetilde{\boldsymbol{v}},$$

and treat the equations in the local approximation, i.e., consider only a small region $\Delta r \ll r$, $\Delta z \ll h$, neglecting for instance terms in $\widetilde{v}_\theta / r$ compared with $\partial_r \widetilde{v}_\theta$,

$$\frac{d}{dt'}\widetilde{v}_r - 2\Omega\widetilde{v}_\theta = -\frac{1}{\rho}\,\partial_r p, \tag{11.32}$$

$$\frac{d}{dt'}\widetilde{v}_\theta + \frac{\kappa^2}{2\Omega}\widetilde{v}_r = -\frac{1}{\rho r}\,\partial_\theta p \tag{11.33}$$

$$\frac{d}{dt'}\widetilde{v}_z = -\frac{1}{\rho}\,\partial_z p + \partial_z \phi_g, \tag{11.34}$$

where

$$\frac{d}{dt'} = \partial_{t'} + \widetilde{\boldsymbol{v}}\cdot\nabla, \quad \partial_{t'} = \partial_t + \Omega\,\partial_\theta,$$

and viscous effects are omitted for simplicity. These have to be supplemented by the equations for density (2.16) and pressure (2.20). In the local approximation we neglect the r-dependence of the gravitional force, $\phi_g = \phi_g(z)$, (11.7). The parameter κ in (11.33),

$$\kappa^2 = \frac{1}{r^3}\frac{d(r^2\Omega)^2}{dr}, \tag{11.35}$$

is called the *epicyclic frequency*, which is the frequency of the oscillation (in a rotating reference frame) a fluid element performs when it is perturbed from its circular orbit. To see this consider (11.32) and (11.33) for a test fluid element, neglecting pressure effects,

$$\partial_{t'}\widetilde{v}_r - 2\Omega\widetilde{v}_\theta = 0, \quad \partial_{t'}\widetilde{v}_\theta + \frac{\kappa^2}{2\Omega}\widetilde{v}_r = 0.$$

With the *Ansatz* $\widetilde{v}_r, \widetilde{v}_\theta \sim e^{-i\varpi t'}$ these yield

$$\varpi^2 = \kappa^2. \tag{11.36}$$

For a Keplerian flow $\Omega = \sqrt{GM}/r^{3/2}$ one finds $\kappa = \Omega$.

To illustrate the difference between the stability properties in rotating and straight, or Cartesian, flows, we consider an energy formulation of the equations (11.32)–(11.34) following Balbus *et al.* (1996). Multiplying (11.32) by $\widetilde{v}_r$ and (11.33) by $\widetilde{v}_\theta$ and averaging over the local region Δr, Δz, we obtain

$$\tfrac{1}{2}\,\partial_{t'}\langle\rho\widetilde{v}_r^2\rangle + \tfrac{1}{2}\langle\nabla\cdot(\rho\widetilde{\boldsymbol{v}}\widetilde{v}_r^2)\rangle = \underbrace{2\Omega\langle\rho\widetilde{v}_r\widetilde{v}_\theta\rangle}\, - \langle\widetilde{v}_r\,\partial_r p\rangle, \tag{11.37}$$

$$\tfrac{1}{2}\,\partial_{t'}\langle\rho\widetilde{v}_\theta^2\rangle + \tfrac{1}{2}\langle\nabla\cdot(\rho\widetilde{\boldsymbol{v}}\widetilde{v}_\theta^2)\rangle = \underbrace{-\frac{\kappa^2}{2\Omega}\langle\rho\widetilde{v}_r\widetilde{v}_\theta\rangle}\, - \frac{1}{r}\langle\widetilde{v}_\theta\,\partial_\theta p\rangle. \tag{11.38}$$

(Note that, in contrast to (11.10), the average is not taken over the vertical structure of the disk.) Variations in density and hence buoyancy effects are neglected, such that v_z decouples and need not be considered. There are two source terms $\propto \langle\rho\widetilde{v}_r\widetilde{v}_\theta\rangle$ in the equations, indicated by underbraces, which for $\kappa^2 > 0$ have opposite signs. If the shear flow drives the instability, we expect the momentum flux to be in the direction opposite to the velocity gradient, i.e., $\langle\rho\widetilde{v}_r\widetilde{v}_\theta\rangle > 0$, since the shear velocity decreases outward in the Keplerian flow, $r\Omega \sim r^{-1/2}$. Hence the Ω-term in (11.37) acts as a source of radial fluctuation energy $\widetilde{v}_r^2$, while the κ^2-term acts as a sink of $\widetilde{v}_\theta^2$. Since these two terms are coupled by the incompressibility condition, instability will not occur if κ^2 is sufficiently large.

Equations (11.37) and (11.38) should be contrasted with the corresponding equations in Cartesian coordinates x, y for a plane shear flow $v_{0y} = V(x)$,

$$\tfrac{1}{2}\,\partial_t\langle\rho\widetilde{v}_x^2\rangle + \tfrac{1}{2}\langle\nabla\cdot(\rho\widetilde{\boldsymbol{v}}\widetilde{v}_x^2)\rangle = -\langle\widetilde{v}_x\,\partial_x p\rangle, \tag{11.39}$$

$$\tfrac{1}{2}\,\partial_t\langle\rho\widetilde{v}_y^2\rangle + \tfrac{1}{2}\langle\nabla\cdot(\rho\widetilde{\boldsymbol{v}}\widetilde{v}_y^2)\rangle = \underbrace{-V'\langle\rho\widetilde{v}_x\widetilde{v}_y\rangle} - \langle\widetilde{v}_y\,\partial_y p\rangle. \tag{11.40}$$

In this case there is only one source term, which is always destabilizing since $-V'\langle\widetilde{v}_x\widetilde{v}_y\rangle > 0$, driving $\widetilde{v}_y^2$, which, in turn, entrains $\widetilde{v}_x^2$ by virtue of the pressure term through the incompressibility condition. Hence the Cartesian flow is unstable (often only nonlinearly), while the Keplerian flow is stable, even nonlinearly, the reason being the presence of the epicyclic term. Only for nearly constant specific angular momentum, $\kappa^2 \simeq 0$, can nonlinear shear-flow instability be expected, since in this case the equations (11.37) and (11.38) become formally identical to those of the Cartesian flow (11.39) and (11.40) when the coordinates are interchanged, $r \to y$, $\theta \to x$. Extensive numerical simulations by Balbus *et al.* (1996) confirmed these expectations, showing that nonlinear instability occurs only for very small positive $\kappa^2 \ll \Omega^2$, which excludes the Keplerian flow, for which $\kappa^2 = \Omega^2$.

11.3.2 Effects of convective turbulence

A further possible source of angular-momentum transport is convection driven by the vertical temperature gradient. Convective turbulence is expected to occur, in particular, in optically thick protostellar disks, where surface cooling may readily generate a superadiabatic temperature gradient (3.50), as was first discussed by Cameron (1978) and Lin and Papaloizou (1980). In a self-consistent picture of the dynamics the turbulence should generate the radial angular-momentum transport, which allows accretion and hence dissipation of (half of) the gravitational energy set free. This would then lead to an increase of the mid-plane temperature and hence steeper vertical gradients, which in turn would drive convection. The uncertain point in this chain, however, is whether convective turbulence does in fact give rise to angular-momentum transport with the correct sign (outward) and of sufficiently high intensity, i.e., whether the turbulence is self-sustaining.

Convective turbulence is a very effective mechanism for transporting heat, which plays a crucial role in the outer opaque radial layers in stars, giving rise to extended convection zones. It is, however, not clear, even rather doubtful, whether this kind of turbulence, which generates strong correlations between vertical velocity and temperature fluctuations, also entails strong (and positive) correlations between the radial and azimuthal fluctuations in velocity in the Reynolds stress which are responsible for the transport of angular momentum.

The problem has enticed numerous investigators. Authors of earlier studies treated the linearized problem by calculating eigenmodes of the convective instability and their growth rates and estimating the resulting parameter α in the phenomenological model of the angular-momentum flux (11.22) by use of a simple mixing-length approximation. Only rather recently did Ryu and Goodman (1992), studying general nonaxisymmetric linear modes, find that the Reynolds stress is predominantly negative, which is incompatible with an outward flux of angular momentum. Fully nonlinear compressible numerical simulations of convectively unstable accretion disks have been performed by Stone and Balbus (1996) using a 3D version of the hydrodynamic code ZEUS (Stone and Norman, 1992a–1992c), which has become a convenient tool in astrophysical hydrodynamics. The main results of these studies are that the average transport of angular momentum is very weak (corresponding to a value of $\alpha \sim 10^{-5}$) and, in addition, has a negative sign, confirming the linear result obtained by Ryu and Goodman, which essentially excludes convective turbulence as the driving agent of accretion.

11.4 Magnetorotational instability

As we have seen in the preceding section, accretion disks are, contrary to intuitive expectation, hydrodynamically either stable or, if they are unstable, the resulting turbulence does not lead to the required transport of angular momentum. How can the free energy of the Keplerian rotation be extracted in such a system? At least for ionized systems, in which the plasma properties are important, this problem is solved by including the coupling to the magnetic field. This amounts to adding the Lorentz force to the r.h.s. of (11.32)–(11.34), i.e., adding $\boldsymbol{B} \cdot \nabla B_j/(4\pi)$ and replacing the pressure by the total pressure $P = p + B^2/(8\pi)$, where the evolution of the magnetic field follows Faraday's law (2.15).

The qualitative difference compared with the hydrodynamic case can already be noticed by considering the equations for $\langle \rho(\widetilde{v}_j^2 + \widetilde{v}_{\mathrm{A}j}^2)\rangle$ generalizing the hydrodynamic equations (11.37) and (11.38). While the underbraced driving term in the equation for r remains unchanged, the driving in the equation for θ becomes

$$\begin{aligned} -\frac{\kappa^2}{2\Omega}\langle \rho\widetilde{v}_r\widetilde{v}_\theta\rangle \rightarrow -\frac{\kappa^2}{2\Omega}\langle \rho\widetilde{v}_r\widetilde{v}_\theta\rangle - \frac{d\Omega}{d\ln r}\langle \rho\widetilde{v}_{\mathrm{A}r}\widetilde{v}_{\mathrm{A}\theta}\rangle \\ = -2\Omega\langle \rho\widetilde{v}_r\widetilde{v}_\theta\rangle - \frac{d\Omega}{d\ln r}\langle \rho(\widetilde{v}_r\widetilde{v}_\theta - \widetilde{v}_{\mathrm{A}r}\widetilde{v}_{\mathrm{A}\theta})\rangle, \end{aligned}$$

where $\langle\rho(\widetilde{v}_r\widetilde{v}_\theta - \widetilde{v}_{Ar}\widetilde{v}_{A\theta})\rangle$ is (essentially) the turbulent stress tensor $R_{r\theta}$, (11.13), the sum of the Reynolds tensor and the Maxwell tensor, which drives the accretion and determines the rate of accretion, (11.15). We see that, in the MHD case, the turbulent stress tensor now couples to the angular-velocity gradient $d\Omega/d\ln r$ instead of the angular-momentum gradient $\propto \kappa^2$ in the hydrodynamic case, and hence has the opposite sign.

11.4.1 Linear instability

Consider an axisymmetric disk threaded by a magnetic field with the components B_θ and B_z (assuming that $B_r = 0$ implies that $\partial_z B_z = 0$). The field should be sufficiently weak that the hydrodynamic equilibrium of the disk, (11.6) and (11.7), is not modified. In the local approximation we can make the Fourier *Ansatz* $\sim e^{i\boldsymbol{k}\cdot\boldsymbol{x}}$, but it is useful, in order to simplify the somewhat bulky formalism, to restrict our consideration to perturbations $\boldsymbol{k} = k\boldsymbol{e}_z$, which represent modes typical of the destabilizing effect of the magnetic field. A straightforward linear analysis yields the dispersion relation (for details, see Balbus and Hawley, 1998)

$$
\begin{aligned}
&(\varpi^2 - k_\parallel^2 v_A^2)[\varpi^4 - \varpi^2 k^2(c_s^2 + v_A^2) + k^2 k_\parallel^2 c_s^2 v_A^2] \\
&\quad = \kappa^2\varpi^4 - \varpi^2\left(\kappa^2(k^2c_s^2 + k_\perp^2 v_A^2) + k_\parallel^2 v_A^2 \frac{d\Omega^2}{d\ln r}\right) - k^2 k_\parallel^2 c_s^2 v_A^2 \frac{d\Omega^2}{d\ln r},
\end{aligned}
\tag{11.41}
$$

where $k_\parallel^2 = (\boldsymbol{k}\cdot\boldsymbol{B})^2/B^2 = k^2B_z^2/B^2$, $k_\perp^2 = (\boldsymbol{k}\times\boldsymbol{B})^2/B^2 = k^2B_\theta^2/B^2$, and $c_s = \sqrt{\gamma p/\rho}$ is the sound speed and $v_A = B/\sqrt{4\pi\rho}$ is the Alfvén speed, where $B^2 = B_\theta^2 + B_z^2$.

In the absence of rotation the r.h.s. of (11.41) vanishes, thus reproducing the dispersion relation (2.91) for a homogeneous MHD plasma with its three different dispersion branches, the (shear) Alfvén wave ϖ_A, the fast magnetosonic wave ϖ_{fast}, and the slow mode ϖ_{slow} with $\varpi_{\text{fast}} \geq \varpi_A \geq \varpi_{\text{slow}}$. As the flow rotation and shear represented by κ^2 and $d\Omega^2/d\ln r$ increase, both ϖ_{fast}^2 and ϖ_A^2 increase, but ϖ_{slow}^2 decreases, finally turning negative, which corresponds to instability.

The physics of this instability becomes particularly clear on considering the linear equations in a reference frame rotating with the angular velocity $\Omega(r_0) = \Omega_0$ at some radius r_0, which means replacing $\widetilde{v}_\theta$ by $\widetilde{u}_\theta = \widetilde{v}_\theta + r(\Omega - \Omega_0) \simeq \widetilde{v}_\theta + x(r\,d\Omega/dr)_0$ in the vicinity of r_0, $x = r - r_0$, $\widetilde{u}_r = \widetilde{v}_r$, $\widetilde{u}_z = \widetilde{v}_z$. We thus obtain from (11.32) and (11.33) supplemented by the Lorentz force

$$
\frac{d}{dt'}\widetilde{u}_r - 2\Omega\widetilde{u}_\theta + x\frac{d\Omega^2}{d\ln r} = -\frac{1}{\rho}\,\partial_r P + \frac{1}{4\pi}\boldsymbol{B}\cdot\nabla\widetilde{b}_r,
\tag{11.42}
$$

$$\frac{d}{dt'}\widetilde{u}_\theta + 2\Omega\widetilde{u}_r = -\frac{1}{\rho r}\,\partial_\theta P + \frac{1}{4\pi}\boldsymbol{B}\cdot\nabla\widetilde{b}_\theta, \tag{11.43}$$

$$\frac{d}{dt'}\widetilde{u}_z = -\frac{1}{\rho}\,\partial_z P + \partial_z\phi_g + \frac{1}{4\pi}\boldsymbol{B}\cdot\nabla\widetilde{b}_z \tag{11.44}$$

omitting the subscript 0, and $\widetilde{b}_{r,\theta,z}$ are the deviations from the equilibrium field. The l.h.s. of (11.43) results from the relation $\widetilde{v}_r\,\partial_r\widetilde{v}_\theta + [\kappa^2/(2\Omega)]\widetilde{v}_r = \widetilde{u}_r\,\partial_r\widetilde{u}_\theta + 2\Omega\widetilde{u}_r$. The term $2\Omega u$ is the Coriolis force, while the term $d\Omega^2/d\ln r$, representing the effect of the velocity shear, is the tidal force. The form (11.42)–(11.44) of the fluid equations for a rotating system is called the *Hill approximation* (Hill, 1878). Neglecting resistive effects, the magnetic field is frozen into the fluid, such that an infinitesimal perturbation can be written in terms of the displacement $\boldsymbol{\xi}$ of a fluid element, $\dot{\boldsymbol{\xi}} = \delta\boldsymbol{u}$. From Faraday's law (2.15) the perturbation of the magnetic field becomes

$$\begin{aligned}\delta\boldsymbol{b} &= i\boldsymbol{k}\times(\boldsymbol{\xi}\times\boldsymbol{B})\\ &= i\boldsymbol{k}\cdot\boldsymbol{B}\boldsymbol{\xi} \quad \text{for } \xi_z = 0,\ \boldsymbol{k} = k\boldsymbol{e}_z.\end{aligned}$$

If, in addition, the equilibrium field is vertical, $\boldsymbol{B} = B\boldsymbol{e}_z$, there is no pressure perturbation, $\delta P = \boldsymbol{B}\cdot\delta\boldsymbol{b}/(4\pi) = 0$; hence the linearized equations (11.42) and (11.43) can be written in the following form:

$$\ddot{\xi}_r - 2\Omega\dot{\xi}_\theta = -\left(\frac{d\Omega^2}{d\ln r} + (k_\parallel v_\mathrm{A})^2\right)\xi_r, \tag{11.45}$$

$$\ddot{\xi}_\theta + 2\Omega\dot{\xi}_r = -(k_\parallel v_\mathrm{A})^2\xi_\theta. \tag{11.46}$$

As discussed by Balbus and Hawley (1998), these equations decribe the oscillation of two neighboring rotating fluid elements a distance $\boldsymbol{\xi}$ apart, which are connected by a spring with the spring constant $(k_\parallel v_\mathrm{A})^2$ resulting from the stiffness of magnetic-field lines. In the absence of a field the fluid elements perform epicyclic oscillations $\varpi^2 = d\Omega^2/d\ln r + 4\Omega^2 = \kappa^2$. For conditions under which the rotation frequency decreases with radius but the angular momentum increases, as in the case of Keplerian rotation, the presence of a magnetic field leads to instability, which can be understood by the following argument. The fluid element at the slightly smaller radius rotates faster, thus increasing ξ_θ, which increases the spring tension. This will brake the motion, leading to a loss of angular momentum, which makes the fluid element recede to ever smaller radii, thus increasing ξ_r. The opposite behavior occurs for the fluid element at a slightly larger radius, which is accelerated and, because of the corresponding increase of the angular momentum, moves ever further outward. This is the basic mechanism of the magnetorotational instability.

Instability occurs for any finite magnetic field, as long as dissipation effects can be neglected. On the other hand, (11.45) indicates that the magnetic field should not be too strong, or too stiff, figuratively speaking, since a reversal of sign on the r.h.s. will quench the instability. Indeed, from the dispersion relation following from (11.45) and (11.46), namely

$$\varpi^4 - \varpi^2[\kappa^2 + 2(k_\parallel v_A)^2] + [(k_\parallel v_A)^2 + d\Omega^2/d\ln r] = 0 \tag{11.47}$$

(which agrees with the incompressible limit $c_s \to \infty$ of (11.41)), one obtains the stability condition for a wavenumber $k_\parallel$ ($= k$ in this restricted model):

$$(k_\parallel v_A)^2 > -\frac{d\Omega^2}{d\ln r}. \tag{11.48}$$

Hence, with growing magnetic-field strength, the spectrum of unstable modes is shifted to longer wavelengths, until the vertical scale height h is reached, such that for

$$v_A^2 > h^2 \frac{d\Omega^2}{d\ln r} \sim c_s^2 \tag{11.49}$$

there are no more unstable modes. For a smaller magnetic field the maximum growth rate is of the order

$$\gamma_{max} \sim \left|\frac{d\Omega}{d\ln r}\right| \sim \Omega, \quad \text{for } k^2 v_A^2 \sim \left|\frac{d\Omega^2}{d\ln r}\right| = 3\Omega^2 \tag{11.50}$$

for Keplerian flow. Thus one finds that the instability is growing very fast, which could rapidly lead to fully developed turbulence. Though the presentation given above is limited to a vertical equilibrium field, the instability is found to be rather insensitive to the orientation of the field; even a predominantly toroidal field, which is expected in differentially rotating systems, gives rise to similar growth rates.

The discovery of the basic phenomenon of the destabilization of a shear flow by a perpendicular magnetic field dates back a long time, to Velikhov (1959) and Chandrasekhar (1960). The existence of the instability was also verified in a laboratory experiment (using mercury, not a plasma discharge) by Donnelly and Ozima (1960). Chandrasekhar notes in his book (1961, p. 389) that the limit of vanishing magnetic field is discontinuous in the sense that the stability condition for $B = 0$ differs from that for an arbitrarily weak field, $B \to 0$. While in the former case Rayleigh's criterion (3.44) requires the angular momentum to increase with radius, in the latter only the angular velocity must increase.

However it was only three decades later that the significance of these stability results for accretion disks became fully appreciated, after the introduction of concept of the magnetorotational instability by Balbus and Hawley (1991).

11.4.2 Nonlinear saturation and magnetoviscosity

We have seen that the magnetorotational instability is a fast-growing instability, which should be active in an accretion disk under rather general conditions. However, the question of nonlinear saturation remains, namely the question of whether the unstable modes grow to large amplitudes leading to genuine MHD turbulence, and, more importantly, giving rise to sufficiently large angular-momentum transport to explain the observed rates of accretion. Here, as in most problems in nonlinear fluid dynamics, only numerical simulations can provide a reliable answer. While the results from a single simulation run have to be judged with some caution, a rather general picture of the nonlinear properties gradually appears from a broad range of computations using different numerical methods, with different boundary and initial conditions as well as geometry.

The first computations by Hawley and Balbus (1991, 1992) were restricted to axisymmetry and performed in the local approximation, i.e., in a small section $L_r = 2L_z \ll r, h$ of a Keplerian disk about the midplane of the disk, where density gradients can be neglected. They showed that the magnetorotational instability leads to a strongly nonlinear behavior with outward transport of angular momentum. In general the nonlinear state consists of coherent flows, radial streamers in the r, z projection, whose vertical structure exhibits a significant dependence on the initial field. For a weak vertical field the flow settles at the lowest vertical mode number, $n_z = 1$. Magnetic reconnection of the radially stretched magnetic field plays an important role,[6] allowing field lines to snap back into a more vertical direction, which reduces but does not quench the growth rate, as kinetic and magnetic energies continue to grow exponentially.[7] In the case of a radial initial field, again periodic radial flows are set up, but with a large vertical mode number, $n_z \gg 1$. However, in this case the poloidal field energy seems to saturate, the saturation level depending on the size of the system, increasing for larger systems, so that again no true saturation occurs. Hence

[6] Formally the ideal MHD equations are solved as is often done in astrophysical fluid-dynamics computations. Hence reconnection processes rely on numerical resistivity due to finite-grid effects, whose physical properties are, strictly speaking, somewhat obscure. However, it now appears that fast reconnection rates occur under very diverse physical conditions, such that precise knowledge of the numerical resistivity might not be crucial.

[7] The continued growth of the magnetic field does not contradict Cowling's anti-dynamo theorem that 2D motions cannot sustain a magnetic field (see, e.g., Moffatt, 1978), since here one is dealing with an open system, in which the mean field is maintained by the boundary conditions.

the imposed approximations, especially that of axisymmetry, appear to be too restrictive to properly describe the nonlinear behavior of the magnetorotational instability.

Goodman and Xu (1994) noticed that the radial streamers represent exact nonlinear solutions of the MHD equations, but that these states are unstable with respect to 3D perturbations, primarily to those of Kelvin–Helmholtz type. Three-dimensional nonlinear simulations have since been performed by a number of groups. Most of these authors treat the problem in the local approximation by solving the Hill system (11.42)–(11.44) with 'shearing-box' boundary conditions (Hawley *et al.*, 1995). To further simplify matters, stratification of density is neglected by focussing on a section of the disk centered about the midplane. If the system is sufficiently large in the radial direction to allow destabilization of the Kelvin–Helmholtz mode, the 2D streamers are, indeed, disrupted, leading to fully developed turbulence with efficient transport of angular momentum. The effective value of the Shakura–Sunyaev parameter $\alpha = R_{r\theta}/c_s^2$, (11.22), is ~ 0.1, to which the contribution of the Maxwell tensor $\langle \widetilde{b}_r \widetilde{b}_\theta \rangle/(4\pi)$ is significantly larger than that of the Reynolds tensor $-\langle \rho \widetilde{u}_r \widetilde{u}_\theta \rangle$, hence α can be called the effective magnetoviscosity,

$$\alpha \simeq -\frac{\langle \widetilde{b}_r \widetilde{b}_\theta \rangle}{4\pi \rho c_s^2}.$$

There is still a certain dependence on the initial vertical field; in particular, for a purely toroidal field α is found to be significantly smaller, $\lesssim 10^{-2}$, which demonstrates the importance of a vertical mean field for the transport of angular momentum even for $B_z \ll B_\theta$. The turbulence is highly anisotropic, being strongest in the vertical direction, reflecting the primary drive toward instability, weaker in the radial direction, and much weaker in the azimuthal direction, since k_θ modes are generated only by the secondary (Goodman–Xu) instability. Spectra are roughly consistent with a Kolmogorov power law, but numerical resolution in these computations is still far too low to allow clear identification of the spectral exponent. Though most of the energy resides in the longest-wavelength modes and also the main contribution to the transport of angular momentum comes from these modes, small-scale fluctuations are, of course, not negligible, just as they are not in ordinary hydrodynamic turbulence, and sophisticated subgrid modeling would be required in order to account for their effect.

In more realistic modeling, vertical stratification of density is included while still maintaining the local approximation, which has been done in several independent studies (Brandenburg *et al.*, 1995; Stone *et al.*, 1996; Matsuzaki and Matsumoto, 1997). Though there is now an additional reservoir of free

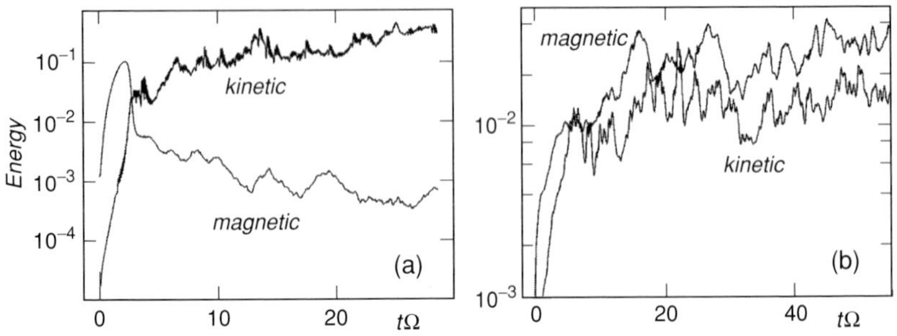

Figure 11.4: Evolution of magnetic and kinetic turbulence energies. (a) Turbulence in a nonrotating shear flow driven by hydrodynamic Kelvin–Helmholtz instability. The magnetic energy seems to decay after an initial amplification phase; hence there is no genuine dynamo action. (b) Turbulence in a rotating shear flow driven by the magnetorotational instability. Dynamo action sustains the magnetic energy against dissipative decay. (From Balbus and Hawley, 1998.)

energy, its effect is altogether rather weak. This is not too surprising, since the buoyancy-driven dynamics characterized by a rate $v_{\rm buoy}/h$ is weaker than the growth rate $\gamma \sim \Omega$ of the shear-flow drive, as long as the fluid motions remain subsonic (only for $v_{\rm buoy} \sim c_{\rm s}$ are the two rates similar). The value of α is found to be somewhat smaller by Brandenburg *et al.* and Stone *et al.*, $\alpha \sim 10^{-2}$, but increases with increasing spatial resolution. It also depends on the value of β for the plasma, being 10^{-2} for $\beta \sim 10$, but 10^{-1} for $\beta \sim 1$ (Matsuzaki and Matsumoto, 1997).

Since the drive toward instability is quenched by strong magnetic fields, (11.49), β in the saturated state remains significantly above unity, typically $\beta = 8\pi p/B^2 \sim$ 10–100, though the magnetic energy exceeds the kinetic energy of the velocity fluctuations. An interesting phenomenon observed in simulations of stratified accretion disks, for which several vertical scale heights are included in the computational domain, is the formation of a strongly magnetized disk corona due to the Parker instability, by virtue of which magnetic flux generated in the high-β central part of the disk is transported upward (Miller and Stone, 2000) in a way similar to the ejection of flux tubes from the solar convection zone into the corona.

A particular aspect of the magnetorotational instability is the amplification of the magnetic field, i.e., dynamo action. Here the azimuthal field is the largest component, being primarily driven by the basic shear flow, followed by the radial component, which is generated by the intermediate flow structures, the radial streamers, while the (fluctuating) vertical component is weakest, resulting only from the secondary 3D instability dynamics. Mean-field dynamo theory (see

Section 4.1.4) shows that generation of large-scale fields requires a mean helicity of the flow $H^K = \langle \widetilde{\boldsymbol{v}} \cdot \widetilde{\boldsymbol{\omega}} \rangle$, which results from the combination of rotation and density stratification, $H^K \propto \boldsymbol{\Omega} \cdot \nabla \rho$. However, even in nonstratified systems, for which the mean helicity vanishes, field amplification is observed numerically, though the field structure is less coherent. Such behavior had already been observed in simulations of homogeneous forced turbulence by Meneguzzi *et al.* (1981), who showed that, for nonhelical forcing of turbulence, only rather intermittent field structures are generated. Saturation of the amplification of magnetic energy can be caused by vertical convection of magnetic field out of the disk or by reconnection and field dissipation. Though convection plays an important role in magnetizing the corona of the disk, simulations indicate that the main saturation mechanism is reconnection and dissipation.

Though the shear-flow turbulence in the disk obviously drives the dynamo, this does not mean that any shear-flow turbulence will lead to dynamo action. Hawley *et al.* (1996) studied the evolution of a seed field immersed in the 3D turbulence excited by a nonrotating shear flow. Here the magnetic-field energy, after an initial phase of amplification, is found to decay, in contrast to the behavior in a rotating turbulent flow, as shown in Fig. 11.4.

12
Interstellar turbulence

The interstellar medium (ISM), which had formerly been better known for the allegoric shapes of its nebulae and dark clouds than for its physical properties, has developed into a fascinating area of astrophyscical research during the past few decades. An important aspect is the turbulent flows observed in many regions of the ISM, with velocities that, at least in the cooler parts, far exceed the speed of sound. These flows seem to play a decisive role in the cloud dynamics, slowing down gravitational contraction and star formation. In contrast to the objects studied in the two previous chapters, which had well-defined physical and geometrical properties and thus allowed a detailed analysis, the ISM is a rather diffuse system, whose modeling is far more uncertain and arbitrary. While the various atomic processes, such as transition probabilities and excitation rates, are well known and also the thermodynamic properties are fairly well understood, we have only a coarse picture of their hydrodynamics, including the effect of magnetic fields. Thus it is, for instance, difficult to apply hydrodynamic-stability theory without very special assumptions regarding geometry and flows. Hence the results we discuss in this chapter are mainly of qualitative nature, in which general arguments, such as equipartition and virialization, play an important role. The precise numerical factors, which are often found in the astrophysical literature, imply special choices of geometry and profiles and should not be taken too literally. In Section 12.1 we give a brief overview of the characteristic properties of the ISM. In Section 12.2 the main observational results on turbulent motions and magnetic fields in molecular clouds are presented. Section 12.3 deals with the role of turbulent and magnetic pressures in supporting the system against gravitational collapse in order to explain the low rate of star formation and the long lifetimes of molecular clouds.

12.1 The main properties of the interstellar medium

The term ISM refers to the gas, dust, and plasma filling the interstellar space in our Galaxy and probably also in other similar galaxies. Although the ISM contains only a small fraction (a few per cent) of the total mass of the Galaxy, it is the site of star formation, and hence its physical and chemical properties play a fundamental role in the *evolution* of the Galaxy. As the short overview in this section can only very cursorily touch on the various aspects of this complex system, we refer the reader interested in a broader introduction to several treatises on the physics of the ISM, by Spitzer (1978), Dyson and Williams (1980), Scheffler and Elsässer (1988), Ruzmaikin *et al.* (1988), and Padmanabhan (2001). Unfortunately, much of the observational and theoretical material has appeared only in short articles collected in the numerous, more or less well prepared, conference proceedings, which gradually assume the role of an alternative kind of scientific journal, as the growing number of references to such articles in the astrophysical literature reveals.

The ISM is very nonuniformly distributed, being concentrated primarily in the Galactic disk, covering regions of rather diverse physical conditions. It consists mainly of gaseous material with the primary constituents hydrogen (90%) and helium (10%), which makes up most of the ISM mass, and a small amount of dust. Hydrogen occurs in ionized (H II), neutral (H I), and molecular (H_2) forms, with temperatures $T \sim 10^4$ K, 10^2 K, and 10 K, respectively. Most of the volume is occupied by hot dilute gas of ionized and atomic hydrogen, while most of the mass is concentrated in dense molecular clouds consisting of cool molecular gas, mostly hydrogen. The main parameters of the different regions of the ISM are summarized in Table 12.1. It can be seen from these numbers that, in spite of the vastly different densities and temperatures, the product of density and temperature is roughly constant, $nT \sim 10^3$–10^4 cm^{-3} K, and the pressure $p = nk_{\rm B}T \sim 10^{-13}$–$10^{-12}$ dyn cm^{-3}. (This pressure equilibrium is, however, relevant only on sufficiently large scales. On small scales much larger values of the density occur, which tend to be gravitationally unstable.)

In addition to these thermal-gas components there is a homogeneous background of high-energy particles, the Galactic cosmic-ray component, consisting mainly of protons with energies from 10^6 eV to 10^{20} eV exhibiting a remarkably uniform and smooth energy spectrum with a mean particle energy of about 3×10^9 eV and an energy density of $nT \sim 1\,{\rm eV\,cm^{-3}} \simeq 10^4\,{\rm cm^{-3}\,K}$. An important effect of cosmic rays is to maintain a finite, though low, degree of ionization of the gas in the dense cool regions, which are too opaque to high-energy photons to be photoionized. High-energy and thermal components are coupled dynamically by a magnetic field B with a mean strength of a few microgauss, which is roughly the equipartition value, $B^2/(8\pi) \sim nk_{\rm B}T$.

Table 12.1: *Parameters of the main components, or phases, coexisting in the ISM*

Component	n (cm^{-3})	T (K)	Volume fraction	Mass fraction
Coronal H II plasma	0.005	10^6	0.5	10^{-3}
Warm H II plasma	0.3	10^4	0.5	0.1
H I clouds	10	10^2	0.05	0.4
H_2 clouds	10^2–10^3	10	0.005	0.5
Cosmic rays	3×10^{-10}	3×10^9 eV	1	0.0
Dust grains			0.5	0.01

A further component, which exists throughout the cooler parts of the ISM, consists of dust grains (diameter 1–10 μm). Though they comprise only 1% of the total mass contained in the ISM (but a major fraction of the heavier elements) and their effect on the dynamics of the gas is therefore weak, they lead to efficient scattering, polarization, and absorption of starlight and thus provide important observational clues.

The inhomogeneity of the ISM, which gives rise, for instance, to the visible structures such as loops and bubbles, arises through differences in the heating and cooling processes. Since the gas is in overall pressure equilibrium, a strong local energy source leads to a hotter and more dilute environment, whereas, where efficient cooling processes occur, gas temperatures are low and, concomitantly, densities are high. Individual regions of rather uniform conditions may have sharp boundaries, such as the transition from the chromosphere to the hot corona in stellar atmospheres, the outer edge of the hot bubble surrounding a star, and the outer shell of a supernova remnant. These boundaries are often not stationary, since the absorption of energy from a local source increases the pressure, which drives a shock wave into the ambient cooler medium, for instance the ionization front driven by the ionizing ultraviolet radiation from a hot massive star.

Cooling is mainly due to line radiation of neutral atoms and molecules, preferentially of heavier elements (called 'metals'), which is lost from the system, since for this low-energy radiation in the far-infrared or radio-frequency range the ISM is optically thin. Hence regions of neutral gas tend to become cool and dense. Evaluating the balance of ionization, recombination, and line cooling indicates the existence of several phases of different densities, and hence temperatures, for the same value of the pressure, which may coexist in the ISM.

The densest parts of the ISM, the giant molecular clouds (GMCs), are the sites of star formation. Gravitational instability leads to local density condensation, until accretion and nuclear ignition make the newborn star a strong source of energy, with jets from protostellar accretion disks and radiation and stellar winds from stars, which thus give 'recycled' material enriched with heavier elements back to the ISM. This process is particular efficient in supernova explosions, in which most of the stellar mass is ejected. Given the occurrence of such violent processes, the ISM cannot be in static equilibrium, but rather is expected to exhibit a strongly dynamic turbulent behavior. Indeed, observations clearly reveal the presence of supersonic random flows, i.e., the kinetic energy greatly exceeds the thermal energy and will therefore, possibly together with the magnetic field, control the conditions for gravitational instability and the rate of star formation. Most of this chapter will therefore be concentrated on turbulence in GMCs.

12.2 Observational results on molecular clouds

Molecular clouds appear visually dark because of the extinction of background starlight, which is caused mainly by dust. On the other hand, molecular clouds are efficient radiators at infrared and radio wavelengths, which provides information about temperatures, bulk motions, and magnetic fields in these dense regions of the ISM. Radiation consists mainly of line radiation emitted by minority molecules such as CO, the second in abundance, OH, CN, H_2O, and even complex carbohydrates, since H_2, the most abundant molecular species, has no strong molecular transitions. (By contrast, atomic hydrogen is conveniently observed in the 21-cm line.)

A major fraction of the molecular gas is organized in GMCs, which have sizes of about 50 pc (1 parsec $\simeq 3 \times 10^{18}$ cm $\simeq 2 \times 10^5$ AU) and masses of 10^4–$10^5 M_\odot$ (for a detailed overview of properties of GMCs see, e.g., Blitz, 1993; McKee, 1999). GMCs are discrete isolated objects with sharp boundaries. The mean density is about $\overline{n} \sim 10^2$ cm^{-3} and the temperature is, rather uniformly, low, $T \sim 10$ K; hence the sound speed is small, $c_s \simeq 0.2$ km s^{-1}. With these numbers a GMC should be strongly gravitationally bound; the Jeans radius $R_J \sim 10^3$ pc, which results from the balance of the pressure force and gravitational attraction (see Section 12.3), greatly exceeds the dimensions of the cloud. As a consequence the cloud would be expected to contract and collapse in a free-fall time, which would be considerably shorter than the observed lifetime. The discrepency becomes even more severe on considering the discrete substructures of a GMC, which may have much higher densities, $n \sim 10^3$ cm^{-3} in the clumps and 10^6 cm^{-3} in the cores, since the mass distribution is very nonuniform. The stability estimates are, however, due to the tacit assumption that the gas is in static equilibrium, which is not the case at all. As we shall

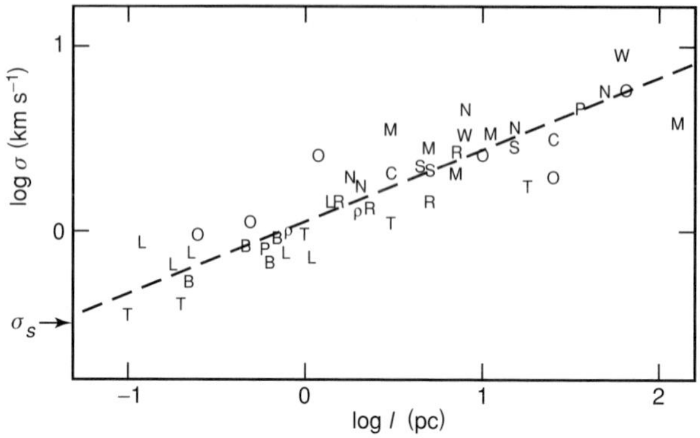

Figure 12.1: A log–log plot of the velocity dispersion versus the linear dimension for a number of different molecular clouds. (From Larson, 1981.)

see, the system is in a violently dynamic state, of which the strong variation in density itself is an indication, reminiscent of the filamentary structure of the density fluctuations found in the simulations of supersonic turbulence discussed in Chapter 9.

12.2.1 Supersonic turbulence

It has been noted for a long time that spectral line widths observed from interstellar clouds are much broader than would be expected from the low temperatures in these gases, which indicates the presence of flows with high Mach numbers. The supersonic character of these flows is mainly due to the low gas temperature, such that even rather modest flow speeds of the order of 1 km s^{-1} greatly exceed the thermal velocities. It was not clear *a priori* whether these flows are regular, resulting from gravitational collapse and rotation, or random. The idea of supersonic random motions had previously been rejected, since they should rapidly be slowed down by the efficient dissipation in shock waves. Nonetheless, the turbulent character has been verified by observations of the velocity dispersion δv, which is found to exhibit a power-law dependence on the scale l. Here one measures the dependence of the width of some molecular emission line, which is proportional to δv, on the size l of the emitting area. Observations and their interpretation have been summarized by Larson (1981), who collected measurements of δv from different regions varying in size by three orders of magnitude, 0.1–100 pc; see Fig. 12.1. The data are fitted by the power law

$$\delta v \sim l^{\alpha}, \quad \alpha \simeq 0.38, \tag{12.1}$$

where the exponent is close to the Kolmogorov value $\frac{1}{3}$. Though Kolmogorov's theory refers to incompressible turbulence, the numerical simulations discussed in Section 9.2.1 have shown that the basic scaling of the velocity fluctuations remains valid also in supersonic turbulence.

Larson's result has not remained uncontested. Solomon *et al.* (1987), evaluating an even larger sample of spectra, claimed that the scaling exponent is $\alpha = 0.5 \pm 0.05$, thus excluding the Kolmogorov value. Instead, it is assumed that clouds are gravitationally bound and that the relation $\delta v \sim l^{1/2}$ arises from the virial theorem, $\delta v^2 \sim GM/l$, which will be discussed in more detail in Section 12.3. The difference between the results may be due to whether observations are taken from only one particular GMC or from a collection of different GMCs, since, in general, mixing data from different turbulent systems with varying rates of input of energy and inertial range widths can indeed lead to a steeper effective spectral power law. On similar lines, Caselli and Myers (1995) found a scaling exponent $\alpha \simeq 0.5$ for observations from low-mass cores and a much smaller value, $\alpha \simeq 0.2$, from high-mass cores, which indicates that special care in the choice of data is required for comparison with homogeneous turbulence theory. On combining results from different ranges of size, $0.02\,\mathrm{pc} < l < 80\,\mathrm{pc}$, and density, $50\,\mathrm{cm}^{-3} < \overline{n}_{\mathrm{H}} < 3 \times 10^4\,\mathrm{cm}^{-3}$, Falgarone *et al.* (1992) found an overall power law of the velocity dispersion with exponent $\alpha \simeq 0.4$, thus essentially confirming Larson's result.

12.2.2 Gravity in molecular clouds

We have seen in Table 12.1 that the ISM is in approximate pressure equilibrium, which implies that GMCs are confined by the ambient gas pressure. The numbers refer, however, only to the thermal pressure, ignoring the turbulent flows, which are supersonic in the cool regions of the ISM and hence constitute an effective pressure $\langle \rho\, \delta v^2 \rangle$ greatly exceeding the thermal pressure. Nevertheless, because of the high density, self-gravity becomes important, and the question of whether these structures are gravitationally bound arises. In this case one expects from the virial theorem (see Section 12.3.1)

$$(\delta v)^2 \sim GM/l \sim G\overline{\rho}l^2, \tag{12.2}$$

where l is the linear dimension of the object, $M \sim \overline{\rho}l^3$ is the enclosed mass, and $\delta v^2 \sim \overline{v^2}$, factors of order unity being neglected. Also the mass is observed to exhibit a size-scaling relation $M \sim l^s$, see Section 12.2.3, and the equipartition relation (12.2) implies that this scaling is connected with the scaling of $\delta v \sim l^\alpha$, $2\alpha = s - 1$. Using the mass data of the same objects as in (12.1), Larson (1981) obtained the scaling $\delta v \sim M^q$, $q \simeq 0.2$, which gives a second relation between

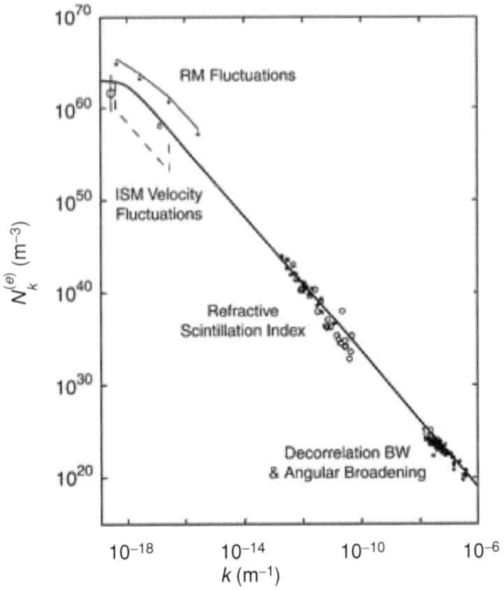

Figure 12.2: The electron-density power spectrum $N_k^{(e)} = |n_e(k)|^2$ in the ISM resulting from various direct or indirect observations (from Armstrong *et al.*, 1995; see their paper for an explanation of the different observational methods indicated in the figure). The solid line gives the Kolmogorov spectrum $\sim k^{-11/3}$ defined without the k-space volume factor k^2.

α and s, $\alpha \simeq 0.2s$. Both relations determine, in principle, α and s, which are roughly consistent with Larson's results $\alpha \simeq 0.38$ and $s \simeq 2$. However, results of this kind should be considered with some caution. Solomon *et al.* (1987) claimed from their evaluation of observational data to find similarly consistent scalings with their value $\alpha \simeq 0.5$.

12.2.3 The density spectrum and mass distribution

Most of the interstellar volume is filled with ionized gas; see Table 12.1. The spectrum of the electron-density fluctuations, which has been observed by different methods in the various spectral ranges, is found to follow a 3D Kolmogorov law (Armstrong *et al.*, 1981 and 1995),

$$N_k^{(e)} = |n_e(k)|^2 \propto k^{-11/3},$$

or, when it is angle-integrated, $k^2 N_k^{(e)} \sim k^{-5/3}$ extending over wavenumbers from 10^{-17} m^{-1} to 10^{-6} m^{-1}, i.e., scales from 10^{-4} AU to 10^2 pc, Fig. 12.2. This agrees with the theory for weakly compressible MHD turbulence proposed by Montgomery *et al.* (1987) and discussed in Section 9.2.2. Attractive as such a uniform scaling law may be, it is difficult to believe that this spectrum extending through an inertial range spanning 12 decades in wavenumber is due to a single cascade or transfer process. Instead it seems to be more appropriate to assume that there are several different spectral ranges k_j with particular sources and sinks joining smoothly, such that each is characterized by a transfer rate ϵ_j, $k^2 N_k^{(e)} \sim \epsilon_j k^{-5/3}$, with such an overall smooth spectrum requiring these transfer rates to be roughly equal. The underlying physics would be that the energy dissipated in one spectral range will serve as free energy driving turbulence in the adjoining one.

In molecular clouds, which occupy only a tiny fraction of the interstellar volume but have much higher densities, the level of ionization is very low. Here density fluctuations, which primarily concern the neutral gas, are driven by supersonic turbulent motions. Indeed, a characteristic feature of molecular clouds is the very inhomogeneous, spotty density distribution with a large contrast between the dense clumps and the voids in between. There is a hierarchy of structures, which, while loosely characterized by the sequence clouds – clumps – cores, in fact exhibits a continuous decrease of scales. Because of the scale-free nature of the ISM – there seems to exist no explicit scale length between the global structures in the Galaxy $\sim$ 1 kpc and some lower microscale $l_d < 0.01$ pc[1] –, a suitable approach is to describe the hierarchical density distribution as a fractal, i.e., a self-similar sequence of levels of elements with decreasing size (see Section 7.1). In this concept the density is confined to the smallest structures $l \sim l_d$ distributed in an intricate and chaotic way as illustrated by the simplest fractal set, the Cantor set.[2] The density measured in a volume of a larger size $l > l_d$ is the mean value over the elemental structures contained in the volume $V_l \sim l^3$. As a consequence of the self-similar character of the fractal mass distribution there is a power-law relation in a GMC,

$$n(l)\, dl \sim l^{-\alpha_l}\, dl, \tag{12.3}$$

$n(l)\, dl$ being the number of dense structures of size l in the interval dl. The exponent is related to the fractal dimension D of these structures,

$$N(l)\, d\ln l \sim l^{-D}\, d\ln l,$$

[1] The radius of the hot bubbles generated by stellar winds $l \sim 10^2$ AU constitutes such a lower limiting scale.

[2] The Cantor set is constructed in the following way. The unit interval is divided into three equal intervals, of which the middle one is discarded. On the second level the remaining two intervals are again divided into three and the middle ones discarded, etc. The Cantor set is the singular set of N intervals which results from this procedure at the level n for $n \to \infty$. With the definition of the (Hausdorff) dimension d, $N \sim l^{-D}$, where l is the length of an interval on the level n, one obtains for the Cantor set $N = 2^n$, $l = (\frac{1}{3})^n$, and hence $D = \ln 2/\ln 3$.

where $N(l)$ is the number of objects per logarithmic interval, $N(l)\,d\ln l = n(l)\,dl$; hence $n(l) \sim l^{-(1+D)}$, with $\alpha_l = 1 + D$. A broad survey of different clouds gives $D = 2.3 \pm 0.3$ (Elmegreen and Falgarone, 1996). The fractal character of the density distribution is expected to be connected with the scaling properties of the turbulence. In incompressible turbulence there is a simple relation between the fractal dimension D of level surfaces of an advected scalar field and the scaling exponent of the velocity increment $\delta v \sim l^{\xi}$, $D = 2 + \xi \simeq 2.33$ (Meneveau and Sreenivasan, 1990). It appears from the results just presented that a similar relation is valid for level surfaces of the density in supersonic turbulence, in spite of the fact that the density fragmentation is mainly due to compression rather than advection.

Observations also suggest that there is a power-law scaling of the size l with the mass M of a structure,

$$M \sim l^{s}. \tag{12.4}$$

The value of the exponent varies somewhat in the literature. Earlier results give $s \simeq 2$ (Larson, 1981; Falgarone *et al.*, 1992), whereas Elmegreen and Falgarone (1996) obtained values in the range $2.4 < s < 3.7$. If the mass fragmentation results from the fractality of the density distribution, one may expect the identity $s = D$. From (12.3) and (12.4) follows the scaling relation for the mass-distribution function $n(M)\,dM$, the number of structures of mass M in the interval dM,

$$n(M)\,dM = n(l)\left(\frac{dM}{dl}\right)^{-1} dM \sim l^{-D-s}\,dM \sim M^{-\alpha_M}\,dM. \tag{12.5}$$

Hence, using (12.4),

$$\alpha_M = 1 + D/s. \tag{12.6}$$

The power-law mass distribution $n(M)$ has been observed directly from various different clouds; it is essentially the distribution of the masses of clumps within a GMC, which gives values in the range $\alpha_M \simeq 1.6$–2, which is consistent with (12.6).

12.2.4 Magnetic fields

It has been mentioned before that dense molecular clouds are only weakly ionized, since their interiors are shielded from high-energy photons and only the pervasive cosmic-ray component generates a certain low level of ionization, $n_e/n_{\rm H} \sim 10^{-5}/[n_{\rm H}\,({\rm cm}^{-3})]^{1/2}$ (Elmegreen, 1979). Hence the electrical conductivity is low. Nevertheless, when it is considered on the huge spatial scales involved in the ISM structures, the magnetic field is expected to be frozen into the gas, a point that will be considered more closely in Section 12.3.2. The question, to be answered by observations, is that of whether the field is sufficiently strong to affect the dynamics and, in particular, to support clouds against gravitational collapse. A simple criterion compares the magnitude of the magnetic-energy density with the turbulent kinetic and gravitational energy densities. If the former is large such that $\mathsf{M}_\mathrm{A} < 1$, field lines are stiff, allowing fast motions only parallel to $\boldsymbol{B}$. In the opposite case, $\mathsf{M}_\mathrm{A} > 1$, the field is mainly advected with the fluid and field lines become highly tangled.

Cosmic magnetic fields are mainly observed by exploiting the polarization of radiation. There are four main methods to measure magnetic fields in the ISM (for a broader review, see, for instance, Heiles *et al.*, 1993, or Crutcher *et al.*, 2002).

(a) Faraday rotation. When a linearly polarized electromagnetic wave propagates along a magnetic field in a plasma, the two oppositely circularly polarized components, into which the wave may be decomposed, attain different phase velocities. This leads to a rotation of the plane of polarization by the angle θ. One has in general $\theta \propto \lambda^2 \int_0^L B_\parallel n_e \, dl$, where λ is the wavelength, L the path length, and $B_\parallel$ the field component parallel to the line of sight. Hence, to determine the magnitude of the mean parallel magnetic field $\overline{B_\parallel}$, one has to know the electron density.

(b) Synchrotron radiation. Electrons gyrate about magnetic-field lines due to the Lorentz force, and the corresponding acceleration causes them to emit electromagnetic radiation, which is linearly polarized perpendicularly to the magnetic field. For relativistic electrons this radiation is called synchrotron radiation. Its intensity depends on the field component perpendicular to the line of sight, on the energy-distribution function of the electrons, and on the wavelength. Since the density of high-energy electrons is usually not known, a measurement of the angle of polarization gives primarily only the direction of the magnetic field in the plane of the sky. Both methods (a) and (b) are more suitable for the ionized regions of the ISM.

(c) Zeeman splitting. If the region of spectral line emission is permeated by a magnetic field, the line of frequency ν_0 is split into three lines with the frequencies $\nu_{\sigma+} = \nu_0 - \nu_\mathrm{Z}$, $\nu_\pi = \nu_0$, and $\nu_{\sigma+} = \nu_0 + \nu_\mathrm{Z}$, where the Zeeman frequency ν_Z is essentially the electron cyclotron frequency $eB/(m_e c)$. The polarizations and the relative intensities of the individual components depend on the direction of the field. In the IMS the frequency splitting is, however, too small compared with the Doppler-broadened line width

$\Delta\nu_D$ to be directly measurable, $\nu_Z/\Delta\nu_D \sim 10^{-4}$, since $\Delta\nu_D/\nu_0 \sim 10^{-5}$ for $v \sim 3\,\mathrm{km\,s^{-1}}$, $\nu_0 \sim 10^9$ Hz, and $\nu_Z \sim 10$ Hz for $B \sim 5\,\mu$G. The Zeeman splitting becomes visible only on subtracting the right-circular polarization from the left-circular one, which gives essentially the derivative of the line profile and provides a measure of $B_\parallel$. The Zeeman effect has been observed only in regions of particularly high field strength.

(d) Dust polarization. Dust grains, which are never exactly spherical, tend to align along the field (with their *minor* axis). Since star light passing through the dusty region is most strongly attenuated when it is polarized along the major axis of the grains, i.e., perpendicularly to the field, the observed light is preferentially polarized parallelly to the field. A combination of Zeeman-splitting and polarization measurements together with some statistical modeling has yielded relatively accurate values of the local field in certain regions, though the spatial resolution is still rather coarse.

There are several review articles summarizing magnetic-field observations in the ISM (e.g., Heiles *et al.*, 1993; Beck, 2001; Crutcher *et al.*, 2002). Here we mention only a few general features. The large-scale field in the Galaxy is rather weak, $B \lesssim 5\,\mu$G, and oriented mainly in the Galactic plane along the spiral arms. In molecular clouds, however, significantly higher values of B have been observed, exceeding $10^2\,\mu$G, and even $10^3\,\mu$G, with random directions with respect to the mean Galactic field. Here the magnitude seems to agree with that required by the virial theorem, which basically implies equipartition of kinetic, magnetic, and gravitational energies. The relation

$$B \propto \sqrt{n}, \tag{12.7}$$

which is found to hold approximately in a density range of more than ten orders of magnitude, see Fig. 12.3, is usually interpreted in the sense that pressure and magnetic-energy density are in equilibrium, assuming that $p \sim n$, since the temperature does not vary strongly, $T \sim 10$ K in molecular clouds. The argument ignores, however, the fact that the *effective* pressure is caused by turbulent supersonic fluid motions instead of thermal motions. Hence, for equipartition, one would expect the relation

$$B \propto \delta v \sqrt{n} \tag{12.8}$$

to hold, where the velocity dispersion δv is a measure of the turbulent velocity at the particular scale. Figure 12.4(b) shows that this relation is, indeed, significantly better satisfied (there is less scatter of data points) than is relation (12.7), Fig. 12.4(a), for the same data set.

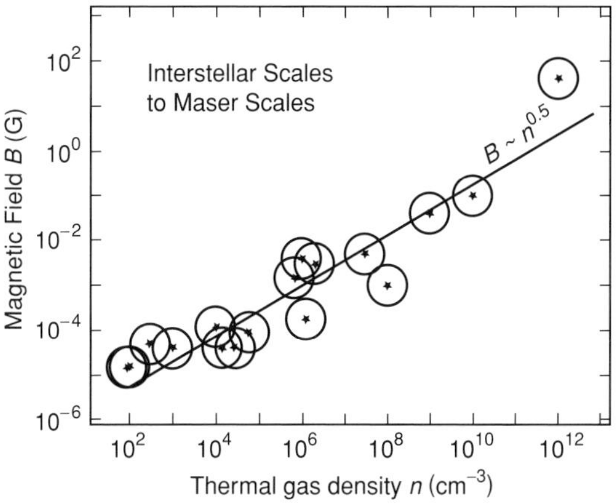

Figure 12.3: Observations of the magnetic field as a function of the gas density. The line gives the statistical fit $B = 0.4\sqrt{n}$ with B in microgauss and n in cm^{-3}. (From Vallée, 1997.)

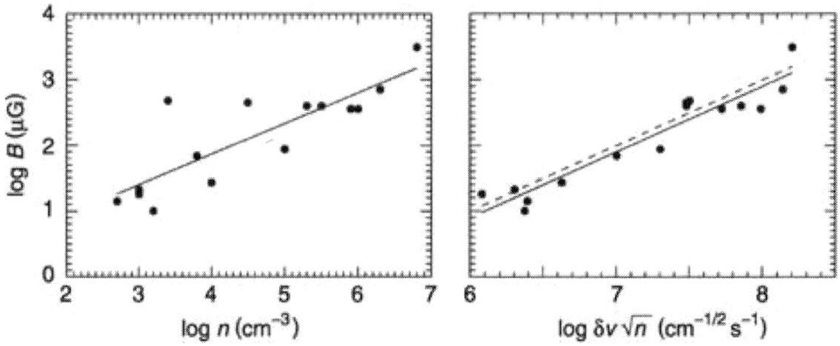

Figure 12.4: The magnitude of $B_{\|}$ versus the gas density n, (a), and $B_{\|}$ versus $\delta v\,\sqrt{n}$, (b), for the same set of data points. (From Basu, 2000, using a data set from Crutcher, 1999.)

12.3 Stability of molecular clouds

In view of their high densities and low temperatures, the question of the gravitational stability of GMCs arises, both globally and locally in the constituent clumps. If the pressure in some volume in space is too low, it cannot support the gas against self-gravitational contraction. To give a rough estimate of the stability limit, assume that we have a sphere of radius r filled with gas of mass density $\rho(r)$. In equilibrium, pressure and gravity balance each other,

$$p = \rho \frac{k_{\mathrm{B}} T}{\mu} \sim \frac{G M \overline{\rho}}{r}, \tag{12.9}$$

where μ is the mass of atomic or molecular particles, $k_B T/\mu \simeq c_s^2/\gamma$. This balance suggests a crude size scaling of the mass of a cloud, or cloud substructure, $M \sim rT$ or $\overline{\rho} \sim T/r^2$. If the pressure is smaller than that required by the equilibrium (12.9), the gas is said to be gravitationally unstable. According to the classical criterion, instability arises if, for a given mass M, the radius becomes smaller than the Jeans radius R_J,

$$r < R_J = b\,GM/c_s^2, \tag{12.10}$$

or if, for a given density, the mass exceeds the Jeans mass $M_J = \frac{4}{3}\pi\overline{\rho}R_J^3$,

$$M > M_J = c\frac{c_s^3}{(G^3\overline{\rho})^{1/2}}, \tag{12.11}$$

where the numerical coefficients b and c will be determined on applying the virial theorem $2E^K = |E^G|$ (see Section 12.3.1) instead of the qualitative balance relation (12.9) and specifying the density distribution $\rho(r)$, $b \simeq \frac{1}{3}$ for $\gamma \simeq 1.5$, and $c \simeq 3$ assuming that we have a constant density. Expressions (12.10) and (12.11) refer to the case of an isolated static gas sphere. In general, however, the contracting gas volume is embedded in a larger ambient gas region of finite pressure, and, in addition, the system is not static but rather is strongly turbulent, so that the sound speed must be replaced by a general velocity spread arising both from macroscopic and from thermal motions. This will be treated in Section 12.3.1, where we study the stability of molecular clouds more closely using the virial theorem, considering, in particular, the effect of the magnetic field. Section 12.3.2 analyzes the decoupling of the gas from the magnetic field by ambipolar diffusion, and Section 12.3.3 discusses the possible mechanisms capable of maintaining the observed level of turbulence.

12.3.1 The virial theorem

The virial theorem is a convenient tool for a qualitative, or semi-quantitative, analysis of the equilibrium and stability properties without the need to specify in detail the configuration and geometry, which is particularly useful for objects in the ISM with their widely varying shapes. Scalar multiplication of the equation of motion

$$\rho\frac{d\boldsymbol{v}}{dt} = -\nabla\left(p + \frac{1}{8\pi}B^2\right) + \frac{1}{4\pi}\boldsymbol{B}\cdot\nabla\boldsymbol{B} - \rho\,\nabla\phi_g$$

by the radius vector $\boldsymbol{r}$ measured from the center of gravity, integration over the volume V of the cloud, and integration by parts gives the *virial theorem* for a magnetized gas (Chandrasekhar and Fermi, 1957; Mestel and Spitzer, 1956; Strittmatter, 1966):

$$\frac{1}{2}\frac{d^2 I}{dt^2} = 2E^K + 3(\overline{p} - p_0)V + E^M - S^M + E^G. \tag{12.12}$$

In the derivation we used the following expressions for the inertial effect:

$$\int_V \boldsymbol{r}\cdot\frac{d\boldsymbol{v}}{dt}\rho\, dV = \int_0^M \boldsymbol{r}\cdot\frac{d^2\boldsymbol{r}}{dt^2}\, dM$$
$$= \frac{1}{2}\frac{d^2}{dt^2}\int_0^M r^2\, dM - \int_0^M v^2\, dM = \frac{1}{2}\frac{d^2 I}{dt^2} - 2E^K;$$

for the thermal pressure contribution

$$-\int_V \boldsymbol{r}\cdot\nabla p\, dV = 3\int_V p\, dV - \oint_{S_V} p\boldsymbol{r}\cdot d\boldsymbol{S} = 3(\overline{p} - p_0)V,$$

where p_0 is the pressure at the surface S_V, i.e., the pressure in the medium surrounding the volume V; for the magnetic contribution:

$$-\frac{1}{8\pi}\int_V \boldsymbol{r}\cdot\nabla B^2\, dV + \frac{1}{4\pi}\int_V \boldsymbol{r}\cdot(\boldsymbol{B}\cdot\nabla\boldsymbol{B})\, dV$$

$$= \frac{1}{8\pi}\int_V B^2\, dV - \frac{1}{4\pi}\oint_{S_V} \boldsymbol{r}\cdot(\tfrac{1}{2}B^2\boldsymbol{I} - \boldsymbol{BB})\cdot d\boldsymbol{S} = E^M - S^M;$$

and for the gravitational contribution:

$$-\int \rho\boldsymbol{r}\cdot\nabla\phi_g = -4\pi\int_0^r \rho\frac{GM(r')}{r'^2}r'^2\, dr' = -a\frac{GM^2}{r} = E^G,$$

where $a = \frac{3}{5}$ for a sphere of homogeneous density. If the r.h.s. of (12.12) is positive, the mass expands; if it is negative, the mass contracts. We assume that the system is in approximate equilibrium, which is satisfied if the evolution, for instance by losing energy by radiation, occurs on a time scale much longer than the dynamic time, on which the system reacts to a deviation from equilibrium. The latter is essentially the free-fall time, $d^2r/dt^2 \equiv r/t_{\rm ff}^2 = GM/r^2$; hence

$$t_{\rm ff} = \frac{r^{3/2}}{(GM)^{1/2}} = \left(\frac{3}{4\pi G\overline{\rho}}\right)^{1/2} \simeq \frac{5\times 10^7}{[n_{\rm H}\,({\rm cm}^{-3})]^{1/2}}\ \text{years}. \tag{12.13}$$

In the virial theorem (12.12) we neglected rotation, which seems to play no major role in the dynamics of molecular clouds. The situation is therefore rather different from that encountered in accretion disks, where gravity is mainly balanced by the centrifugal force, (11.6), while the pressure gradient is important only in the axial direction, (11.7).

For a weak magnetic field the gas is supported by the internal pressure $\overline{P}$ against gravity and the external pressure p_0. Hence we have from (12.12)

$$3(\overline{P} - p_0)V = aGM^2/r, \tag{12.14}$$

where $\overline{P}$ defined by $\overline{P} \equiv \rho\sigma_v^2$ includes the effect of the turbulence and σ_v is the velocity dispersion due both to turbulent motions and to thermal motions, $\sigma_v^2 \simeq \delta v^2 + c_s^2$. For given σ_v the equilibrium relation can be satisfied only for M up to a certain limit. To calculate the critical mass we write (12.14) in the form

$$p_0 = \frac{3M\sigma_v^2}{4\pi r^3} - \frac{3}{20\pi}\frac{GM^2}{r^4}, \tag{12.15}$$

using $M = \frac{4}{3}\pi r^3 \rho$ and $a = \frac{3}{5}$. By varying (12.15) with respect to r, we obtain the critical mass[3]

$$M_{\mathrm{BE}} \simeq 1.8 \frac{\sigma_v^6}{(G^3 p_0)^{3/2}}, \tag{12.16}$$

which is called the Bonnor–Ebert mass (Bonnor, 1956; Ebert, 1955). Hence, in the presence of an external pressure p_0, the critical mass depends more strongly on the velocity dispersion than it does in the case of an isolated sphere of gas (12.11). As discussed before, observations indicate that σ_v varies with the size of the cloud or clump under consideration, $\sigma_v^2 \simeq \delta v^2 \propto r^{2/3}$, (12.1). For larger cloud volumes we have, say, $\delta v \sim 1\,\mathrm{km\,s^{-1}}$, which gives $M_{\mathrm{BE}} \sim 5 \times 10^4 M_\odot$ using the value of the ambient pressure $p_0 \simeq 3 \times 10^{-13}\,\mathrm{dyn\,cm^{-3}}$, whereas for the smallest clumps, the star-forming cores, the turbulent velocity dispersion is low, $\sigma_v \sim c_s \simeq 0.2\,\mathrm{km\,s^{-1}}$, such that $M_{\mathrm{BE}} \sim 3M_\odot$.

Now consider the effect of a magnetic field. The average field B in a GMC is higher than the external Galactic field B_0, and even more so in a small constituent clump, so that the magnetic contribution in (12.12), $E^M - S^M \simeq E^M > 0$, is stabilizing the cloud against gravitational collapse. Because of conservation of flux,

$$\Phi = \int \boldsymbol{B} \cdot d\boldsymbol{S} \sim \overline{B} r^2 = \text{constant},$$

gravitational and magnetic terms have the same r-dependence:

$$E^M \sim \overline{B}^2 r^3 \sim \Phi^2/r \sim GM^2/r.$$

Consequently, if the gravitational energy exceeds the magnetic energy at one particular radius, it will do so at all smaller radii, i.e., in this case the magnetic

[3] To calculate the extremum y_m of a function $y(x)$ given implicitly $f(x, y) = 0$, one solves $dy/dx = -\partial_x f/\partial_y f = 0$, i.e., $\partial_x f = 0$, and inserts the solution x_m into $f = 0$.

field cannot prevent gravitational collapse. The virial theorem may be written in the approximate form

$$3(\overline{p} - p_0) - a\frac{GM^2}{r}\left(1 - \frac{M_\Phi^2}{M^2}\right) = 0, \tag{12.17}$$

introducing the magnetic 'mass' M_Φ by defining $E^M = aGM_\Phi^2/r$. Using the same variational procedure as in the derivation of (12.16), an approximate expression is obtained from (12.17) for the critical mass in the presence of a magnetic field,

$$M_{\rm crit} \simeq M_{\rm BE} + M_\Phi. \tag{12.18}$$

This picture is, however, oversimplified since, even in the presence of a strong field, gravitational collapse may still occur along the field, where only pressure can balance gravity. Hence, for $\overline{p} \ll \overline{B}^2$, a thin disk is established. For a more detailed analysis the magnetic configuration must be specified. The simplest case is a homogeneous field pervading the gas. In general, however, the field in the gas will not be constant, not even purely poloidal, but rather will develop a toroidal component due to rotation with respect to the surrounding medium. Nonetheless, expression (12.17) gives the right order of magnitude of the stabilizing effect of the magnetic field. The fact that molecular clouds, or their constituent clumps, are not disk-shaped seems to indicate that the magnetic field is not the dominant stabilizing factor, $M_\Phi < M_{\rm BE}$ (McKee, 1989). On the same lines, Padoan and Nordlund (1999) showed that, in numerical simulations of supersonic MHD turbulence, super-Alfvénic systems, $\mathrm{M_A} \gg 1$, i.e., systems without a strong mean magnetic field, give better agreement with the known properties of molecular clouds than do sub-Alfvénic systems, $\mathrm{M_A} \lesssim 1$, which had previously been favored (Myers and Goodman, 1988; Mouschovias and Psaltis, 1995), for which the observed motions were interpreted as hydromagnetic waves (Arons and Max, 1975; Zweibel and Josafatsson, 1983).

Turbulent magnetic-field fluctuations $\delta\boldsymbol{B}$ exert the same stabilizing effect as a mean magnetic field, $E^M \propto \overline{B}^2 + \overline{\delta B^2}$. Though observations do not yet provide a sufficiently well-resolved picture of the local field distribution, the general tendency seems to be clear. On large scales the turbulence in GMCs is essentially isotropic, since the magnetic field exhibits random directions and field strengths often greatly exceed the mean galactic field. In small clumps, however, where the field is compressed along with the density, the field threading the clump may be considered as a static field about which only small fluctuations occur.

12.3.2 Ambipolar diffusion

In the preceding discussion of the stabilizing effects of magnetic fields it has been assumed that the field is tightly coupled, or frozen in, to the gas. Since ionization is very low in molecular clouds, one might have certain doubts about the validity of this assumption. The electrical conductivity σ is, indeed, low, or the magnetic diffusivity $\eta = c^2/(4\pi\sigma) = (c^2/\omega_{pe}^2)\nu_{ei} \propto T^{-3/2}$ high by usual plasma standards, $\eta \sim 10^{12}\,\mathrm{cm^2\,s^{-1}}$ for $T = 10\,\mathrm{K}$, but even for these values the frozen-in condition would still be extremely well satisfied considered on the astronomical scales of the ISM, for which the time taken for the magnetic field to diffuse across, say, 1 pc would be $\tau_\eta = L^2/\eta \sim 10^{17}$ years. Hence, if resistivity due to electron–ion collisions were the only nonideal process in Ohm's law, conservation of flux would be perfect and the magnetic Reynolds number huge, $\mathrm{Rm} \sim 10^{11}$ assuming that the fluid velocity is $1\,\mathrm{km\,s^{-1}}$.

This, however, means only that the field is tightly coupled to the *ions* which, in turn, have to drag along the neutral species by ion–neutral-species collisions. As the Lorentz force acts only on the ions, we are left with the approximate system of equations (Mestel and Spitzer, 1956)

$$\rho_n \frac{d\boldsymbol{v}_n}{dt} \simeq \rho_n \nu_{ni} \boldsymbol{v}_d - \rho_n \nabla\phi_g, \tag{12.19}$$

$$\rho_i \frac{d\boldsymbol{v}_i}{dt} \simeq -\rho_n \nu_{ni} \boldsymbol{v}_d + \frac{1}{c}\boldsymbol{j} \times \boldsymbol{B} \simeq 0, \tag{12.20}$$

since for $\rho_i \ll \rho_n$ the inertial term for ions is negligible, and $\nu_{ni} = n_i\langle\sigma v\rangle$ is the collision frequency for the scattering of neutral species by ions. Equation (12.20) then gives the ion–neutral-species relative velocity, which is called the ambipolar drift,

$$\boldsymbol{v}_d = \boldsymbol{v}_i - \boldsymbol{v}_n = \frac{\boldsymbol{j} \times \boldsymbol{B}}{c\rho_n \nu_{ni}}, \tag{12.21}$$

and hence Ohm's law, expressed in terms of the fluid velocity $\boldsymbol{v}_n$,

$$\boldsymbol{E} = -\frac{\boldsymbol{v}_i \times \boldsymbol{B}}{c} = -\frac{\boldsymbol{v}_n \times \boldsymbol{B}}{c} + \frac{B^2}{c^2 \rho_n \nu_{ni}} \boldsymbol{j}_\perp, \tag{12.22}$$

which defines an effective magnetic diffusivity

$$\eta_{\mathrm{amb}} = \frac{B^2}{4\pi \nu_{ni} \rho_n}. \tag{12.23}$$

This process, called ambipolar diffusion in a weakly ionized gas, is much more important than resistive diffusion. For typical GCM parameters $n_n \simeq n_{\mathrm{H}_2} \sim 10^3\,\mathrm{cm^{-3}}$, $n_i \sim 10^{-3}\,\mathrm{cm^{-3}}$, $\langle\sigma v\rangle \simeq 1.5 \times 10^{-9}\,\mathrm{cm^3\,s^{-1}}$ (Nakano, 1984), and

$B \sim 30\,\mu\text{G}$ we find $\eta_{\text{amb}} \sim 10^{22}\,\text{cm}^2\,\text{s}^{-1}$, ten orders of magnitude greater than the corresponding value for resistive diffusion. The time required for magnetic diffusion across 1 pc is only 10^7 years, which is comparable to the lifetime of GMCs, and the effective magnetic Reynolds number is quite low, $\text{Rm}_{\text{eff}} = vL/\eta_{\text{amb}} \sim 10$. Since a value of $\text{Rm} \lesssim 1$ implies the decoupling of the fluid from the magnetic field, the stabilizing effect of the latter is lifted for smaller-sized clumps with $L \lesssim 0.1\,\text{pc}$ (Zweibel, 1999 and 2002). Thus ambipolar diffusion plays a crucial role in allowing cores to collapse and form (low-mass) stars, since also the stabilizing turbulent velocity dispersion is small at these scales.

12.3.3 Generation of turbulence in molecular clouds

A major problem for understanding molecular clouds had been to explain the observed long lifetimes and low star-formation rates. If these clouds were gas volumina in static equilibrium, the Jeans mass (12.11) for characteristic parameters $n \sim 10^3\,\text{cm}^{-3}$ and $T = 10\,\text{K}$, i.e., $c_s \simeq 0.2\,\text{km}\,\text{s}^{-1}$, would be only $10 M_\odot$, while a GMC contains up to $10^5 M_\odot$ and should hence collapse in a free-fall time (12.13) of $t_{\text{ff}} \simeq 3 \times 10^6$ years. However, lifetimes of GMCs are considerably longer, typically 3×10^7 years, after which time these clouds seem to disintegrate diffusively (Blitz and Shu, 1980) with only a small fraction of the cloud mass formed into stars. Obviously, molecular clouds must be maintained in a state of approximate equilibrium, and there seems to be rather general agreement that supersonic turbulent motions in combination with magnetic fields balance that gravitational attraction and that gravitational instability occurs only in the small cores and filaments of high density generated by the turbulence itself. Only in those clouds which are not strongly turbulent does the gas seem to collapse in a time of the order of the free-fall time and is the star-formation rate high.

The problem thus reduces to the question of what mechanism generates and maintains this turbulence. Since supersonic turbulence decays fast, as numerical simulations have shown (e.g., Mac Low *et al.*, 1998), essentially during one eddy-turnover time $\tau \sim L/\delta v(L) \sim 3 \times 10^6$ years for $L \sim 10\,\text{pc}$ and $\delta v \sim 3\,\text{km}\,\text{s}^{-1}$, which is much shorter than the lifetime of the cloud, the turbulence has to be regenerated continuously. There are probably several mechanisms active at different scales. Large-scale turbulent eddies in a GMC may arise due to instability of Galactic shear motions, though this is a relatively inefficient drive. Most of the power seems to be provided by internal processes, which are driven by gravitational and nuclear energy set free by formation and burning of hot massive stars (OB stars), which constitute a turbulence drive on relatively small sub-parsec scales by protostellar jets and stellar winds. The huge amounts of energy ejected from supernovae are certainly the most efficient agent to stir up the interstellar gas. These are, however, rare events and therefore cannot be primarily responsible for the observed motions in most molecular clouds.

Outlook

MHD turbulence has come of age, after a long period of life as a somewhat exotic branch of turbulence theory, where interesting new concepts were discussed, but contact with real-world phenomena was tenuous. Whereas formerly the MHD dynamics in a reversed-field pinch, and the hypothetical dynamo effect in the liquid-metal flows in a fast-breeder reactor, were often considered as the only potential applications, interest has now broadened, being shifted more to astrophysical systems, for which improved measurements of magnetic fields and turbulent flows combined with extended statistical evaluations point to the crucial role of high-Reynolds-number turbulence. Three specific areas have been treated in this volume.

Strong impact has come from the computational side. Owing to the progress in computing power during the past decade with the introduction of massively parallel computers, fully 3D numerical simulations could be performed with sufficiently high spatial resolution to allow a first glimpse of the inertial-range behavior. Further advances in this direction are to be expected during the next, say, five years, when simulations with up to 2000^3 modes, or grid points, will become feasible, which should provide reasonably accurate values of the most important scaling exponents of homogeneous MHD turbulence, though it is, of course, not guaranteed that these will correspond to the asymptotic high-Reynolds-number behavior. In any case the picture will be significantly more complex than that in hydrodynamic turbulence because of the dependence on magnetic helicity and alignment, and also on the way of driving, i.e., on the ratio of kinetic and magnetic energies.

Because of these dependences, self-consistent modeling of real-world MHD turbulent systems seems to be even more important than it is in hydrodynamic turbulence, which implies accounting for the globally inhomogeneous structures of these systems. Numerical studies of the accretion-disk problem, for instance, should at least include the cross-disk stratification in order to allow competition of pressure-driven and shear-flow-driven dynamics. Moreover, the turbulence dynamics may also depend on the transport processes, such as radiative transport in simulations of solar convection. Obviously a fully realistic numerical treatment will not be possible, not even in the more distant future; hence the development of specific-problem-adapted subgrid-scale approximations is particularly important.

References

Agullo, O., Müller, W.-C., Knaepen, B. and Carati, D. (2001). Large-eddy simulation of decaying magnetohydrodynamic turbulence with dynamic subgrid modeling, *Phys. Plasmas* **8**, 3502–5.

Anselmet, F., Gagne, Y., Hopfinger, E. J. and Antonia, R. A. (1984). High-order velocity structure functions in turbulent shear flows, *J. Fluid Mech.* **140**, 63–89.

Antonia, R. A., Ould-Rouis, M., Anselmet, F. and Zhu, Y. (1997). Analogy between predictions of Kolmogorov and Yaglom, *J. Fluid Mech.* **332**, 395–409.

Armstrong, J. W., Cordes, J. M. and Rickett, B. J. (1981). Density power spectrum in the local interstellar medium, *Nature* **291**, 561–4.

Armstrong, J. W., Rickett, B. J. and Spranger, S. R. (1995). Electron density power spectrum in the local interstellar medium, *Astrophys. J.* **443**, 209–21.

Arons, J. and Max, C. E. (1975). Hydromagnetic waves in molecular clouds, *Astrophys. J.* **196**, L77–82.

Balbus, S. A. and Hawley, J. F. (1991). A powerful local shear instability in weakly magnetized disks, *Astrophys. J.* **376**, 214–22.

Balbus, S. A. and Hawley, J. F. (1998). Instability, turbulence, and enhanced transport in accretion disks, *Rev. Mod. Phys.* **70**, 1–53.

Balbus, S. A., Hawley, J. F. and Stone, J. M. (1996). Nonlinear stability, hydrodynamical turbulence, and transport in disks, *Astrophys. J.* **467**, 76–86.

Balsara, D. S. (2001). Divergence-free adaptive mesh refinement for magnetohydrodynamics, *J. Comput. Phys.* **174**, 614–48.

Basu, S. (2000). Magnetic fields and the triaxiality of molecular cloud cores, *Astrophys. J.* **540**, L103–6.

Batchelor, G. K. (1950). On the spontaneous magnetic field in a conducting liquid in turbulent motion, *Proc. Roy. Soc. London* A **201**, 405–16.

Batchelor, G. K. (1951). Pressure fluctuations in isotropic turbulence, *Proc. Cambridge Phil. Soc.* **47**, 359–74.

Batchelor, G. K. (1959). Small-scale variation of convected quantities like temperature in turbulent fluid. Part 1. General discussion and the case of small conductivity, *J. Fluid Mech.* **5**, 113–34.

Batchelor, G. K. (1967). *An Introduction to Fluid Dynamics* (Cambridge University Press, Cambridge).

Batchelor, G. K. (1969). Computation of the energy spectrum in homogeneous two-dimensional turbulence, *Phys. Fluids* **12**, II 233–9.

Batchelor, G. K., Howells, I. D. and Townsend, A. (1959). Small-scale variation of convected quantities like temperature in a turbulent fluid. Part 2. The case of large conductivity, *J. Fluid Mech.* **5**, 135–9.

Batchelor, G. K. and Proudman, I. (1956). The large-scale structure of homogeneous turbulence, *Phil. Trans. Roy. Soc.* **268**, 369–405.

Bauer, F., Betancourt, O. and Garabedian, P. (1978). *A Computational Method in Plasma Physics* (Springer, New York).

Beck, R. (2001). Galactic and extragalactic magnetic fields, *Space Sci. Rev.* **99**, 243–60.

Benzi, R., Ciliberto, S., Tripiccione, R., Baudet, C., Massaioli, F. and Succi, S. (1993). Extended self-similarity in turbulent flows, *Phys. Rev.* E **48**, R29–32.

Benzi, R., Paladin, G., Parisi, G. and Vulpiani, A. (1984). On the multifractal nature of fully developed turbulence and chaotic systems, *J. Phys.* A **17**, 3521–31.

Berger, M. A. and Field, G. B. (1984). The topological properties of magnetic helicity, *J. Fluid Mech.* **147**, 133–48.

Berger, M. J. and Colella, P. (1989). Local adaptive mesh refinement for shock hydrodynamics, *J. Comput. Phys.* **82**, 64–84.

Bertin, G. (1982). Effects of local current gradients on magnetic reconnection, *Phys. Rev.* A **25**, 1786–9.

Belcher, J. W. and Davis, L. (1971). Large-amplitude Alfvén waves in the interplanetary medium, *J. Geophys. Res.* **76**, 3534–63.

Biskamp, D. (1993a). *Nonlinear Magnetohydrodynamics* (Cambridge University Press, Cambridge).

Biskamp, D. (1993b). Geometric properties of level surfaces in MHD turbulence, *Europhys. Lett.* **21**, 563–7.

Biskamp, D. (1995). Scaling properties in MHD turbulence, *Chaos, Solitons and Fractals* **5**, 1779–93.

Biskamp, D. (2000). *Magnetic Reconnection in Plasmas* (Cambridge University Press, Cambridge).

Biskamp, D. and Bremer, U. (1994). Dynamics and statistics in inverse cascade processes in 2D magnetohydrodynamic turbulence, *Phys. Rev. Lett.* **72**, 3469–72.

Biskamp, D., Hallatschek, K. and Schwarz, E. (2001). Scaling laws in two-dimensional turbulent convection, *Phys. Rev.* E **63**, 045302(R).

Biskamp, D. and Müller, W.-C. (2000a). Decay laws for three-dimensional magnetohydrodynamic turbulence, *Phys. Rev. Lett.* **83**, 2195–8.

Biskamp, D. and Müller, W.-C. (2000b). Scaling properties of three-dimensional isotropic magnetohydrodynamic turbulence, *Phys. Plasmas* **7**, 4889–900.

Biskamp, D. and Schwarz, E. (2001). On two-dimensional magnetohydrodynamic turbulence, *Phys. Plasmas* **8**, 3282–92.

Biskamp, D. and Welter, H. (1989). Dynamics of decaying two-dimensional magnetohydrodynamic turbulence, *Phys. Fluids* B **1**, 1964–79.

Blitz, L. (1993). In *Protostars and Planets* III, eds. E. H. Levy and J. I. Lunine (The University of Arizona Press, Tuscon), pp. 125–61.

Blitz, L. and Shu, F. H. (1980). The origin and lifetime of giant molecular cloud complexes, *Astrophys. J.* **238**, 148–57.

Boffetta, G., Celani, A. and Vergassola, M. (2000). Inverse energy cascade in two-dimensional turbulence: Deviations from Gaussian behavior, *Phys. Rev.* E **61**, R29–32.

Boldyrev, S. (2001). Kolmogorov–Burgers model for star-forming turbulence, to be published.

Boldyrev, S., Nordlund, Å. and Padoan, P. (2001). Scaling relations of supersonic turbulence in star-forming molecular clouds, to be published.

Bolgiano, R. (1959). Turbulent spectra in a stably stratified atmosphere, *J. Geophys. Res.* **64**, 2226–9.
Bolgiano, R. (1962). Structure of turbulence in stratified media, *J. Geophys. Res.* **67**, 3015–23.
Bonnor, W. B. (1956). Boyle's law and gravitational instability, *Mon. Notes Roy. Astron. Soc.* **116**, 351–9.
Borue, V. (1993). Spectral exponents of enstrophy cascade in stationary two-dimensional homogeneous turbulence, *Phys. Rev. Lett.* **71**, 3967–70.
Borue, V. (1994). Inverse energy cascade in stationary two-dimensional homogeneous turbulence, *Phys. Rev. Lett.* **72**, 1475–8.
Borue, V. and Orszag, S. A. (1996). Kolmogorov's refined similarity hypothesis for hyperviscous turbulence, *Phys. Rev.* E **53**, R21–24.
Bouchaud, J. P., Mézard, M. and Parisi, G. (1995). Scaling and intermittency in Burgers turbulence, *Phys. Rev.* E **52**, 3656–74.
Brachet, M. E., Meneguzzi, M., Politano, H. and Sulem, P. L. (1988). The dynamics of freely decaying two-dimensional turbulence, *J. Fluid Mech.* **194**, 333–49.
Brachet, M. E., Meneguzzi, M., Vincent, A., Politano, H. and Sulem, P. L. (1992). Numerical evidence of smooth self-similar dynamics and possibility of subsequent collapse for three-dimensional ideal flows, *Phys. Fluids* A **4**, 2845–54.
Brandenburg, A. (2001). The inverse cascade and nonlinear α-effect in simulations of isotropic helical hydromagnetic turbulence, *Astrophys. J.* **550**, 824–40.
Brandenburg, A., Jennings, R. L., Nordlund, Å., Rieutord, M., Stein, R. F. and Tuominen, I. (1996). Magnetic structures in a dynamo simulation, *J. Fluid Mech.* **306**, 325–52.
Brandenburg, A., Nordlund, Å., Pulkkinen, P., Stein, R. F. and Tuominen, I. (1990). 3D simulation of turbulent cyclonic magnetoconvection, *Astron. Astrophys.* **232**, 277–91.
Brandenburg, A., Nordlund, Å., Stein, R. F. and Torkelsson, U. (1995). Dynamo-generated turbulence and large-scale magnetic fields in a Keplerian shear flow, *Astrophys. J.* **446**, 741–54.
Burlaga, L. F. (1991). Intermittent turbulence in the solar wind, *J. Geophys. Res.* **96**, 5847–51.
Burlaga, L. F. (1993). Intermittent turbulence in large-scale velocity fluctuations at 1 AU near solar maximum, *J. Geophys. Res.* **98**, 17 467–73.
Camargo, S. J. and Tasso, H. (1992). Renormalization group in magnetohydrodynamic turbulence, *Phys. Fluids* B **4**, 1199–212.
Cameron, A. G. W. (1978). Physics of the primitive solar accretion disk, *Moon Planets* **18**, 5–40.
Cannizo, J. K., Shafter, A. W. and Wheeler, J. C. (1988). On the outburst recurrence time for the accretion disk limit cycle mechanism in dwarf novae, *Astrophys. J.* **33**, 227–35.
Canuto, C., Hussaini, M. Y., Quarteroni, A. and Zang, T. A. (1988). *Spectral Methods in Fluid Dynamics* (Springer, New York).
Canuto, V. M. (1994). Large-eddy simulation of turbulence: a subgrid-scale model including shear, vorticity, rotation, and buoyancy, *Astrophys. J.* **428**, 729–52.
Carbone, V. (1993). Cascade model for intermittency in fully developed magnetohydrodynamic turbulence, *Phys. Rev. Lett.* **71**, 1546–8.
Carbone, V. (1994). Scaling exponents of the velocity structure functions in the interplanetary medium, *Ann. Geophys.* **12**, 585–90.

Carnevale, G. F., McWilliams, J. C., Pomeau, Y., Weiss, J. B. and Young, W. R. (1992). Rates, pathways, and end states of nonlinear evolution in decaying two-dimensional turbulence, *Phys. Fluids* A **4**, 1314–16.

Caselli, P. and Myers, P. C. (1995). The line width–size relation in massive cloud cores, *Astrophys. J.* **446**, 665–8.

Castaing, B., Gunaratne, G., Heslot, F., Kadanoff, L., Libchaber, A., Thomae, S., Wu, X. Z., Zaleski, S. and Zanetti, G. (1989). Scaling of hard thermal turbulence in Rayleigh–Bénard convection, *J. Fluid Mech.* **201**, 1–30.

Celani, A. and Biskamp, D. (1999). Bridge relations in Navier–Stokes turbulence, *Europhys. Lett.* **46**, 332–8.

Celani, A., Lanotte, A., Mazzino, M. and Vergassola, M. (2000). Universality and saturation of intermittency in passive scalar turbulence, *Phys. Rev. Lett.* **84**, 2385–8.

Chandrasekhar, S. (1960). The stability of non-dissipative Couette flow in hydromagnetics, *Proc. Nat. Acad. Sci.* **46**, 253–7.

Chandrasekhar, S. (1961). *Hydrodynamic and Hydromagnetic Stability* (Clarendon Press, Oxford).

Chandrasekhar, S. and Fermi, E. (1957). Problems of gravitational instability in the presence of a magnetic field, *Astrophys. J.* **118**, 116–41.

Chasnov, J. R. (1997). On the decay of two-dimensional homogeneous turbulence, *Phys. Fluids* **9**, 171–80.

Chen, S. Y. and Cao, N. Z. (1997). Anomalous scaling and structure instability in three-dimensional passive scalar turbulence, *Phys. Rev. Lett.* **78**, 3459–62.

Chen, S. Y., Doolan, G., Herring, J. R., Kraichnan, R. H., Orszag, S. A. and She, Z. S. (1993). Far-dissipation range of turbulence, *Phys. Rev. Lett.* **70**, 3051–4.

Chilla, F., Ciliberto, S., Innocenti, C. and Pampaloni, E. (1993). Boundary layer and scaling properties in turbulent thermal convection, *Nuovo Cimento* **15**, 1229–49.

Ching, E. S. C. (2000). Intermittency of temperature field in turbulent convection, *Phys. Rev.* E **61**, R33–6.

Cho, J. and Vishniac, E. T. (2000). The anisotropy of magnetohydrodynamic Alfvénic turbulence, *Astrophys. J.* **539**, 273–82.

Cioni, S., Ciliberto, S. and Sommeria, J. (1995). Temperature structure functions in turbulent convection at low Prandtl number, *Europhys. Lett.* **32**, 413–18.

Colella, P. and Woodward, P. R. (1984). The piecewise parabolic method (PPM) for gas-dynamical simulations, *J. Comput. Phys.* **54**, 174–201.

Couder, Y. and Basdevant, C. (1986). Experimental and numerical study of vortex couples in two-dimensional flows, *J. Fluid Mech.* **173**, 225–51.

Crutcher, R. (1999). Magnetic fields in molecular clouds: Observations confront theory, *Astrophys. J.* **520**, 706–13.

Crutcher, R., Heiles, C. and Troland, T. (2002). Observations of interstellar magnetic fields, to be published.

Dahlburg, R. B. and Picone, J. M. (1989). Evolution of the Orszag–Tang vortex system in a compressible medium. I. Initial average subsonic flow, *Phys. Fluids* B **1**, 2153–71.

Dai, W. and Woodward, P. R. (1994). An approximate Riemann solver for ideal magnetohydrodynamics, *J. Comput. Phys.* **111**, 354–72.

Dai, W. and Woodward, P. R. (1998). On the divergence-free condition and conservation laws in numerical simulations for supersonic magnetohydrodynamic flows, *Astrophys. J.* **494**, 317–35.

Das, C., Kida, S. and Goto, S. (2001). Overall selfsimilar decay of two-dimensional turbulence, *J. Phys. Soc. Japan* **70**, 966–76.

De Hoffmann, F. and Teller, E. (1950). Magnetohydrodynamic shocks, *Phys. Rev.* **80**, 692–703.
Donnelly, R. J. and Ozima, M. (1960). Hydromagnetic stability of flow between rotating cylinders, *Phys. Rev. Lett.* **4**, 497–8.
Dobrowolny, M., Mangeney, A. and Veltri, P. (1980). Fully developed anisotropic hydromagnetic turbulence in interplanetary space, *Phys. Rev. Lett.* **45**, 144–7.
Drazin, P. G. and Reid, W. G. (1981). *Hydrodynamic Stability* (Cambridge University Press, Cambridge).
Dubrulle, B. (1994). Intermittency in fully developed turbulence: Log-Poisson statistics and generalized scale covariance, *Phys. Rev. Lett.* **73**, 959–62.
Dyson, J. E. and Williams, D. A. (1980). *The Physics of the Interstellar Medium* (Institute of Physics, London).
Ebert, R. (1955). Über die Verdichtung in H I Gebieten, *Z. Astrophys.* **37**, 217–32.
Edenstrasser, J. W. (1980). The only three classes of symmetric MHD equilibria, *J. Plasma Phys.* **24**, 515–18.
Elmegreen, B. G. (1979). Magnetic diffusion and ionization fractions in dense molecular clouds: the role of charged grains, *Astrophys. J.* **232**, 729–39.
Elmegreen, B. G. and Falgarone, E. (1996). A fractal origin for the mass spectrum in interstellar clouds, *Astrophys. J.* **471**, 816–21.
Elsässer, W. M. (1950). The hydromagnetic equations, *Phys. Rev.* **79**, 183.
Elsässer, K. and Schamel, H. (1976). Energy spectra of turbulent sound waves, *Z. Phys.* B **23**, 89–95.
Falgarone, E., Puget, J.-L. and Pérault, M. (1992). The small-scale density and velocity structure of quiescent molecular clouds, *Astron. Astrophys.* **257**, 715–30.
Falkovich, G. (1994). Bottleneck phenomenon in developed turbulence, *Phys. Fluids* **6**, 1411–14.
Feller, W. (1966). *An Introduction to Probability Theory and its Applications* (Wiley, New York), vol. II.
Finn, J. M. and Antonsen, T. M. (1985). Magnetic helicity: What is it and what is it good for?, *Comments Plasma Phys. Controlled Fusion* **26**, 111–26.
Fournier, J. D., Sulem, P. L. and Pouquet, A. (1982). Infrared properties of forced magnetohydrodynamic turbulence, *J. Phys.* A **15**, 1393–420.
Frank, J., King, A. and Raine, D. (1992). *Accretion Power in Astrophysics*, second edition (Cambridge University Press, Cambridge).
Friedel, H., Grauer, R. and Marliani, C. (1997). Adaptive mesh refinement for singular current sheets in incompressible magnetohydrodynamic flows, *J. Comput. Phys.* **134**, 190–8.
Frisch, U. (1995). *Turbulence. The Legacy of A. N. Kolmogorov* (Cambridge University Press, Cambridge).
Frisch, U., Sulem, P. M. and Nelkin, M. (1978). A simple dynamical model of intermittent fully developed turbulence, *J. Fluid Mech.* **87**, 719–36.
Frisch, U., Pouquet, A., Léorat, J. and Mazure, A. (1975). Possibility of an inverse cascade of magnetic helicity in magnetohydrodynamic turbulence, *J. Fluid Mech.* **68**, 769–78.
Fulachier, L. and Dumas, R. (1976). Spectral analogy between temperature and velocity fluctuations in a turbulent boundary layer, *J. Fluid Mech.* **77**, 257–77.
Furth, H. P., Killeen, J. and Rosenbluth, M. N. (1963). Finite-resistivity instabilities of a sheet pinch, *Phys. Fluids* **6**, 459–84.
Fyfe, D. and Montgomery, D. (1976). High-beta turbulence in two-dimensional magnetohydrodynamics, *J. Plasma Phys.* **16**, 181–91.

Gagne, Y. (1987). *Etude expérimentale de l'intermittence et des singularités dans le plan complex en turbulence développée*, Thèse de docteur des Sciences Physiques, Institut Polytechnique Universitaire de Grenoble.
Gagne, Y., Marchand, M. and Castaing, B. (1994). Conditional velocity pdf in 3D turbulence, *J. Phys. II, Paris* **4**, 1–8.
Galtier, S. and Pouquet, A. (1998). Solar flare statistics with a one-dimensional MHD model, *Solar Phys.* **179**, 141–65.
George, W. K., Beuther, P. D. and Arndt, R. E. A. (1984). Pressure spectra in turbulent free shear flows, *J. Fluid Mech.* **148**, 155–91.
Germano, M. (1992). Turbulence: the filtering approach, *J. Fluid Mech.* **238**, 325–36.
Gharib, M. and Derango, P. (1989). A liquid film (soap film) tunnel to study two-dimensional laminar and turbulent shear flows, *Physica* D **37**, 406–16.
Ghosh, S., Hossain, M. and Matthaeus, W. H. (1993). The application of spectral methods in simulating compressible fluid and magnetofluid turbulence, *Comput. Phys. Commun.* **74**, 18–40.
Gilman, P. A. and Glatzmaier, G. A. (1981). Compressible convection in a rotating spherial shell. I. Anelastic equations, *Astrophys. J. Suppl.* **45**, 335–49.
Goldreich, P. and Sridhar, S. (1995). Toward a theory of interstellar turbulence. II. Strong Alfvénic turbulence, *Astrophys. J.* **438**, 763–75.
Goldstein, M. L., Roberts, D. A. and Matthaeus, W. M. (1995). Magnetohydrodynamic turbulence in the solar wind, *Ann. Rev. Astron. Astrophys.* **33**, 283–325.
Gollub, J. P., Clarke, J., Gharib, M., Lane, B. and Mesquita, O. N. (1991). Fluctuations and transport in a stirred fluid with a mean gradient, *Phys. Rev. Lett.* **67**, 3507–10.
Gomez, T., Politano, H. and Pouquet, A. (1999). On the validity of a nonlocal approach for MHD turbulence, *Phys. Fluids* **11**, 2298–309.
Goodman, J. and Xu, G. (1994). Parasitic instabilities in magnetized differentially rotating disks, *Astrophys. J.* **432**, 213–23.
Gotoh, T. and Kraichnan, R. H. (1993). Statistics of decaying Burgers' turbulence, *Phys. Fluids* A **5**, 445–57.
Gotoh, T. and Fukayama, D. (2001). Pressure spectrum in homogeneous turbulence, *Phys. Rev. Lett.* **86**, 3775–8.
Grappin, R., Frisch, U., Léorat, J. and Pouquet, A. (1982). Alfvénic fluctuations as asymptotic states of MHD turbulence, *Astron. Astrophys.* **105**, 6–14.
Grappin, R, Pouquet, A. and Léorat, J. (1983). Dependence of MHD turbulence spectra on the velocity–magnetic field correlation, *Astron. Astrophys.* **126**, 51–8.
Grauer, R. and Sideris, T. C. (1991). Numerical computation of 3D incompressible ideal fluids with swirl, *Phys. Rev. Lett.* **67**, 3511–14.
Grauer, R., Krug, J. and Marliani, C. (1994). Scaling of high-order structure functions in magnetohydrodynamic turbulence, *Phys. Lett.* A **195**, 335–8.
Grauer, R., Marliani, C. and Germaschewski, K. (1998). Adaptive mesh refinement for singular solutions of the incompressible Euler equations, *Phys. Rev. Lett.* **80**, 4177–80.
Grauer, R. and Marliani, C. (2000). Current-sheet formation in 3D ideal incompressible magnetohydrodynamics, *Phys. Rev. Lett.* **84**, 4850–3.
Hatori, T. (1984). Kolmogorov-style argument for decaying homogeneous MHD turbulence, *J. Phys. Soc. Japan* **53**, 2539–45.
Hawley, J. F. and Balbus, S. A. (1991). A powerful local shear instability in weakly magnetized disks. II Nonlinear evolution, *Astrophys. J.* **376**, 223–33.
Hawley, J. F. and Balbus, S. A. (1992). A powerful local shear instability in weakly magnetized disks. III Long-term evolution in a shearing sheet, *Astrophys. J.* **400**, 595–609.

Hawley, J. F., Gammie, C. F. and Balbus, S. A. (1995). Local three-dimensional magnetohydrodynamic simulations of accretion disks, *Astrophys. J.* **440**, 742–63.

Hawley, J. F., Gammie, C. F. and Balbus, S. A. (1996). Local three-dimensional magnetohydrodynamic simulations of an accretion disk hydromagnetic dynamo, *Astrophys. J.* **464**, 690–703.

Hawley, J. F. and Stone, J. M. (1995). MOCCT: A numerical technique for astrophysical MHD, *Comput. Phys. Commun.* **89**, 127–48.

Heiles, C., Goodman, A. A., McKee, C. F. and Zweibel, E. G. (1993). Magnetic fields in star-forming regions: Observations, in *Protostars and Planets III*, eds. E. H. Levy and J. I. Lunine (University of Arizona Press, Tuscon), pp. 279–326.

Helmholtz, H. von (1868). Über diskontinuierliche Flüssigkeitsbewegungen, *Monats. Königl. Preuβ. Akad. Wiss. Berlin* **23**, 215–28.

Hénon, M (1976). A two-dimensional map with a strange attractor, *Commun. Math. Phys.* **50**, 69.

Hill, G. W. (1878). *Am. J. Math.* **1**, 5.

Hollweg, J. V. (1974). Transverse Alfvén waves in the solar wind, *J. Geophys. Res.* **79**, 1539–41.

Hopfinger, E. J., Browand, F. K. and Gagne, Y. (1982). Turbulence and waves in a rotating tank, *J. Fluid Mech.* **125**, 505–34.

Horbury, T. S. (1999). Waves and turbulence in the solar wind – an overview, in *Plasma Turbulence and Energetic Particles in Astrophysics*, eds. M. Ostrowski and R. Schlickeiser, (Jagiellonian University, Kraków), pp. 115–34.

Horbury, T. S. and Balogh, A. (1997). Structure function measurements of intermittent MHD turbulent cascade, *Nonlin. Proc. Geophys.* **4**, 185–99.

Hosokawa, I. and Yamamoto, K. (1992). Evidence against the Kolmogorov refined similarity hypothesis, *Phys. Fluids* A **4**, 457–9.

Hurlburt, N. E., Toomre, J., Massaguer, J. M. and Zahn, J.-P. (1994). Penetration below a convective zone, *Astrophys. J.* **421**, 245–60.

Iroshnikov, P. S. (1964). Turbulence of a conducting fluid in a strong magnetic field, *Sov. Astron.* **7**, 566–71.

Jeffrey, A. and Taniuti, T. (1964). *Nonlinear Wave Propagation* (Academic Press, New York).

Kadomtsev, B. B. and Petviashvili, V. I. (1973). Acoustic turbulence, *Sov. Phys. Doklady* **18**, 115–16.

Kadomtsev, B. B. and Pogutse, O. P. (1974). Nonlinear helical perturbations of a plasma in a tokamak, *Sov. J. Plasma Phys.* **1**, 389–91.

Kármán, T. von and Howarth, L. (1938). On the statistical theory of isotropic turbulence, *Proc. Roy. Soc.* A **164**, 192–215.

Kelvin, Lord (1871). Hydrokinetic solutions and observations, *Phil. Mag.* **42**, 362–77.

Kerr, R. M. (1993). Evidence of a singularity of the three-dimensional incompressible Euler equations, *Phys. Fluids* A **5**, 1725–46.

Kida, S. (1979). Asymptotic properties of Burgers' turbulence, *J. Fluid Mech.* **93**, 337–77.

Kida, S., Yanase, S. and Mizushima, J. (1991). Statistical properties of MHD turbulence and turbulent dynamo, *Phys. Fluids* A **3**, 457–65.

Kida, S. and Orszag, S. A. (1992). Energy and spectral dynamics in decaying compressible turbulence, *J. Sci. Computing* **7**, 1–34.

Kinney, R., McWilliams, J. C. and Tajima, T. (1995). Coherent structures and turbulent cascades in two-dimensional incompressible magnetohydrodynamic turbulence, *Phys. Plasmas* **2**, 3623–39.

Kolmogorov, A. N. (1941a). The local structure of turbulence in incompressible viscous fluid for very large Reynolds numbers, *Dokl. Akad. Nauk. SSSR* **30**, 299–303 (reprinted in *Proc. Roy. Soc.* A **434**, 9–13 (1991)).

Kolmogorov, A. N. (1941b). Dissipation of energy in locally isotropic turbulence, *Dokl. Akad. Nauk SSSR* **32**, 19–21 (reprinted in *Proc. Roy. Soc.* A **434**, 15–17 (1991)).

Kolmogorov, A. N. (1962). A refinement of previous hypotheses concerning the local structure of turbulence in a viscous incompressible fluid at high Reynolds number, *J. Fluid Mech.* **13**, 82–5.

Kraichnan, R. H. (1959). The structure of isotropic turbulence at very high Reynolds numbers, *J. Fluid Mech.* **5**, 497–543.

Kraichnan, R. H. (1965a). Lagrangian-history closure approximation for turbulence, *Phys. Fluids* **8**, 575–98.

Kraichnan, R. H. (1965b). Inertial range spectrum in hydromagnetic turbulence, *Phys. Fluids* **8**, 1385–7.

Kraichnan, R. H. (1966). Isotropic turbulence and intertial-range structure, *Phys. Fluids* **9**, 1728–52.

Kraichnan, R. H. (1967). Inertial ranges in two-dimensional turbulence, *Phys. Fluids* **10**, 1417–23.

Kraichnan, R. H. (1968). Lagrangian-history statistical theory for Burgers equation, *Phys. Fluids* **11**, 265–77.

Kraichnan, R. H. (1971). Inertial-range transfer in two- and three-dimensional turbulence, *J. Fluid Mech.* **47**, 525–35.

Kraichnan, R. H. (1973). Helical turbulence and absolute equilibrium, *J. Fluid Mech.* **59**, 745–52.

Kraichnan, R. H. (1994). Anomalous scaling of a randomly advected passive scalar, *Phys. Rev. Lett.* **72**, 1016–19.

Kraichnan, R. H. (1997). Passive scalar: Scaling exponents and realizability, *Phys. Rev. Lett.* **78**, 4922–5.

Kraichnan, R. H. and Montgomery, D. (1980). Two-dimensional turbulence, *Rep. Prog. Phys.* **43**, 547–619.

Krause, F. K. and Rädler, K. H. (1980). *Mean-Field Magnetohydrodynamics and Dynamo Theory* (Pergamon Press, Oxford).

Landau, L. D. and Lifshitz, E. M. (1959). *Fluid Mechanics* (Pergamon Press, London).

Larson, R. B. (1981). Turbulence and star formation in molecular clouds, *Mon. Notes Roy. Astron. Soc.* **194**, 809–26.

Leamon, R. J., Smith, C. W., Ness, N. F. and Matthaeus, W. H. (1998). Observational constraints on the dynamics of the interplanetary magnetic field dissipation range, *J. Geophys. Res.* **103**, 4775–87.

Legras, B., Santangelo, P. and Benzi, R. (1988). High-resolution numerical experiments for forced two-dimensional turbulence, *Europhys. Lett.* **5**, 37–42.

Lesieur, M. (1997). *Turbulence in Fluids*, third revised and enlarged edition (Kluwer Academic Publishers, Dordrecht).

Lesieur, M. and Schertzer, D. (1978). Amortissement auto-similaire d'une turbulence à grand nombre de Reynolds, *J. Mécanique* **17**, 609–46.

Leslie, D. C. (1973). *Developments in the Theory of Turbulence* (Clarendon Press, Oxford).

LeVeque, R. J., Mihalas, D., Dorfi, E. A. and Müller, E. (1998). *Computational Methods for Astrophysical Fluid Flow* (Springer, Berlin).

Lifshitz, A. E. (1989). *Magnetohydrodynamics and Spectral Theory* (Kluwer Academic Publishers, Dortrecht).

Lin, D. N. C. and Papaloizou, J. (1980). On the structure and evolution of the primordial solar nebula, *Mon. Notes Roy. Astron. Soc.* **191**, 37–48.
Lin, D. N. C. and Papaloizou, J. C. B. (1996). Theory of accretion disks II: Application to observed systems, *Ann. Rev. Astron. Astrophys.* **34**, 703–47.
Loève, M. (1963). *Probability Theory* (D. van Nostrand Company, Princeton), p. 156.
Lohse, D. and Müller-Groeling, A. (1995). Bottleneck effects in turbulence: Scaling phenomena in r versus p space, *Phys. Rev. Lett.* **74**, 1747–50.
Lumley, J. L. (1970). *Stochastic Tools in Turbulence* (Academic Press, New York).
L'vov, V. S. (1991). Spectra of velocity and temperature fluctuations with constant entropy flux of fully developed free-convective turbulence, *Phys. Rev. Lett.* **67**, 687–70.
Mac Low, M.-M., Klessen, R. S. and Burkert, A. (1998). Kinetic energy decay rates of supersonic and super-Alfvénic turbulence in star-forming clouds, *Phys. Rev. Lett.* **80**, 2754–7.
Maltrud, M. E. and Vallis, G. K. (1991). Energy spectra and coherent structures in forced two-dimensional and beta-plane turbulence, *J. Fluid Mech.* **228**, 321–42.
Marsch, E. (1991). MHD turbulence in the solar wind, in *Physics of the Inner Heliosphere*, eds. R. Schwenn and E. Marsch (Springer, Berlin), pp. 159–241.
Marsch, E. and Tu, C.-Y. (1990a). On the radial evolution of MHD turbulence in the inner heliosphere, *J. Geophys. Res.* **95**, 8211–29.
Marsch, E. and Tu, C.-Y. (1990b). Spectral and spatial evolution of compressible turbulence in the inner solar wind, *J. Geophys. Res.* **95**, 11 945–56.
Marsch, E. and Tu, C.-Y. (1993). Structure functions and intermittency of velocity fluctuations in the inner solar wind, *Ann. Geophys.* **11**, 227–38.
Marsch, E. and Tu, C.-Y. (1997). Intermittency, non-Gaussian statistics and fractal scaling of MHD fluctuations in the solar wind, *Nonlin. Proc. Geophys.* **4**, 101–24.
Matsuzaki, T. and Matsumoto, R. (1997). Three-dimensional MHD simulations of the Parker instability in a differentially rotating disk, *Accretion Phenomena and Related Outflows*, eds. D. T. Wickramasinghe, L. Ferrario, and G. V. Bicknell (Astronomical Society of the Pacific, San Francisco), pp. 766–7.
Matthaeus, W. H., Goldstein, M. L. and Smith, C. (1982). Evaluation of magnetic helicity in homogeneous turbulence, *Phys. Rev. Lett.* **48**, 1256–9.
McKee, C. F. (1989). Photoionization-regulated star formation and the structure of molecular clouds, *Astrophys. J.* **345**, 782–801.
McKee, C. F. (1999). In *The Origin of Stars and Planetary Systems*, eds. C. J. Lada and N. D. Kylafis (Kluwer Academic Publishers, Dortrecht), pp. 29–66.
McMurtry, P. A., Riley, J. J. and Metcalfe, R. W. (1989). Effects of heat release on the large-scale structure of turbulent mixing layers, *J. Fluid Mech.* **146**, 21–43.
McWilliams, J. C. (1984). The emergence of isolated coherent vortices in turbulent flow, *J. Fluid Mech.* **146**, 21–43.
McWilliams, J. C. (1990). The vortices of two-dimensional turbulence, *J. Fluid Mech.* **219**, 361–85.
Meneguzzi, M., Frisch, U. and Pouquet, A. (1981). Helical and nonhelical turbulent dynamos, *Phys. Rev. Lett.* **47**, 1060–4.
Meneveau, C. and Katz, J. (2000). Scale-invariance and turbulence models for large-eddy simulation, *Ann. Rev. Fluid Mech.* **32**, 1–32.
Meneveau, C. and Sreenivasan, K. R. (1987). Simple multifractal cascade model for fully developed turbulence, *Phys. Rev. Lett.* **59**, 1424–7.
Meneveau, C. and Sreenivasan, K. R. (1990). Interface dimension in intermittent turbulence, *Phys. Rev.* A **41**, 2246–8.

Mestel, L. and Spitzer, L., Jr (1956). Star formation in magnetic dust clouds, *Mon. Notes Roy. Astron. Soc.* **116**, 503–14.
Meyer-Hofmeister, E. and Spruit, H., eds. (1997). *Accretion Disks – New Aspects*, Lecture Notes in Physics 487 (Springer, Berlin).
Milano, L. J., Matthaeus, W. H., Dmitruk, P. and Montgomery, D. (2001). Local anisotropy in incompressible magnetohydrodynamic turbulence, *Phys. Plasmas* **8**, 2673–81.
Miller, K. A. and Stone, J. M. (1997). Magnetohydrodynamic simulations of stellar magnetosphere-accretion disk interaction, *Astrophys. J.* **489**, 890–902.
Miller, K. A. and Stone, J. M. (2000). The formation and structure of a strongly magnetized corona above a weakly magnetized accretion disk, *Astrophys. J.* **534**, 398–419.
Millionshtchikov, M. (1941). On the theory of homogenous isotropic turbulence, *Dokl. Akad. Nauk. SSSR* **32**, 615–18.
Moffatt, K. (1978). *Magnetic Field Generation in Electrically Conducting Fluids* (Cambridge University Press, Cambridge).
Monin, A. S. and Yaglom, A. M. (1975). *Statistical Fluid Mechanics: Mechanics of Turbulence* (The MIT Press, Cambridge, Massachusetts), vol. 2.
Montgomery, D., Brown, M. R. and Matthaeus, W. H. (1987). Density fluctuation spectra in magnetohydrodynamic turbulence, *J. Geophys. Res.* **92**, 282–4.
Montgomery, D. and Matthaeus, W. H. (1995). Anisotropic model of energy transfer in interstellar turbulence, *Astrophys. J.* **447**, 706–7.
Mouschovias, T. C. and Psaltis, D. (1995). Hydromagnetic waves and linewidth–size relation in interstellar clouds, *Astrophys. J.* **444**, L105–8.
Müller, W.-C. and Biskamp, D. (2000). Scaling properties of three-dimensional magnetohydrodynamic turbulence, *Phys. Rev. Lett.* **84**, 475–8.
Müller, W.-C., Biskamp, D. and Grappin, R. (2002). Statistical anisotropy of magnetohydrodynamic turbulence, submitted to *Phys. Rev. Lett.*
Müller, W.-C. and Carati, D. (2001). Dynamic gradient-diffusion subgrid models for incompressible magnetohydrodynamic turbulence, *Phys. Plasmas* **9**, 824–34.
Mydlarski, L. and Warhaft, Z. (1998). Paasive-scalar statistics in high-Péclet number grid-generated wind tunnel turbulence, *J. Fluid Mech.* **320**, 331–68.
Myers, P. C. and Goodman, A. A. (1988). Evidence for magnetic and virial equilibrium in molecular clouds, *Astrophys. J.* **326**, L27–30.
Nakano, T. (1984). Contraction of magnetic interstellar clouds, *Fund. Cosmic Phys.* **9**, 139–232.
Nelkin, M. (1995). Inertial range scaling of intense events in turbulence, *Phys. Rev.* E **52**, R4610–11.
Nordlund, Å., Brandenburg, A., Jennings, R. L., Rieutord, M., Ruokolainen, J., Stein, R. F. and Tuominen, I. (1992). Dynamo action in stratified convection with overshoot, *Astrophys. J.* **392**, 647–52.
Nordlund, Å and Padoan, P. (1999). The density p.d.f. of supersonic random flows, in *Interstellar Turbulence*, eds. J. Franco and A. Carramiñana (Camibridge University Press, Cambridge), pp. 218–22.
Novikov, E. A. (1994). Infinitely divisible distributions in turbulence, *Phys. Rev.* E **50**, R3303–5.
Obukhov, A. M. (1941). Energy distribution in the spectrum of a turbulent flow, *Izv. Akad. Nauk SSSR, Ser. Geofiz.*, No. 4–5, 453–66.
Obukhov, A. M. (1962). Some specific features of atmospheric turbulence, *J. Fluid Mech.* **13**, 77–81.

Ogura, Y. (1963). A consequence of the zero fourth cumulant approximation in the decay of isotropic turbulence, *J. Fluid Mech.* **16**, 33–40.

Olla, P. (1991). Renormalization group analysis of two-dimensional incompressible turbulence, *Phys. Rev. Lett.* **67**, 2465–8.

Orszag, S. A. (1970). Analytic theories of turbulence, *J. Fluid Mech.* **41**, 363–86.

Orszag, S. A. (1977). In *Fluid Dynamics*, eds. R. Balian and J. J. Peube (Gordon and Breach, New York), pp. 235–374.

Orszag, S. A. and Kells, L. C. (1980). Transition to turbulence in plane Poiseuille and plane Couette flow, *J. Fluid Mech.* **96**, 159–205.

Orszag, S. A. and Tang, C. M. (1979). Small-scale structure of two-dimensional magnetohydrodynamic tubulence, *J. Fluid Mech.* **90**, 129–43.

Padmanabhan, T. (2001). *Theoretical Astrophysics* (Cambridge University Press, Cambridge), vol. II.

Padoan, P. and Nordlund, Å. (1999). A super-Alfénic model of dark clouds, *Astrophys. J.* **526**, 279–94.

Paladin, G. and Vulpiani, A. (1987). Anomalous scaling laws in multifractal objects, *Phys. Rep.* **156**, 147–225.

Papaloizou, J. C. B. and Lin, D. N. C. (1995). Theory of accretion disks I: Angular momentum tranport, *Ann. Rev. Astron. Astrophys.* **33**, 505–40.

Paret, J. and Tabeling, P. (1998). Intermittency in the two-dimensional inverse energy cascade: Experimental observations, *Phys. Fluids* **10**, 3126–36.

Parker, E. N. (1958). Dynamics of interplanetary gas and magnetic fields, *Astrophys. J.* **128**, 664–76.

Parker, E. N. (1963). The solar flare phenomenon and the theory of magnetic reconnection and annihilation of magnetic fields, *Astrophys. J. Suppl. Ser.* **8**, 177–211.

Passot, T. and Pouquet, A. (1987). Numerical simulation of compressible homogeneous flows in the turbulent regime, *J. Fluid Mech.* **181**, 441–66.

Passot, T., Pouquet, A. and Woodward, P. (1988). The plausibility of Kolmogorov-type spectra in molecular clouds, *Astron. Astrophys.* **197**, 228–34.

Passot, T. and Vázquez-Semadeni, E. (1998). Density probability distribution in one-dimensional polytropic gas dynamics, *Phys. Rev.* E **58**, 4501–10.

Pelz, R. B. (1997). Locally self-similar, finite-time collapse in a high-symmetry vortex filament model, *Phys. Rev.* E **55**, 1617–25.

Petschek, H. E. (1964). Magnetic field annihilation, in *ASS/NASA Symposium on the Physics of Solar Flares*, ed. W. N. Hess (NASA, Washington), pp. 425–37.

Picone, J. M. and Boris, J. P. (1988). Vorticity generation by shock propagation through bubbles in a gas, *J. Fluid Mech.* **189**, 23–51.

Pneuman, G. W. and Kopp, R. A. (1971). Gas–magnetic field interactions in the solar corona, *Solar Phys.* **18**, 258–70.

Politano, H. and Pouquet, A. (1995). Model of intermittency in magnetohydrodynamic turbulence, *Phys. Rev.* E **52**, 636–41.

Politano, H. and Pouquet, A. (1998a). Von Kármán–Howarth equation for magnetohydrodynamics and its consequences on third-order longitudinal structure and correlation functions, *Phys. Rev* E **57**, R21–4.

Politano, H. and Pouquet, A. (1998b). Dynamic length scales for turbulent magnetized flows, *Geophys. Res. Lett.* **25**, 273–6.

Politano, H., Pouquet, A. and Carbone, V. (1998). Determination of anomalous exponents of structure functions in two-dimensional magnetohydrodynamic turbulence, *Europhys. Lett.* **43**, 516–21.

Politano, H., Pouquet, A. and Sulem, P. L. (1989). Inertial ranges and resistive instabilities in two-dimensional magnetohydrodynamic turbulence, *Phys. Fluids* B **1**, 2330–9.

Pomeau, Y. and Manneville, P. (1980). Intermittent transition to turbulence in dissipative dynamical systems, *Commun. Math. Phys.* **74**, 189–97.

Porter, D., Pouquet, A., Syntine, I. and Woodward, P. (1999). Turbulence in compressible flows, *Physica* A **263**, 263–70.

Porter, D. H., Pouquet, A. and Woodward, P. R. (1992a). Three-dimensional supersonic homogeneous turbulence: a numerical study, *Phys. Rev. Lett.* **68**, 3156–9.

Porter, D. H., Pouquet, A. and Woodward, P. R. (1992b). A numerical study of supersonic turbulence, *Theor. Comput. Fluid Dyn.* **4**, 13–49.

Porter, D. H., Pouquet, A. and Woodward, P. R. (1994). Kolmogorov-like spectra in decaying three-dimensional supersonic flows, *Phys. Fluids* **6**, 2133–42.

Porter, D. H., Pouquet, A. and Woodward, P. (1998a). Intermittency in compressible flows, in *Advances in Turbulence VII*, ed. U. Frisch (Kluwer Academic Publishers, Dordrecht), pp. 255–8.

Porter, D. H., Woodward, P. R. and Pouquet, A. (1998b). Inertial-range structures in decaying compressible turbulent flows, *Phys. Fluids* **10**, 237–45.

Pouquet, A. (1978). On two-dimensional magnetohydrodynamic turbulence, *J. Fluid Mech.* **88**, 1–16.

Pouquet, A., Frisch, U. and Léorat, J. (1976). Strong MHD helical turbulence and the nonlinear dynamo effect, *J. Fluid Mech.* **77**, 321–54.

Pouquet, A., Lesieur, M., André, J. C. and Basdevant, C. (1975). Evolution of high-Reynolds number two-dimensional turbulence, *J. Fluid Mech.* **72**, 305–19.

Pouquet, A., Sulem, P. L. and Meneguzzi, M. (1988). Influence of velocity–magnetic field correlations on decaying magnetohydrodynamic turbulence with neutral X-points, *Phys. Fluids* **31**, 2635–43.

Prandtl, L. (1925). Bericht über Untersuchungen zur ausgebildeten Turbulenz, *Z. angew. Math. Mech.* **5**, 136–9.

Praskovsky, A. A. (1992). Experimental verification of the Kolmogorov refined similarity hypothesis, *Phys. Fluids* A **4**, 2589–91.

Pringle, J. E. (1981). Accretion disks in astophysics, *Ann. Rev. Astron. Astrophys.* **19**, 137–62.

Proudman, I. and Reid, W. H. (1954). On the decay of a normally distributed and homogeneous turbulent velocity field, *Phil. Trans. Roy. Soc.* A **247**, 163–89.

Pumir, A., Schraiman, B. I. and Siggia, E. D. (1991). Exponential tails and random advection, *Phys. Rev. Lett.* **3**, 2838–41.

Pumir, A. and Siggia, E. D. (1992a). Finite-time singularities in the axisymmetric three-dimensional Euler equations, *Phys. Rev. Lett.* **68**, 1511–14.

Pumir, A. and Siggia, E. D. (1992b). Development of singular solutions to the axisymmetric Euler equations, *Phys. Fluids* A **4**, 1472–91.

Rasmussen, H. O. (1999). A new proof of Kolmogorov's 4/5-law, *Phys. Fluids* **11**, 3495–8.

Rayleigh, Lord (1883). On the crispations of fluid resting on a vibrating support, *Phil. Mag.* **16**, 50–8.

Rayleigh, Lord (1916). On the dynamics of revolving fluids, *Proc. Roy. Soc. London* A **93**, 148–54.

Richardson, L. F. (1922). Atmospheric diffusion shown on a distance–neighbour graph, *Proc. R. Soc. London* A **110**, 709–37.

Richtmyer, R. D. and Morton, K. W. (1967). *Difference Methods for Initial-Value Problems* (Interscience Publishers, New York).

Roberts, D. A. and Goldstein, M. L. (1991). Turbulence and waves in the solar wind, *Rev. Geophys. Suppl.* **29**, 932–43.
Roberts, P. H. and Soward, A. M. (1992). Dynamo theory, *Ann. Rev. Fluid mech.* **24**, 459–515.
Ruiz-Chavarria, G., Baudet, C. and Ciliberto, S. (1996). Scaling laws and disspation scale of a passive scalar in fully developed turbulence, *Physica* D **99**, 369–80.
Russell, T. C., ed. (1995). Physics of collisionless shocks, *Advances in Space Research*, Vol. 15 (Elsevier, Oxford).
Rutgers, M. A. (1998). Forced 2D turbulence: Experimental evidence of simultaneous inverse energy and forward enstrophy cascades, *Phys. Rev. Lett.* **81**, 2244–7.
Rutherford, P. H. (1973). Nonlinear growth of the tearing mode, *Phys. Fluids* **16**, 1903–8.
Ruzmaikin, A. A., Feynman, J., Goldstein, B. E., Smith, E. J. and Balogh, A. (1995). Intermittent turbulence in the solar wind from the south polar hole, *J. Geophys. Res.* **100**, 3395–3403.
Ruzmaikin, A. A., Shukurov, A. M. and Sokoloff, D. D. (1988). *Magnetic Fields in Galaxies* (Kluwer Academic Publishers, Dordrecht).
Ryu, D. and Goodman, J. (1992). Convective instability in differentially rotating disks, *Astrophys. J.* **388**, 438–50.
Saddoughi, S. G. and Veeravalli, S. V. (1994). Local isotropic turbulence in turbulent boundary layers at high Reynolds number, *J. Fluid Mech.* **268**, 333–72.
Sagaut, P. (2001). *Large-Eddy Simulations for Incompressible Flows* (Springer, Berlin).
Santangelo, P., Benzi, R. and Legras, B. (1989). The generation of vortices in high-resolution two-dimensional decaying turbulence and the influence of initial conditions on the breaking of selfsimilarity, *Phys. Fluids* A **1**, 1027–34.
Scalo, J., Vázquez-Semadini, E., Chappell, D. and Passot, T. (1998). On the probability density function of galactic gas. I. Numerical simulations and the significance of the polytropic index, *Astrophys. J.* **504**, 835–53.
Scheffler, H. and Elsässer, H. (1988). *Physics of the Galaxy and Interstellar Matter* (Springer, Berlin).
Schuster, H. G. (1988). *Derministic Chaos* (VCH Verlagsgesellschaft, Weinheim).
Shakura, N. I. and Sunyaev, R. A. (1973). Black holes in binary systems. Observational appearance, *Astron. Astrophys.* **24**, 337–55.
She, Z. S., Chen, S., Doolen, G., Kraichnan, R. H. and Orszag, S. A. (1993). Reynolds number dependence of isotropic Navier–Stokes turbulence, *Phys. Rev. Lett.* **70**, 3251–4.
She, Z. S. and Jackson, E. (1993). On the universal form of energy spectra in fully developed turbulence, *Phys. Fluids* A **5**, 1526–8.
She, Z.-S. and Lévêque, E. (1994). Universal scaling laws in fully developed turbulence, *Phys. Rev. Lett.* **72**, 336–9.
Shebalin, J. V., Matthaeus, W. H. and Montgomery, D. (1983). Anisotropy in MHD turbulence due to a mean magnetic field, *J. Plasma Phys.* **29**, 525–47.
Shraiman, B. I. and Siggia, E. D. (2000). Scalar turbulence, *Nature* **405**, 639–46.
Siggia, E. D. (1994). High-Rayleigh-number convection, *Ann. Rev. Fluid Mech.* **26**, 137–68.
Smagorinsky, J. (1963). General circulation experiments with the primitive equations, *Mon. Weath. Rev.* **91**, No. 3, pp. 99–164.
Smith, L. and Yakhot, V. (1993). Bose condensation and small-scale structure generation in a random force driven 2D turbulence, *Phys. Rev. Lett.* **71**, 352–5.

Smith, M. D., Mac Low, M.-M. and Zuev, J. M. (2000). The shock waves in decaying supersonic turbulence, *Astron. Astrophys.* **356**, 287–300.

Solomon, P. M., Rivolo, A. R., Barrett, J. and Yahil, A. (1987). Mass, luminosity, and line width relations of galactic molecular clouds, *Astrophys. J.* **319**, 730–41.

Sommeria, J. (1986). Experimental study of the two-dimensional inverse energy cascade in a square box, *J. Fluid Mech.* **170**, 139–68.

Spitzer, L. (1978). *Physical Processes in the Interstellar Medium* (Wiley, New York).

Sreenivasan, K. R. (1995). On the universality of the Kolmogorov constant, *Phys. Fluids* **7**, 2778–84.

Sreenivasan, K. R. (1996). The passive-scalar spectrum and the Obukhov–Corrsin constant, *Phys. Fluids* **8**, 189–96.

Sridhar, S. and Goldreich, P. (1994). Toward a theory of interstellar turbulence. I. Weak Alfvénic turbulence, *Astrophys. J.* **432**, 612–21.

Steenbeck, M., Krause, F. and Rädler, K. H. (1966). Berechnung der mittleren Lorentz Feldstärke für ein elektrisch leitendes Medium in turbulenter, durch Coriolis Kräfte beeinflußter Bewegung, *Z. Naturforschung* **21**a, 369–76.

Stein, R. F. and Nordlund, Å. (1989). Topology of convection beneath the solar surface, *Astrophys. J.* **342**, L95–8.

Stix, T. H. (1962). *Theory of Plasma Waves* (McGraw-Hill, New York).

Stone, J. M. and Balbus, S. A. (1996). Angular momentum transport in accretion disks via convection, *Astrophys. J.* **464**, 364–72.

Stone, J. M., Hawley, J. F., Gammie, C. F. and Balbus, S. A. (1996). Three-dimensional magnetohydrodynamic simulations of vertically stratified accretion disks, *Astrophys. J.* **463**, 656–73.

Stone, J. M. and Norman, M. L. (1992a). ZEUS-2D: A radiation magnetohydrodynamics code for astrophysical flows in two space dimensions. I The hydrodynamic algorithms and tests, *Astrophys. J. Suppl. Ser.* **80**, 753–90.

Stone, J. M. and Norman, M. L. (1992b). ZEUS-2D: A radiation magnetohydrodynamics code for astrophysical flows in two space dimensions. II The magnetohydrodynamic algorithms and tests, *Astrophys. J. Suppl. Ser.* **80**, 791–818.

Stone, J. M. and Norman, M. L. (1992c). ZEUS-2D: A radiation magnetohydrodynamics code for astrophysical flows in two space dimensions. III The radiation hydrodynamic algorithms and tests, *Astrophys. J. Suppl. Ser.* **80**, 819–45.

Strauss, H. R. (1976). Nonlinear three-dimensional magnetohydrodynamics in noncircular tokamaks, *Phys. Fluids* **19**, 134–40.

Stribling, T. and Matthaeus, W. H. (1991). Relaxation processes in a low-order three-dimensional magnetohydrodynamic model, *Phys. Fluids* B **3**, 1848–64.

Strittmatter, P. A. (1966). Gravitational collapse in the presence of a magnetic field, *Mon. Notes Roy. Astron. Soc.* **132**, 359–78.

Sulem, P. L., Frisch, U., Pouquet, A. and Meneguzzi, M. (1985). On the exponential flattening of current sheets near neutral X-points in two-dimensional ideal MHD flow, *J. Plasma Phys.* **33**, 191–8.

Sweet, P. A. (1958). The production of high-energy particles in solar flares, *Nuovo Cimento Suppl.* **8**, Ser. X, 188–96.

Taylor, G. I. (1935). Statistical theory of turbulence, I–IV, *Proc. Roy. Soc. London* A **151**, 421–78.

Taylor, G. I. (1950). The instability of liquid surfaces when accelerated in a direction perpendicular to their planes, *Proc. Roy. Soc. London* A **201**, 192–6.

Taylor, G. I. and Green, A. E. (1937). Mechanism of the production of small eddies from large ones, their planes, *Proc. Roy. Soc. London* A **158**, 499–521.

Taylor, J. B. (1974). Relaxation of toroidal plasma and generation of reversed magnetic fields, *Phys. Rev. Lett.* **33**, 1139–41.

Theobald, M. L., Fox, P. A. and Sofia, S. (1994). A subgrid-scale resistivity for magnetohydrodynamics, *Phys. Plasmas* **1**, 3016–32.

Thoroddsen, S. T. and Van Atta, C. W. (1992). Experimental evidence supporting Kolmogorov's refined similarity hypothesis, *Phys. Fluids* A **4**, 2592–4.

Toomre, J., Brummell, N. and Cattaneo, F. (1990). Three-dimensional compressible convection at low Prandtl numbers, *Comput. Phys. Commun.* **59**, 105–17.

Tu, C.-Y. and Marsch, E. (1995). MHD structures, waves and turbulence in the solar wind: Observations and theories, *Space Sci. Rev.* **73**, 1–210.

Tu, C.-Y. and Marsch, E. (1996). An extended structure function model and its application to the analysis of solar wind intermittency properties, *Ann. Geophys.* **14**, 270–85.

Vallée, J. P. (1997). Observations of the magnetic fields inside and outside the Milky Way, starting with globules ($\sim$1 parsec), filaments, clouds, superbubbles, spiral arms, galaxies, superclusters, and ending with the cosmological universe's background surface (at $\sim$8 teraparsec), *Fund. Cosmic Phys.* **19**, 1–89.

Vázquez-Semadini, E. and Passot, T. (1999). Turbulence as an organizing agent in the interstellar medium, in *Interstellar Turbulence*, eds. J. Franco and A. Carramiñana (Cambridge University Press, Cambridge), pp. 223–31.

Velikhov, E. P. (1959). Stability of an ideally conducting liquid flowing between cylinders rotating in a magnetic field, *Sov. Phys. JETP* **36**, 1389–404.

Velli, M., Grappin, R. and Mangeney, A. (1989). Turbulent cascade of incompressible unidirectional Alfvén waves in the interplanetary medium, *Phys. Rev. Lett.* **63**, 1807–10.

Vincent, A. and Meneguzzi, M. (1991). The spatial structure and statistical properties of homogeneous turbulence, *J. Fluid Mech.* **225**, 1–25.

Wang, L. P., Chen, S. Y., Brasseur, J. G. and Wyngaard, J. C. (1996). Examination of hypotheses in the Kolmogorov refined turbulence theory through high-resolution simulations. Part 1. Velocity field, *J. Fluid Mech.* **309**, 113–56.

Wahrhaft, Z. (2000). Passive scalars in turbulent flows, *Annual Rev. Fluid Mech.* **32**, 203–40.

Wahrhaft, Z. and Lumney, J. L. (1978). An experimental study of the decay of temperature fluctuations in grid-generated turbulence, *J. Fluid Mech.* **88**, 659-84.

Weber, E. J. and Davis, L. (1967). Angular momentum in the solar wind, *Astrophys. J.* **148**, 217–27.

Wickramasinghe, D. T., Ferrario, L. and Bicknell, G. V., eds. (1997). *Accretion Phenomena and Related Outflows*, APS Conference Series, Vol. 121 (American Physical Society, San Francisco).

Woltjer, L. (1958). A theorem on force-free magnetic fields, *Proc. Nat. Acad. Sci. USA* **44**, 489–92.

Wu, X. Z., Kadanoff, L., Libchaber, A. and Sano, M. (1990). Frequency power spectrum of temperature fluctuations in free convection, *Phys. Rev. Lett.* **64**, 2140–3.

Yaglom, A. M. (1949). Local structure of the temperature field in a turbulent flow, *Dokl. Akad. Nauk SSSR* **69**, 743–6.

Yaglom, A. M. (1966). Effect of fluctuations in energy dissipation rate on the form of turbulence characteristics in the inertial subrange, *Dokl. Akad. Nauk SSSR* **166**, 49–52.

Yakhot, V. and Orszag, S. A. (1986). Renormalization group analysis of turbulence I. Basic theory, *J. Sci. Comput.* **1**, 3–51.
Yakhot, A., Orszag, S. A., Yakhot, V. and Israeli, M. (1989). Renormalization group formulation of large-eddy simulations, *J. Sci. Comput.* **4**, 139–58.
Yanase, S. (1997). New one-dimensional model equations of magnetohydrodynamic turbulence, *Phys. Plasmas* **4**, 1010–17.
Yoshizawa, A. (1990). Self-consistent turbulent dynamo modeling of reversed-field pinches and planetary magnetic fields, *Phys. Fluids* B **2**, 1589–1600.
Yoshizawa, A. (1991). Subgrid-scale modeling of magnetohydrodynamic turbulence, *J. Phys. Soc. Japan.* **60**, 9–12.
Zakharov, V. E. and Sagdeev, R. Z. (1970). Spectrum of acoustic turbulence, *Sov. Phys. Doklady* **15**, 439–41.
Zhou, Ye and Vahala, G. (1991). Aspects of subgrid modelling and large-eddy simulation of magnetohydrodynamic turbulence, *J. Plasma Phys.* **45**, 239–49.
Zweibel, E. (1999). Magnetohydrodynamics problems in the interstellar medium, *Phys. Plasmas* **6**, 1725–31.
Zweibel, E. (2002). Ambipolar drift in a turbulent medium, *Astrophys. J.* **567**, 962–70.
Zweibel, E. and Josafatsson, K. (1983). Hydromagnetic wave dissipation in molecular clouds, *Astrophys. J.* **270**, 511–18.

Index